ENVIRONMENTAL IMPACT OF MINING

ENVIRONMENTAL IMPACT OF MINING

C. G. DOWN, Ph.D.

and

J. STOCKS, B.Sc., A.R.S.M., C.Eng.

Department of Mineral Resources Engineering,
Royal School of Mines, London

A HALSTED PRESS BOOK

JOHN WILEY & SONS

NEW YORK—TORONTO

PUBLISHED IN THE U.S.A. AND CANADA BY
HALSTED PRESS
A DIVISION OF JOHN WILEY & SONS, INC., NEW YORK

Library of Congress Cataloging in Publication Data

Down, Christopher Gordon.
   Environmental impact of mining.

   "A Halsted Press book."
   Includes bibliographies and index.
   1. Mineral industries—Environmental aspects.
I. Stocks, John, joint author. II. Title.
TD195.M5D68 1977   333.7   77–23129
ISBN 0–470–99086–4

WITH 81 TABLES AND 108 ILLUSTRATIONS

© APPLIED SCIENCE PUBLISHERS LTD 1977

Printed in Great Britain by Galliard (Printers) Ltd Great Yarmouth

Preface

The conflict between mining activity and its external environment has intensified in recent years, until in the developed countries most significant mining proposals are greeted with strenuous opposition from some sections of the community. For some years, therefore, mining companies, legislators and others have been concerned to increase their knowledge of mining's environmental effects and the ways of minimising them.

It is surprising that despite an immense literature upon individual facets of the problem, no book which examines the whole range of impacts has hitherto appeared. Partly this must be due to the difficulty of treating an interdisciplinary subject, composed of numerous specialisms and with wide ramifications, within the compass of a single publication. We have written this book because we believe that many mining impacts are interdependent, and hence should be considered in a unified way. Selected bibliographies have been provided to assist in the more detailed study of the various problems.

The rapid march of metrication has not yet reached all sectors of the mining industry. We therefore give both metric and non-metric units, with conversion factors being listed in the Appendix.

C. G. DOWN
J. STOCKS
Royal School of Mines

Contents

Acknowledgements

Much of the information, and many of the ideas, in this book originated while we were examining for the Department of the Environment the environmental impact of openpit metal mines and stone quarries. Without this generous support and our helpful discussions with officers of the Minerals Division about this research project, preparation of this book would have been much harder. We are indebted also to the many individuals in the minerals industry, in central and local government, and with environmental interest groups, who so willingly assisted in this work. Likewise, our ideas owe much to discussion with our colleagues at the Royal School of Mines, and especially the Professor of Mining, Robert N. Pryor. However, the responsibility for the facts and opinions expressed is, of course, ours alone.

Additionally, we are grateful to the undermentioned for permission to use their copyright material:

R. T. Arguile/Institution of Municipal Engineers (Fig. 78 and Table 63)
C. H. Frame (Fig. 19)
Dr R. P. Gemmell/Greater Manchester Council (Fig. 90 and Table 67)
Professor G. T. Goodman/The Royal Society (Table 65)
Institution of Mining and Metallurgy (Figs. 16, 17, 18, 104)
Institution of Mining Engineers (Fig. 93)
Dr J. F. Levy (Figs. 84, 89)
Dr M. H. Martin (Fig. 24)
National Coal Board (Figs. 2, 3, 15, 30, 41, 88, 100)
Quarry Management & Products (Fig. 92)
Sand & Gravel Association (Fig. 70)
Sheffield Evening Telegraph (Fig. 98)
South Wales Institute of Engineers (Fig. 91)
Professor B. G. Wixson (Fig. 25)

1

Mining and the Environment

1.1. INTRODUCTION

Mining is second only to agriculture as the world's oldest and most important industry. The dependence of primitive societies upon mined products is illustrated by the nomenclature of those epochs: Stone Age, Bronze Age and Iron Age, a sequence which also shows the increasing complexity of society's relationship with mining. To produce a flint axe required that a suitable deposit of flint be located, a method of mining be evolved and the resultant crude flint be processed to produce a usable finished product. Stated broadly these requirements are precisely the same for mineral production today. In the Bronze Age this general sequence also applied but the nature of the processing was necessarily of greater complexity. The use of the term Iron Age does not imply that flint or bronze were redundant, merely that a third mineral product had come to the fore in particular usages.[4,5] Thus, over the centuries, man's use of minerals has been characterised by the following features:

(1) Increased variety of minerals used, for a greater range of purposes.
(2) Increased sophistication of the methods of locating, winning and processing those minerals.

Arising from these trends have been several consequences of the most fundamental nature for individuals, nations and their inter-relationships. These are:

(1) Development of specialist skills, such as those of the geologist, miner, mineral processer, smelter, etc. Development of corporate bodies also owes much to the experience of early mining enterprises.
(2) The growth and concentration of particular sectors of the mineral industries into geographical areas where geological and other factors are most favourable for the development; early examples include the Harz metal district of Germany, or the Newcastle coal field of England. Recent examples are discussed in Section 1.2.
(3) Because of the localised nature of many minerals, as opposed to the widespread demand for them, trading of minerals between and within nations has been common since prehistoric times. More than

anything else, trading of mineral products first prompted different nations to meet each other; the Phoenician trade with Spain, or the Roman conquest of Britain, exemplify both peaceful and warlike actions for which mineral wealth (copper and tin, in these cases), actual or supposed, provided the motivation. Indeed Ovid, when writing of mining, stated that 'destructive iron came forth, and gold, more destructive than iron; then war came forth'. A more recent example is the current dispute over the sovereignty of the Spanish Sahara, where valuable phosphate deposits are the motivation.

(4) Absolute increase in the volume of demand for minerals, because of the greater variety of minerals being used and because of increases in both population and standard of living, have resulted in increased pressure to locate new mineral deposits and—latterly—concern over the finite resources of the earth.

(5) The existence of hitherto freely available mineral resources has encouraged developed countries (especially but not exclusively) to become increasingly dependent upon them.

(6) Increased scale and intensity of exploitation of mineral resources, increased population, and social developments have resulted in concern over the side effects of mining and processing; mainly in terms of aesthetic and pollution effects, but also various social effects exemplified by mining villages with inhabitants of a uniform social status and single employer.

This book is concerned with the impact that mining has upon the external environment and the ways by which this impact may best be reduced or eliminated. However, as the above six points demonstrate, the effect of mining upon the land it uses or influences is only one small aspect of the whole relationship between mining and society, and these various aspects are to a great extent interdependent. Action upon one component usually influences the others. At the crudest level detrimental environmental impact could be eliminated by ceasing to mine, but this solution (although seriously advocated by some persons in certain circumstances) ignores the fact that the developed countries are now wholly dependent upon a continuing free flow of minerals, without which society—and most of its members—would simply cease to exist. Every material thing in modern society is either directly a mineral product or else (as in the case of wood or foodstuffs) produced only with the aid of mineral derivatives such as steel, fertilisers or energy. Our present era may not actually be termed the aluminium, uranium or plastic age but we are even more dependent upon such things than was stone age man on his flint axe. Therefore, it is important to understand that the technical, biological and engineering components of environmental impact which are discussed in this book bear a vital relationship to social, legal and political considerations.

TABLE 1

WORLD PRODUCTION OF SELECTED MINERALS[1]

(Note that many of the figures are approximate, due to lack of adequate statistics. In particular, statistics for U.S.S.R. production are not always available.)

Mineral	Mt								
	1941	1945	1949	1953	1957	1961	1965	1969	1973
Aluminium (as bauxite)	6·0	3·9	8·4	13·7	20·1	28·5	36·3	53·0	70·0
Asbestos	0·6	0·6	0·8	1·1	1·8	2·5	2·8	3·3	5·5
China clay	2·0	2·0	3·0	4·0	5·5	7·0	9·5	12·0	16·0
Chrome ore	1·6	1·1	2·1	3·5	4·6	4·1	4·9	5·3	6·8
Coal	1745	1344	1600	1926	2300	2440	2760	2826	2245
Copper metal	2·5	2·1	2·2	2·8	3·5	4·2	4·9	5·8	6·4
Fluorspar	0·4	0·4	0·6	1·2	1·7	2·0	2·8	3·6	5·0
Gypsum	12·0	10·0	17·0	21·6	33·0	41·0	44·0	49·0	60·0
Iron ores	228	157	220	336	429	499	619	697	864
Lead, metal in ore	1·7	1·2	1·6	1·9	2·3	2·4	2·7	3·1	3·5
Manganese ore	5·7	4·0	6·2	10·9	13·4	13·8	18·0	18·6	21·3
Nickel, metal in ore	0·15	0·15	0·15	0·20	0·30	0·37	0·42	0·50	0·68
Phosphate rock	10·1	10·8	19·6	24·7	32·0	45·0	63·0	80·0	97·0
Potash (K_2O)	3·1	2·2	3·9	6·6	8·1	9·9	13·8	16·9	19·0
Pyrites	—	—	9·0	10·9	15·7	17·9	19·8	20·9	22·0
Salt	37·0	33·0	41·5	51·4	70·0	83·0	108·0	131·0	150·0
Tin, metal in ore	0·24	0·09	0·16	0·18	0·18	0·16	0·18	0·20	0·24
Zinc, metal in ore	2·1	1·6	1·9	2·6	3·1	3·3	4·1	4·8	5·5

1.2. MINERAL PRODUCTION

The minerals of concern in this book are those solid rocks or ores which are usually obtained from the earth's crust by conventional earthmoving techniques. Occasionally, other methods are used. Excluded from consideration are liquid or gaseous minerals such as oil and natural gas, which present such markedly different problems that they are incompatible with the present study.

Table 1 demonstrates the growth in world production of some of the main bulk minerals that has taken place since 1941. After a temporary falling-off in production due to the cessation of war demands, these minerals have nearly all been the subject of steadily increasing production. The magnitude of the increase differs considerably. In no case does the production increase relate directly to world population growth, which was from 2000 million persons in the mid-1920s to 3000 million persons in 1963. Consumption of minerals has increased at a greater rate than has population, and this is to be explained by increasing individual and national expectations of developed countries, plus the willingness of less affluent nations to permit indigenous minerals to be exploited for more or less completely foreign consumption. Further impetus to particular sectors has arisen from the substitution of one mineral product for another—notably aluminium for copper and steel.

The minerals listed in Table 1 are important in international trade.

Fig. 1. The physical proximity of gravel pit and new road highlights the interdependence of the source of a mineral and its consumer. It emphasises the illogicality of complaints about mining activity while continuing to demand the mineral.

Excluded are a group of minerals in which international movements seldom occur, except in isolated cases. These are the low-value, bulk civil engineering rocks—granites, limestones, sandstones, sand and gravel—plus clays used for brick and tile manufacture, and chalk and pure limestones used for the production of cement or chemicals. As raw materials, these are seldom traded because they are of very low value (transport costs even over short distances usually forming a major part of the selling price) and because most countries have adequate indigenous exploitable resources. The main exception to this is cement which is sold internationally but, here too, great efforts are being made by many countries to utilise local deposits for cement and lime production.

This situation is the opposite of that which obtains for many other minerals, particularly metals. In these cases, the resources (or, rather, the resources which are currently exploited) are not ubiquitous but often markedly localised. Table 2 demonstrates that for many important commodities more than half of world production is concentrated in very few countries. This is, of course, of great political significance but also means that in certain areas of the world are concentrated specific environmental problems associated with particular minerals.

In environmental terms the significance of the impact relates to population density and public awareness. On *a priori* grounds it might be expected that a direct relationship between these factors would exist but this is not the case. Of the most densely populated countries listed in Table 2 only Japan and the U.K. (268 and 94 persons/km^2, 693 and 243 persons/square mile) have a truly comprehensive framework of environmental legislation. Jamaica, Italy, Spain, China and Malaysia (100, 94, 94, 77 and 51 persons/km^2, 259, 243, 243, 200 and 133 persons/square mile) are also densely peopled but do not have a highly-developed public consciousness of mining pollution if legislation is taken as the measure of this awareness. In fact, contrary to expectation, it is some of the more sparsely-settled countries which have the most important mining and environment legislation, and which have devoted the greatest effort to research and development of control technology. These countries include the U.S.S.R., U.S.A., Canada and South Africa (10–12 persons/km^2, 27–30 persons/square mile). Yet, other countries in this density range (Chile, Bolivia, Peru, Morocco, Zambia) have devoted relatively little effort to such work. Therefore, a third factor—affluence—is implicated as important when considering environmental impact. The only countries which have made effective efforts to control the undesired impacts of mining are those with relatively high per capita income, and population density has had only a marginal effect. Many poorer countries, in which the primary mineral industries are proportionately of greater economic importance than in other countries, are understandably reluctant to place non-essential restrictions upon their main earners of wealth and foreign exchange.

TABLE 2
CONCENTRATION OF PRODUCTION OF CERTAIN MINERALS IN 1970[1]

Mineral (per cent of world production)

Country	Aluminium (bauxite)	Asbestos	China clay	Chrome ore	Coal	Copper (metal in ore)	Fluor-spar	Gold	Iron ore	Lead	Molybdenum	Nickel (metal in ore)	Phosphate rock	Tin (metal in ore)	Zinc (metal in ore)
Jamaica	21	—	—	—	—	—	—	—	—	—	—	—	—	—	—
Australia	16	—	—	—	—	—	—	—	7	15	—	—	—	—	—
Surinam	10	—	—	—	—	—	—	—	—	—	—	—	—	—	—
U.S.S.R.	9	23	—	30	19	14	—	—	—	12	—	17	—	—	11
Canada	—	37	—	—	—	—	—	6	26	12	19	43	—	—	21
U.K.	—	—	c. 40	—	—	—	—	—	—	—	—	—	—	—	—
U.S.A.	—	—	c. 55	—	19	25	—	4	12	15	61	—	41	—	10
China	—	—	—	—	12	—	—	—	—	—	—	—	—	—	—
S. Africa	—	—	—	23	—	—	—	80	—	—	—	—	—	—	—
Zambia	—	—	—	—	—	11	—	—	—	—	—	—	—	—	—
Chile	—	—	—	—	—	11	—	—	—	—	—	—	—	—	—
Mexico	—	—	—	—	—	—	23	—	—	—	—	—	—	—	—
France	—	—	—	—	—	—	13	—	7	—	—	—	—	—	—
Italy	—	—	—	—	—	—	8	—	—	—	—	—	—	—	—
Spain	—	—	—	—	—	—	8	—	—	—	—	—	—	—	—
Morocco	—	—	—	—	—	—	—	—	—	—	—	—	13	—	—
Malaysia	—	—	—	—	—	—	—	—	—	—	—	—	—	36	—
Bolivia	—	—	—	—	—	—	—	—	—	—	—	—	—	14	—
Peru	—	—	—	—	—	—	—	—	—	—	—	—	—	—	5
Japan	—	—	—	—	—	—	—	—	—	—	—	—	—	—	5
Total	56	60	c. 95	53	50	61	52	90	52	54	80	60	54	50	52

1.3. HISTORY OF ENVIRONMENTAL PROBLEMS

The belief that environmental awareness is a phenomenon of the 1950s and later is entirely erroneous, for problems have been experienced, complained about and—at least temporarily—solved for at least seven centuries. An early example of pollution caused by burning coal comes from the year 1257 when the English queen Eleanor was obliged to leave Nottingham town because of the smoke nuisance. Fifty years later occurs the first example of environmental legislation: in the reign of Edward I the use of coal in London for industrial and domestic purposes caused so great a smoke nuisance that the nobility, strongly backed by the populace at large, successfully agitated against its use. In 1306, a royal proclamation forbade the burning of coal but the law proved impossible to enforce with any effectiveness.[2]

The world's first mining textbook, *De Re Metallica*, by Georgius Agricola in 1556, contains an excellent description of the destruction caused by mining in Germany:[3] '. . . the strongest argument of the detractors is that the fields are devastated by mining operations . . . the woods and groves are cut down, for there is need of an endless amount of wood for timbers, machines and the smelting of metals. And when the woods and groves are felled, then are exterminated the beasts and birds, very many of which furnish a pleasant and agreeable food for man. Further, when the ores are washed, the water which has been used poisons the brooks and streams, and either destroys the fish or drives them away . . . Thus it is said, it is clear to all that there is greater detriment from mining than the value of the metals which the mining produces'.

Agricola also instanced an early piece of Italian legislation which forbade metal mining in fertile fields, vineyards and olive groves. With certain allowances for modern problems of noise, vibration, and transport, the above condemnation of mining could almost have been written today. Agricola continued to discuss the need for reclamation: '. . . as the miners dig almost exclusively in mountains otherwise unproductive, and in valleys invested in gloom, they do either slight damage to the fields or none at all. Lastly, where woods and glades have been cut down, they may be sown with grain . . . These new fields soon produce rich crops'.

The reference to cutting down forests for fuel is noteworthy, for this was a widespread problem when wood charcoal was the only fuel which could be used for smelting; in southern England the Sussex iron industry was effectively extinguished by its destruction of forests and lack of fuel.

An early example of reclamation is a Somerset colliery lease of 1791 which required that when the colliery closed the shaft was to be filled up and 'sown with Rye Grass seeds'. This specification of the work to be done is surprisingly precise; even today many British mines and quarries have legal reclamation requirements less stringent than this.

Many private landowners were, through self-interest, similarly enlightened. In the ironstone fields of the Midlands, leases granted after 1850 usually specified that worked-out areas should be topsoiled and returned to agriculture. No such requirements (or power to make them) appeared in British legislation until 1947 and even then were frequently not imposed, a disparity between private and public practice which is only now being removed.

By the 1920s many industries were making spasmodic attempts at reclamation, particularly coal mining in the U.S.A. and U.K., but it was very seldom that these were successful and they were certainly not widespread in any industry. Other problems, notably those of pollution, were seldom tackled before World War II except in special circumstances. It should also be stressed that many problems did not exist, at least in severe form, until the age of modern technology. For example, hand-worked ironstone quarries were often reinstated, but the introduction of the steam navvy and alteration of working methods prevented reclamation.

1.4. CONCLUSIONS

Although the detrimental impact of mining upon the land, air and water has been a topic of concern for many years it is only in the present century that the problems have become so acute that remedial measures have been instituted on any scale. The reasons for this intensification of the nuisances are that society has required massively increased quantities of mineral products, a demand which has only been met by dramatic technological developments enabling new, less accessible resources to be won at the expense of much greater environmental impact.

This trend is typified by the copper mining industry. At the start of the nineteenth century the largest copper mine in the world was on Parys Mountain, North Wales, producing in excess of 9000 tonnes per annum of ore with a grade of about 6% copper. Import of foreign ores began in the 1820s and by 1845 nearly 60 000 tonnes of ore were being imported annually, virtually destroying the home industry. Most such ores were obtained from vein deposits, with the valuable content fairly readily distinguished from the valueless host rock. Many ores were so rich that they could be smelted in the condition that they were excavated from the mine. British ores rarely contained more than 10% of recoverable copper, whereas Cuban ores contained 27%, Australian ores 40% and Chilean ores up to 60%. A marked shift in the location of copper mining therefore took place: in 1830, Europe produced 69% of world copper, but only 9% in 1930. North America, and to a lesser extent South America, Australia, and Africa became the major sources. However, such high-grade ores were rapidly exhausted and since the 1920s attention has been increasingly

directed at disseminated ores of much lower grade, usually under 2%
copper and in some cases today under 0·5% recoverable metal. Therefore,
a mine today may have to excavate and process about 40 times more ore
than in former years merely to produce the same amount of metal.
However, economics and demand dictate that mining is carried out on a
very much greater scale (and usually by open pit rather than underground
methods) than the above comparison would suggest. Thus, more and
larger mines are required, covering larger areas, moving greater tonnages
of material, and generating larger quantities of waste from more sophisticated processing. Hence, the potential for nuisance, pollution and
damage is greatly increased.

The corollary is that the technological developments which have enabled
the world's mineral demands to be satisfied from ever less accessible
resources have also provided the means of containing the environmental
damage. The post-war derelict land reclamation effort has only been
feasible because much of the technology was already developed for the
current requirements of industry. The other essentials are the research
effort to enable the technology to be deployed to the best advantage, and
a mechanism through which the public can decide upon the environmental
controls it wants to impose, as well as being made aware of the consequences
of its wishes.

In essence the basic technology of environmental control already exists.
Research is improving this technology as well as specifying more precisely
the objectives and means of the controls applied. The great defects at
present are in the public response to the problems of mining, the ignorance
of the long-term effects of some pollutants and of the effects of some of the
more drastic controls which are suggested. The environmental engineer
has a responsibility in all these spheres.

REFERENCES

1. *Statistical Summary of the Mineral Industry* (various dates), Institute of
   Geological Sciences, H.M.S.O., London.
2. Galloway, R. L. (1882). *A History of Coal Mining in Great Britain*, Macmillan,
   London.
3. Agricola, G. (1556). *De Re Metallica*, translated by H. C. Hoover and L. H.
   Hoover (1950), Dover Publications, New York.
4. Partridge, M. I. (1975). Mining as the precursor to civilization, *Royal School
   of Mines Journal*, No. 24, 40–4.
5. Rickard, T. A. (1932). *Man and Metals*, Vols. 1 and 2, McGraw-Hill, New
   York and London.

2

Range and Importance of Environmental Problems

2.1. INTRODUCTION

Mining may be defined as the removal of minerals from the earth's crust for use in the service of man. It is thus axiomatic that all mining activity effects some change in the natural environment and hence may be said to have an environmental impact. The extent of this impact can range from scarcely perceptible to highly obtrusive and the nature of the impact can similarly vary widely depending upon the mineral worked, the method of mining and the characteristics of the mine site and its surroundings.

In order to assess and remedy environmental disruption attributable to mining, a logical sequence would be:

(a) Express the nature and extent of the impact with reference to a rational and consistent system of measurement.
(b) Define acceptable standards and criteria in a comparable manner.
(c) Compare the measured (or predicted) impact with the relevant standards.
(d) Implement remedial measures, if necessary, to reduce the impact to conform with the accepted standards.

As will become apparent in ensuing chapters, there are relatively few occasions on which it is practicable to follow the idealised sequence (a)–(d). The principal difficulties encountered are:

  (i) For many types of environmental problem there are no consistent and accepted techniques available to quantify the extent of the impact. This is particularly the case when loss of amenity occurs. Assessment is frequently qualitative rather than quantitative and thus contains a strong element of subjectivity.
 (ii) Well defined and generally accepted criteria exist for only a minority of environmental problems. Some types of impact, such as visual intrusion, seem inherently unsuited to measurement and standard-isation since they are inevitably concerned with subjective reaction. Other cases, such as the quality of drinking water, are ostensibly eminently suited to measurement and objective assessment, and yet

10

there exists a confusing plethora of differing international and national standards.

(iii) There may be no known remedy capable of reducing a particular problem to generally acceptable levels. For instance the long term dereliction and pollution associated with open pit mining of base metal ores is well established and yet no adequate remedial techniques have yet been devised. In such cases it seems necessary to accept the environmental disruption which occurs or to prohibit mining and thus deprive society of some of the raw materials which form the basis of our civilisation.

It is in the last two decades that public concern has escalated to the extent that environmental issues have assumed fundamental importance to the minerals industry. This concern over the quality of life has largely been apparent in the developed countries and almost certainly can be related to the increasing affluence of the majority of people in these countries which has led to a growing interest in leisure and amenity considerations.

The majority, although not necessarily all, of the environmental problems associated with mining diminish with distance. Consequently the most severe environmental hazards commonly arise adjacent to the mining operation and those most affected are likely to be employees. It is necessary to define 'environment' when considering the effects of mining and it is common practice to consider the health and welfare of employees separately from the general well being of the community, even though frequently both are affected by the same matters. This book is concerned with the relationship between mining and the general community and hence excludes effects which solely pertain to the health and safety of employees.

2.2. NATURE OF PROBLEMS

The wide variety of environmental problems which can arise from mining could be categorised in a number of ways. The system adopted below is based on the effect of the problem upon man. In this system, many types of environmental impact can fall into more than one category, depending upon the degree of severity with which they occur.

2.2.1. Direct Hazard to the Safety of Man

Cases of death or serious injury to the general public directly attributable to mining activity are very few in number. In most mining operations there exist potential hazards to employees and in almost all countries there is stringent legislation and control to safeguard mine workers. This legislation commonly protects the general public, either directly or indirectly, from

FIG. 2.   The tips at Merthyr Vale Colliery, Aberfan, soon after the disaster showing (left of centre) the path of the tip slide.

FIG. 3. Aberfan, viewed from the same position as in Fig. 2, five years after the disaster. Note the reclamation of the tip site (by removal of the tip complex) and the drainage ditch across the hillside.

many of the more hazardous aspects of mining. For instance the transport, storage and use of explosives is invariably subject to very strict control, thus affording protection against accidental explosions to both employees and the general public.

Nonetheless public safety can be endangered on occasions through negligence or unusual or unforeseen circumstances. The principal hazards are:

(a) Sudden failure of reservoirs, waste dumps, tailings lagoons, open pit sidewalls or similar artefacts—the tragic potential of such catastrophes has been graphically illustrated in recent times by the Aberfan disaster in South Wales (Figs. 2 and 3) and the failure of tailings dams in the El Cobre district of Chile.

(b) Major subsidence associated with underground mining—subsidence is an inevitable concomitant of certain types of underground mining methods. Engineers normally plan either to let down the surface in a controlled manner or to confine subsidence to areas debarred to the general public. In many countries the principal risk probably lies with very old, abandoned workings whose stability, or even existence, may be unknown.

(c) Shafts, inclines, adits and open pits—adequate fencing or other security measures are normally required around any mine features which could be the scene of an accident. Hence, as in (b), the main risk probably occurs in abandoned operations where there is no continuing maintenance of security aspects. Very old workings which predate present legislation can cause a particular hazard since no safety measures may have been taken at the time of abandonment and natural revegetation may now camouflage potentially dangerous mine openings. Open pits fill up with water to the level of the water table once pumping ceases and this body of water, surrounded by steeply sloping banks, can be dangerous if access by the general public is possible.

(d) The release of toxic effluents—many mine wastes, particularly in metalliferous mining, are toxic if released in sufficient concentrations. Measures are normally adopted either to contain such substances or to dilute them to sub-toxic levels before release. Danger can arise from accidental spillage or from unforeseen events such as exceptionally heavy rainfall which alters the planned pattern of effluent discharge.

(e) The use of explosives—accidental initiation of explosives is very rare but the general public is acutely conscious of the destructive potential of modern explosives. Techniques for controlling flyrock from blasting have greatly improved in the last decade but dangerous instances still occur occasionally.

(f) Transport—mineral transport, and particularly road traffic, poses the same threat to public safety as transport arising from other sources. Most commonly this becomes an issue of importance when a mine is established in a rural area and the volume of traffic generated considerably exceeds previous levels. The authors are unaware of any statistics which single out road accidents specifically associated with mineral workings and there seems to be no reason *a priori* to assume that the impact of minerals traffic differs significantly from that of other traffic.

2.2.2. Indirect Hazard to the Well Being of Man

Because of the obvious and immediate nature of the threat, comprehensive measures are normally adopted to eliminate or minimise features of a mining operation which pose a direct danger to man. Much less easy to detect and remedy are those aspects which, if persistent over a period of time, can affect the health and safety of the general public. In spite of increasing research efforts in the last few years, insufficient is known of the long term effects of exposure to some aspects of mining, the more important of which seem to be:

(a) Air and water pollution—certain types of mining, and particularly for base metals, almost invariably result in a gradual release to the natural environment of toxic substances, of which the most important are probably metal ions and chemical reagents. This release mechanism normally continues for long periods after the mine has ceased to operate. As discussed in Section 2.2.1, it is unusual for lethal concentrations to be released from a mine. However, the long term effects of exposure to sub-lethal doses are uncertain, especially where a complex combination of metallic ions is concerned. There does seem to be some potential threat to the health, if not the life, of humans and there is the possibility that sub-lethal concentrations will be subjected to food chain magnification with consequent danger to human beings.

(b) Noise and dust—long term exposure to high noise levels can cause permanent hearing damage and similar exposure to dust can damage lung tissue and, in extreme cases, cause premature death. These are most important problems in protecting employees but there is no evidence that the general public is ever subjected to levels of noise or dust sufficient to cause a health hazard.

2.2.3. Damage to Property, Crops and Livestock

Many aspects of mining have the potential to cause damage to buildings, animals or crops. Such damage, when it occurs, may be regarded as a direct financial loss to a member or a section of the community although it is almost inevitable that, at the levels necessary to cause damage, nuisance and loss of amenity will also occur. Problems which affect crops or livestock may also pose an indirect threat to man by reducing or polluting sources of food.

It is an essential difficulty of this type of problem that it is frequently impracticable to prove unequivocally the source of the damage. For example, vibrations from blasting can cause cracks in buildings but so also can foundation settlement and many other factors dissociated from mining. It is thus a common complaint from the general public that mining companies are unwilling to accept responsibility for the adverse effects of their operations, whilst the companies frequently claim that they are subjected to unreasonable demands for compensation. Even where damage is admitted, the extent of compensation justified is often contentious.

The aspects of mining which commonly occasion complaints of this nature are:

(a) Air pollution—both airborne dust and gaseous emissions can cause damage, the extent of which depends upon their composition and concentration. Allegations of dust damage are most common when the dust has toxic constituents, such as metallic ions. However, even non-toxic dust can contaminate if present in sufficient concentrations and this can be an issue of importance if the mine is close to population centres. Stack emissions of sulphur dioxide are the most widespread gaseous pollutant, and are normally associated with the smelting of base metal concentrates. Their principal effect is to kill or inhibit the growth of vegetation.

(b) Water pollution—the wide range of liquid effluents from mining can pollute sources of water. The effect can range from the dangerous toxicity discussed previously to minor turbidity affecting aquatic plants and animals at the lower end of the food chain. Periodic flooding of normal water flow may cause land pollution as a secondary effect.

(c) Subsidence—the removal of material from the earth's crust inevitably leads to some stress readjustment and resulting ground movement. The effects of this may be limited by selecting an underground mining method in which the workings are supported by rock left *in situ* or by filling completed workings with waste material. Furthermore, in massive, localised orebodies the extent of the surface zone affected is limited. However, in areally extensive deposits, such as coal seams or sedimentary orebodies, it is difficult to avoid some surface subsidence without leaving a substantial proportion of the mineral unmined. Where deposits of this type underlie surface structures there exists a potential for subsidence damage. This may range from the appearance of sinkholes and the complete collapse of buildings with consequent hazard to life, to minor cracking and architectural damage.

(d) Ground vibrations and air blast—blasting operations give rise to ground vibrations and air blast waves which can cause structural damage. The potential diminishes rapidly with distance from the site and commonly is limited to the cracking of windows and plaster. Vibrations from static machinery do not normally contain sufficient energy to affect property beyond the boundaries of the mine site but heavy mobile plant moving on external roads close to buildings can cause similar vibrational levels to blasting.

2.2.4. Nuisance and Loss of Amenity

There is little dispute that hazards to the health and safety of man, from whatever source, should be removed as far as is practicable. Similarly it is generally accepted that anyone suffering financial loss from the activities of another should receive compensation. Consequently the types of mining impact discussed in Sections 2.2.1, 2.2.2 and 2.2.3 have long received priority consideration from mining companies and the community in general. Nuisance and loss of amenity do not threaten man's existence and seldom cause a readily quantifiable pecuniary loss; consequently they have in the past received less attention. It is in these areas that much, though not all, of the growing environmental awareness of the last two decades has concentrated.

Nuisance and loss of amenity are concerned with the quality of life and disruption of normal human activity. There is thus a strong element of subjectivity in their assessment which is inimical to quantification. At very severe levels of impact there is usually unanimity. Thus a continuous noise level sufficient to prevent speech communication would be generally regarded as unacceptable. However, in order to determine whether or not a nuisance exists, it is necessary to define the boundary between the acceptable and the unacceptable. This may depend on a wide range of factors including the nature of the impact, the location, the time of day, the duration of the impact and, most important of all, the susceptibilities of the individual.

The difficulties of deriving standards and criteria which receive general acceptance are such that for some types of impact, such as visual intrusion, there have been few serious attempts and each individual situation is dealt with *ad hoc*. Even where standards are common, as in the case of noise, the problems of adequate assessment are well illustrated by the enormous range of regulations in force in different countries.

The types of impact which can cause nuisance or loss of amenity are:

(a) Visual intrusion—every aspect of mining from the excavation to the surface buildings and waste disposal areas can cause visual disruption. The extent to which the mine site is visible depends principally upon the size of the operation and the nature of the surrounding topography. Visual intrusion is related mainly to the degree of visibility and the nature of the local landscape, which itself is a matter of personal value judgement. It may in general be thought advisable to reduce visibility to a minimum but this could be an oversimplification since a proportion of the general public finds interest in viewing major earthworks. Thus the massive Bingham Canyon open pit copper mine is claimed to be the second largest tourist attraction in the State of Utah.

(b) Noise and vibration—both noise and vibration can be readily measured and therefore, in contrast to visual impact, are easily quantified. Nuisance potential seems to be related to a number of factors, principally intensity, frequency, duration, time of day, type of locality and individual susceptibility. Some standards in use relate to one factor only, usually intensity, and probably for this reason are often regarded as unsatisfactory. More complex standards incorporating several of the above factors are nonetheless frequently the subject of contention.

(c) Air and water pollution—even at relatively low levels air and water pollution can interfere with wildlife and natural vegetation, discolour the atmosphere and water courses, and cause minor contamination to homes, laundry, etc.

(d) Dereliction—in certain types of mining it is difficult to avoid creating land areas of either total dereliction or very limited land use potential. Surface excavations, solid waste disposal dumps, tailings dams and subsided areas can sterilise land to the extent that it is seldom possible to find new land uses. Conversely, underground mining may create new space capable of a wide range of after uses. In both cases the importance is directly related to location. At one extreme, in arid, sparsely-populated regions land uses are very limited and questions of dereliction or new space are of low priority, yet close to urban centres or in other areas of high land value they may assume great importance.

2.2.5. Resource Utilisation

Problems of resource utilisation differ substantially from those discussed in the preceding sections in that they have almost no present, immediate impact but are concerned with the future supply of raw materials for the long term continuation of civilised society. Prior to the Industrial Revolution, consumption of minerals was almost static at a very low level by present standards. In the last two centuries mineral consumption has grown continuously. This growth has arisen in part from a corresponding growth in world population and in part from increasing *per capita* consumption associated with the rising affluence which has occurred particularly in the developed countries. It is axiomatic that a finite world cannot support exponential growth, or for that matter any other kind of growth, indefinitely. Consequently fears have arisen that, if present trends continue, mineral sources will become depleted to the extent that civilised society can no longer function.

A number of factors combine to make it impossible to predict when, if ever, total depletion of any particular mineral will occur. The more important of these are:

(a) Future consumption is uncertain. Extrapolation of present trends, which commonly forms the basis for prediction, can be misleading since it cannot take account of future changes in society or technology.

(b) Total world ore reserves are unknown. Since exploration is expensive, there is little economic incentive to prove the existence of orebodies beyond the short and medium term horizons of the next 30 years. Furthermore, changes in economic climate and technological development necessitate continual reassessment of what constitutes an economic orebody. Many of today's major mines are working mineral deposits which 20 years ago would have been classed as too chemically complex or low grade for economic exploitation.

(c) Substitution from one mineral to another can occur. As a mineral becomes relatively scarce and therefore expensive, there is a natural tendency for secondary industry to seek acceptable alternatives. Thus aluminium has in recent years partially replaced more expensive copper and tin in a number of applications.

(d) Recycling may increase in importance in the future. Few minerals, apart from those from which energy is derived, are consumed in the sense that their essential nature is changed so that they are no longer usable. At present there is little recycling because it is normally less costly to satisfy demand from primary sources. However, any future tendency towards depletion may be expected to escalate mineral prices and this could alter the economics to the point where it is viable to satisfy a significant proportion of new demand by recycling.

In view of the above factors, continuing discourse on the likelihood of mineral resource depletion seems likely, with neither side able to develop an overwhelmingly conclusive case. Briefly stated, the influences tending to lead to depletion are:

(i) Increasing world population which increases total world demand for minerals commensurately.

(ii) Increasing *per capita* consumption as society strives to add to the material benefit of the individual.

(iii) The finite, although unknown, limit to minerals production from a finite world (it is occasionally postulated that future requirements will be met from other worlds, but this presupposes enormous technological developments, not least of which would be a cheap and almost limitless source of usable energy).

Those influences tending to counteract resource depletion are:

(i) The search for and discovery of new mineral deposits.

(ii) Technological advances and economic changes which permit the working of hitherto unexploitable deposits.

(iii) The assumed ability of society to adapt to changing circumstances. This may lead to zero or negative population growth and similar trends in mineral consumption, although there is as yet no evidence of such fundamental change occurring.

(iv) Substitution from scarce to more abundant minerals.

(v) Increased recycling of minerals, especially metals, thus reducing the need for production from new primary sources.

(vi) The discovery of positive uses for waste materials, which could partially satisfy mineral demand.

Although both technology and economics will play an important part in determining future patterns of supply and demand, it seems probable that political and sociological forces, which are beyond the scope of this book, will predominate.

2.3. FACTORS INFLUENCING THE NATURE AND EXTENT OF ENVIRONMENTAL IMPACT

There is very wide variation in both the type and the severity of environmental impact from mining. This variation can largely be related to four main factors: the size of the operation, its location, the mineral or minerals extracted and the method of mining in use.

2.3.1. The Size of the Operation

It is reasonable to postulate that some relationship must exist between the scale of working and the severity of environmental impact. Indeed, it seems self-evident that a decision to increase output at a given site can only lead to an increase in environmental impact. In contrasting high and low production rates at otherwise identical sites, the following factors tend to aggravate the impact of the larger mine:

(i) Larger fixed plant and buildings are required.
(ii) Larger and/or greater numbers of mobile plant are used.
(iii) Solid and/or slurried waste is produced at a higher rate.
(iv) Over a given period of time the surface area affected by mining (whether open pit or underground) is greater.

The influence of (i)–(iv) would normally be to increase the potential for some, if not all, of the main impacts including visual intrusion, air and water pollution, noise and vibration, and dereliction. However, as discussed previously, there are no known techniques for measuring some of these impacts and for these it is therefore impossible to quantify any increase. Even for those impacts, such as noise, which can be measured it is extremely difficult to obtain data from which general conclusions can be drawn. Information from different sites cannot normally be compared since it is almost impossible to separate the influence of scale of working from other important factors such as location. Therefore in order to derive a relationship between size and the intensity of any quantifiable impact, it seems necessary to select sites at which a significant change in output is planned and to take measurements before and after the change occurs. As far as the authors are aware, the validity of this approach has not yet been tested by any actual site investigations.

Under some circumstances an increase in output can be beneficial in reducing the overall environmental impact. This may happen when:

(i) Total demand is fixed—the effect of obtaining higher output of mineral from some sites, even if at increased environmental impact, is to decrease the total number of sites required to satisfy market demand. It might be claimed that this would decrease the overall impact of mining in a country or region, although this is not amenable to proof and would be a matter for conjecture.[1]
(ii) Total mineral reserves at a site are fixed—in this situation any increase in output reduces the life of the mine. It is again a matter for speculation whether a higher environmental impact for a shorter time is preferable to a lower impact for a longer time.
(iii) Mine waste rock is required for amenity uses—at some sites it is desirable that amenity banks be constructed to screen the operation or that noise baffles are built. The higher level of waste production associated with a larger output might permit such features to be more extensive and more rapidly constructed than in a smaller operation.

It is therefore apparent that no simple relationship exists between scale of working and environmental impact and that assessment must, in the foreseeable future, be subjective. This is true even for a change in output at one site and, when mines at different locations are compared, other factors such as local geography can dwarf considerations of size. In a survey of over 50 U.K. stone quarries[2] it was concluded that output and environmental impact are not connected by any simple mathematical relation. The authors observed that, because of individual site factors, some quarries producing less than 250 000 tonnes p.a. were more obtrusive than others an order of magnitude larger.

2.3.2. Geographical and Locational Factors

Individual site factors are of extreme importance in controlling both the nature and extent of environmental impact from mining. It is perhaps self-evident that, in an extreme case, a mine producing 50 000 tonnes p.a. located in a major city is likely to occasion more public complaints and disquiet than one producing 50 million tonnes p.a. on an inaccessible uninhabited island.

The more important site factors are:

(i) Population density—the extent to which any environmental impact may become a problem is very dependent upon the numbers of people affected by the impact. This is well illustrated by some long established quarries in rural areas in Britain which have recently experienced housing development close to the site perimeter. Not surprisingly operators in such situations find that there is a marked increase in the number of complaints received. Some low-population rural regions are readily accessible to major urban centres and are important leisure and recreation areas. Hence both the local indigenous population density and accessibility of the area can strongly influence environmental issues.

(ii) Topography—the nature of the local topography and the location of the mine within it are key factors. In flat country, ground or low level features are visible for short distances only and water courses flow slowly with consequently low capacity for carrying solids in suspension. Conversely, tall features can be visible over a large area and noise and airborne dust may carry for quite long distances. In hilly or undulating countryside the position of a feature within the landscape is most important. Hilltop installations are visually obtrusive and noise and airborne dust may travel well, whilst those in valleys are often visible only over short distances from a limited number of vantage points, and surrounding hills can form effective barriers to localise noise and dust. The carrying capacity of streams is proportional to the cube of their velocity and hence fast flowing upland watercourses can carry large volumes of sediment to be deposited in flatter lowland areas.

Opposition to mining is more intense in regions of high scenic value or unusual conditions favourable to rare flora and fauna.

(iii) Climate—precipitation, temperature, humidity, wind and other climatic factors strongly affect the mechanisms by which pollution is transported from a mine site to the surrounding environment. Their main influence is therefore upon the intensity of pollution and the distances over which the mine's impact is discernible. Atmospheric effects control the transmission of gaseous effluents, dust, noise and air blast whilst precipitation is of crucial importance in the dissemination of liquid effluents.

(iv) Economic and social factors—the attitude of the general public to mining is conditioned in part by the state of the local economy and the nature of the community. Mine operators generally report fewer complaints from communities or sections of the public which are economically dependent upon the mine. This normally applies mainly to nuisance or loss of amenity. It was noticeable that, when open pit copper mining in the Snowdonia National Park in North Wales was under consideration, some local communities were strongly in favour because of the employment and wealth which would be created, whilst much of the opposition came from sections of the public who would not have benefited directly from mining. In the extreme case, in some underdeveloped countries the need to create employment and raise living standards is accorded such priority that amenity factors are scarcely considered and this has, on occasions, led to accusations that rich countries 'export' some of their pollution to poorer countries which cannot afford to turn down new developments, even if they are environmentally disruptive.

However, where hazard to health or safety is alleged, economic dependence does not prevent strong local reaction. This was exemplified by the Aberfan disaster in South Wales where, despite the importance of the coal mining industry, there was local pressure to remove all waste tips from the area following the collapse of the Aberfan tip.

It is commonly asserted by mine operators that complaints seldom originate from sections of the community with relatively low socio-economic status, such as manual or semi-skilled workers and their families. The membership of many conservation and environmental pressure groups is certainly dominated by teachers, businessmen, 'white-collar' workers and members of the 'middle' and 'upper' rather than 'working' classes. It is, however, unclear whether this situation represents genuinely different attitudes towards environmental issues in different sections of the community, or whether it is simply that the more highly educated members of the public are naturally the most articulate.

Sociological research to clarify the picture might yield most interesting results with implications far beyond the mining industry.

2.3.3. Method of Mining

The large majority of minerals are produced by traditional open pit or underground methods, although other techniques are important for particular minerals. The type of mining method used has a major influence upon both the nature and extent of the environmental impact and the principal advantages and disadvantages are summarised below:

(a) Open pit—generally is cheaper than underground mining and thus permits the working of lower grade orebodies thereby increasing economically available reserves of mineral. Open pit methods frequently achieve a higher percentage recovery of valuable mineral than underground extraction techniques, improving overall resource utilisation. The low costs may also allow extensive environmental control measures which would be economically untenable with more expensive mining methods. However, many environmental impacts are at a maximum in open pit mining, including visual intrusion, noise and airblast, vibration, air and water pollution and dereliction. The size of the excavation and the quantities of solid waste frequently cause major problems.

(b) Open cast—this method is applicable to stratified deposits close to the surface and has the same general advantages and disadvantages as open pit mining. The mineral yield per unit area of land is relatively low and hence large areas are often worked. Whilst this is a potential disadvantage, it limits the time during which any piece of land is being worked and offers the possibility of rolling restoration (discussed further in Chapter 10).

(c) Underground mining—normally has a lower impact than surface working of the same mineral. Some aspects, such as slurried waste disposal, are common to both methods but underground mining normally reduces the levels of visual disruption, noise and airblast, vibration, air and water pollution and can almost eliminate long term dereliction. The high costs of underground mining cannot usually support the removal of large volumes of waste rock and the smaller tonnages produced are often stored underground thus greatly reducing the waste disposal problem. With some underground methods, a proportion of the slurried wastes from the treatment plant are pumped underground to fill the excavated areas and the quantity of tailings requiring surface disposal is thus reduced. Subsidence is a potential problem of all underground methods and, if uncontrolled, can lead to widespread surface damage and even dereliction.

(d) Dredging—although not in widespread use, dredging is a major method of working for some minerals, principally tin and to a lesser extent gold. It is mainly applicable to alluvial deposits and therefore used in low-lying relatively flat country. Dredging has a high potential for water pollution through turbidity even if no toxic materials are present. The most important impact is the potential for widespread dereliction which is well illustrated in some major tin producing countries such as Malaysia, Thailand and Indonesia.

(e) Other methods—for specific minerals or particular circumstances other methods of mining are used. Salt and other soluble minerals are sometimes solution mined by circulating water from the surface through the deposit.[7] Apart from possible interference with groundwater, the major disadvantage of this method is the difficulty in controlling the extent of underground cavities and thus avoiding surface collapse.

High pressure water is used to excavate some weathered or unconsolidated minerals such as china clay. Water pollution, particularly turbidity, can be a special problem of this technique.

Marine mining is currently limited to a few minerals close to shore including beach sands and gemstones. There is growing interest in exploiting the potential of the seabed and this could reduce or eliminate many of the problems of land-based mining whilst possibly creating marine pollution.[5,6]

There is similar interest in the *in situ* leaching of base metal ores. This necessitates shattering the ore body *in situ*, percolating acidic liquors through the ore and collecting the pregnant solution. In the U.S.A. nuclear and conventional explosives have been used to fracture the ore and both present environmental problems.[3,4] Escape of polluted water from the mining zone is a further potential hazard.

2.3.4. Mineral Characteristics

The nature of the mineral can exercise a significant influence upon environmental impact. The number of minerals mined is far too large to permit discussion of more than broad categories.

(i) Industrial minerals—in terms of tonnage, if not value, industrial minerals are the most important class. In the U.S.A. the annual production of crushed stone alone exceeds one billion tonnes. The majority of industrial minerals are of low value and this implies surface working of high grade deposits close to market centres to minimise transport costs. Nonetheless a minority of production is obtained by other methods and some industrial minerals command high prices. Sophisticated processing is exceptional and often comminution and classification is the only treatment prior to marketing. Environmentally, industrial mineral workings often have a high potential impact because of open quarrying close to urban centres. However, few minerals of this class are toxic and the use of chemical reagents is limited. Consequently the typical problems are visual intrusion, dereliction, noise, vibration, air blast and dust rather than serious water pollution.

(ii) Metalliferous minerals—the high price of most metals enables the mining of low grade ores by a variety of methods and in situations where complete access to the ore requires the removal of large volumes of overburden and country rock. Typically the valuable minerals are disseminated throughout the host rock and their recovery requires both fine grinding and sophisticated processing. Therefore commonly in excess of 90% of the ore is rejected as a finely ground slurry contaminated with chemical reagents, and the disposal of these tailings creates severe problems of public safety, air and water pollution, and land sterilisation. The common chemical form of many metalliferous minerals is the sulphide and processing, particularly smelting, can liberate large quantities of sulphur dioxide to pollute the atmosphere.

Additionally, especially from surface mining, large quantities of barren or sub-marginal grade ore may need disposal. These pose problems of stability and visual intrusion and may give rise to airborne dust and water pollution from natural or induced leaching if the rock contains trace quantities of minerals. Metalliferous mining can thus cause the full range of environmental problems and water pollution hazard is particularly marked for base metals which are themselves toxic.

(iii) Coal—both surface and underground methods are used to extract large quantities of coal in many parts of the world. The yield of coal is approximately 13 000 tonnes per hectare per metre (4800 tons/acre/yard) of seam thickness and thus, except in very thick seams, coal workings can quickly extend over large areas. In underground mining this implies subsidence potential over a wide area and in surface mining the possibility of widespread dereliction which can be effectively counteracted by rolling restoration techniques. Although coal is not itself toxic or polluting, it is often associated with pyrite which can cause severe water pollution as evidenced in many old strip mining areas in the U.S.A. In addition to the normal problems of solid waste disposal, colliery waste tips contain a proportion of carbonaceous material and are liable to spontaneous combustion which, once started, is very difficult to eradicate.

(iv) Other minerals—the range of problems associated with these is too diverse to permit simple categorisation.

2.4. CONCLUSIONS

The nature and extent of the environmental problems arising from the mining industry are very wide in their scope. In the ensuing chapters, individual problems are discussed in greater detail. It is impossible to give completely comprehensive coverage of an industry as large and varied as mining and hence attention is focused on the more important and widespread issues.

REFERENCES

1. *Aggregates: the way ahead* (1976). Report of the Advisory Committee on Aggregates, Department of the Environment, H.M.S.O., London, p. 54.
2. Down, C. G. and Stocks, J. (1976). *The environmental impact of large stone quarries and open-pit non-ferrous metal mines in Britain*, Department of the Environment, H.M.S.O., London (in press).

3. Rabb, D. D. (1971). Penetration of leach solution into rocks fractured by a nuclear explosion, *Trans. A.I.M.E.*, **250**, 139–41.
4. AEC and KCC will jointly study potential of nuclear blasting to mine copper (1973). *Engineering & Mining Journal*, **174**(4), 26.
5. Archer, A. A. (1974). Progress and prospects of marine mining, *Mining Magazine*, **126**(3), 150–63.
6. Medford, R. D. (1969). Marine mining in Britain, *Mining Magazine*, **121**(5), 369–81.
7. Martinez, J. D. (1971). Environmental significance of salt, *Amer. Assoc. Petrol. Geol. Bull.*, **55**(6), 810–25.

3
Visual Impact

3.1. INTRODUCTION

There is widespread adverse public reaction to the appearance of most mineral workings, in many developed countries if not elsewhere. This is illustrated in Britain, for example, by the almost universal insistence upon landscaping being undertaken at new mineral workings.

The aesthetic aspect of mining and the environment is difficult to discuss in view of the overwhelming importance of subjective factors and the lack of an objective base from which to work. An additional complication is that opinion on the visual acceptability of a mineral operation varies markedly, depending on the location, as well as on many non-visual criteria. For example, the standard of landscaping practice in Britain tends to be judged according to what can be termed the eighteenth century romantic and picturesque traditions: gently rolling countryside, hedgerows, small copses, etc., all upon a very human scale. Such a tradition does not exist in other countries, nor can it be considered as appropriate in mountain, moorland or other types of landscape, in Britain or elsewhere.

If landscaping of mineral workings was solely a matter of falling in with the aesthetic tastes of the locality, the problem might be simplified, but this is very far from being the case. Antiquarian interest has a strong influence upon whether or not particular structures are acceptable at particular locations.[8] Silbury Hill, one of England's largest prehistoric earth mounds, is in geometric form, and setting, identical in every respect to many colliery waste tips, and yet the former is considered acceptable whereas the latter are not. Moreover the grandeur of a few mineral workings partially transcends many aesthetic objections. Thus Bingham Canyon copper mine, reputedly the largest excavation of its type in the world, is the second largest tourist attraction in the state of Utah.

In short, the aesthetic design of mineral workings is controlled to a large extent by sociological factors outside the control of the engineer. The role of the landscaping practitioner is to interpret subjective opinions in terms of practical landscape possibilities, and it is this aspect which is discussed in this chapter.

28

3.2. LANDSCAPE ANALYSIS

The measurement of a chemical pollutant and the assessment of its significance in the environment are tasks which, at least broadly, are capable of being achieved with widespread agreement amongst environmental scientists. Landscape analysis, though an equally important aspect of environmental control, is by comparison a most rudimentary art.

In precisely parallel manner to the study of a chemical pollutant, it would be ideal if a baseline measure of a landscape could be obtained, against which designs of mineral operations could be compared to determine that which caused the least visual impact, and against which the visual changes caused by mining could be monitored throughout the life of the mine. Many attempts have been made to achieve measurements which can be used in this way, and for cost/benefit analysis. Work on these topics has been reviewed by Jacobs.[4]

The approach most usually adopted is to assign to different components of test landscapes a rank based upon a points system. There is great variability in the extent to which different components are recognised and sub-divided. The ranks of the components are then treated by addition or other transformation, to arrive at a numerical 'value' for that landscape. The meaning of this 'value' is then interpreted by testing subjective public response to sample landscapes.

Such techniques have obvious drawbacks and it is hard to believe that any widely accepted or useful system will be developed in the short or medium term. For many years to come the extent and significance of the visual impact of mineral workings will remain a matter of purely subjective judgement.

3.3. SOURCES OF VISUAL IMPACT

Almost every facet of mineral working can create an undesirable visual impact, but it is possible to class the sources into five main groups.

3.3.1. Surface Excavations
The extent to which surface excavations intrude into the landscape bears no direct or universal relationship to size. The nature of the excavation, the surrounding land-forms, and the relationships between the two, are the main factors.

Fresh rock exposures created by surface mining are often highly visible. In many cases—chalk, limestone, etc.—there is a colour contrast of light rock against darker background, although the converse seldom occurs. Marked colour differences, such as the reddish hues caused by iron minerals, may also result in prominence. It is likely that faces are less conspicuous if the surrounding area is well-endowed with natural rock exposures.

Serious visual impact is caused by the excavation intersecting the skyline. The resulting gap is almost invariably highly prominent.

3.3.2. Waste Disposal

Conventional techniques of tipping solid wastes, and of impounding slurries, often result in considerable visual intrusion. Remedies are usually not applicable until cessation of dumping, so that the impact persists throughout the life of the tip. The magnitude of the problem varies according to the mineral being worked, but is usually greatest at open pit non-ferrous metal mines and underground coal mines, where backfilling cannot be progressively carried out and is, indeed, seldom attempted at any stage.

A particular problem of mining non-ferrous metals and some other minerals such as china-clay is the disposal of tailings. Some impoundments are constructed initially to serve the life of the mine, but it is more common for dam walls to be progressively raised over an extended time period. Such methods result in long-term visual intrusion, as does the impounded pool of discoloured water with a sand or silt shore, scattered with the remains of dead vegetation.

A special case of visual impact from waste disposal is that of leach dumps which are sometimes found at copper mines. Dumps of acid-bleached rock are of course totally devoid of vegetation and are one of the least attractive features of such mines.

3.3.3. Fixed Plant

The fixed plant installed at a mineral operation varies according to the mineral. At hard-rock quarries, it typically comprises crushing, conveying, screening and bunkering equipment, and such items as coating plants, chimneys, lime or cement kilns, ready-mix concrete plants, thickeners and road and rail loadout facilities. Metal mines require a wide range of grinding, classifying and processing circuits. Underground mines of all types may have headframes and haulage or winding equipment. Often these various components are disposed over a wide area to suit operational needs (such as minimising haulage distances, avoiding sterilisation of reserves, etc.), and the physical constraints of the site. Very frequently it is fixed surface plant which provides the greatest visual impact.

This is particularly so in the case of long-term operations where processing plant remains *in situ* for many years. Piecemeal replacement of various sections of plant and the addition of new facilities means that an uncoordinated appearance develops with the passage of time, regardless of any aesthetic merits the plant may have possessed at the time of its construction.

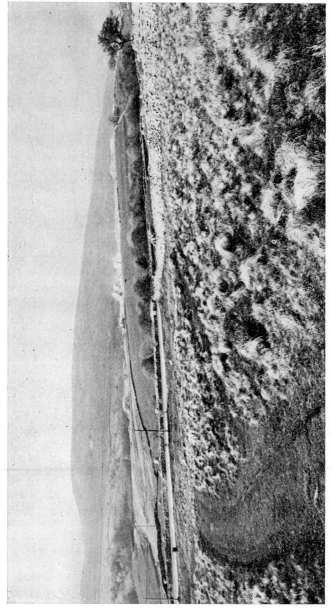

Fig. 4. Distant view of a quarry processing plant in a remote moorland area devoted mainly to tourism. Visual impact problems are acute in such localities.

Conversely, plant erected at short-term operations—sand and gravel pits, coal strip mines, etc.—may have an undesirable impact because of its temporary nature. Often such plant, especially at smaller sites, is semi-portable and attention to its appearance is economically and practically difficult.

Metalliferous mining illustrates slightly different problems. Although there is a history of housing much of the plant within one building or group of buildings—in contrast to many stone quarries at which much plant was not enclosed—the scale of the structures is usually much greater than for most other types of mineral workings.

Fig. 5.   This Yorkshire colliery demonstrates the contrast in visual intrusion between the conventional headframe and the tower-mounted winder.

It is very common for fixed installations to assume the colour of the mineral being processed, due to accumulation of dust and dirt. This can increase the visual impact.

A further important factor is the height of the plant. Many processes depend upon gravity flow, with conveyors to raise the ore up to the next stage of processing. Inevitably such plant is high. Chimneys may also be needed. Headframes at underground mines are also conspicuous and the modern tendency to use tower-mounted Koepe winders, rather than ground-mounted hoists, increases both the size and the height of the headgear (*see* Fig. 5).

3.3.4. Mobile Plant
Mobile machinery can give rise to visual intrusion due to the bright colours used when painting it. Such plant is usually rendered conspicuous for

safety reasons. Occasional nuisance may be caused by vehicles working in prominent locations.

3.3.5. Air and Water Pollutants

In addition to their importance a physical and chemical pollutants (Chapters 4 and 5), aerial emissions and liquid effluents can also cause aesthetic nuisance. Discoloured or turbid streams, dust plumes, etc., can be unsightly and act as long-range visual indications of the presence of mineral workings.

3.4. LANDSCAPE PLANNING

Planning of landscaping at mineral workings has essentially the same features whether the objective is to reduce the visual impact of an existing operation, or to minimise the impact of a new mine although, of course, the scope and flexibility of the remedial measures may be less in the former case. A landscaping plan is normally intended to enable one or more of the following objectives to be attained:

(i) minimum undesired visual impact throughout the life of the operation;

(ii) maximum benefit in respect of other environmental impacts such as noise or dust pollution;

(iii) economical and effective rehabilitation of the closed mine site to a productive after-use.

Landscaping cannot be considered in isolation from the numerous other environmental, technical and economic factors which are of importance in mineral workings. It is essential to take into account every relevant factor —and not merely aesthetic considerations—in the design and execution of landscaping works.[1]

One of the more important factors is time. Surface mineral workings are dynamic and a landscaping scheme which fails to take into account the ultimate as well as the existing extent of the excavation is likely to be unsatisfactory. Even underground mines often undergo major changes in surface plant during their lifetimes so that a landscaping plan which makes provision for plant extension is likely to be more successful.

Overall the process leading up to the production of a landscaping plan can be considered in two stages:

(i) general survey and information gathering in which the main features of the scheme evolve;

(ii) detailed planning. Stage (i) inevitably reveals conflicts between the various requirements of landscaping and economic mine operation, and these must be reconciled to give the most satisfactory overall result. Choices also arise between various ways of achieving the objectives of the scheme.

3.4.1. The General Survey

The general survey must take account of the broad features of the actual or proposed mining operation in relation to its surroundings.[13] The most important components of the survey are:

(i) Landscape Character

The type of landscape in which the operation is set is one of the basic determinants of the landscaping scheme. The scenery may be treeless, wooded, or copses with hedgerows; rugged, rolling or flat with or without natural rock exposures. Settlements, townships or isolated houses may exist. Two particular aspects of landscape character are the topography and the ecology.

(ii) Topography, Contours and Levels

The elevation of the mine site in relation to its surroundings has an important bearing upon its conspicuousness. Also the abruptness with which changes in level occur influences the design of artificial screening mounds as well as the extent to which they can be effective.

(iii) Ecology

A survey of the local ecology—that is, the components of the main plant and animal communities—can provide much valuable information. The types of vegetation which exist naturally in the area are likely also to be the types which survive best in plantings for landscaping purposes (unless planting into radically different soil types is anticipated). The survey thus gives preliminary indication of the possibilities for screening the mine site with vegetation. Treeless landscapes, for example, are unlikely to harmonise with landscaping plans requiring extensive tree plantings—and, moreover, the fact that the natural landscape is treeless, indicates that, for climatic or soil reasons, artificial tree plantings are unlikely to thrive without extensive maintenance.

An additional use of an ecological study is that it may indicate locations of particular scientific value, the preservation of which are desirable in planning the mine. These may range from small occurrences of rare species to major game migration routes or feeding grounds.

(iv) Hydrology

Sub-surface water influences the working of the pit and, particularly, the rehabilitation/landscape possibilities at closure. Surface water flows, which are often susceptible to pollution, are usually difficult to divert and thus tend to limit flexibility of mining and landscaping planning, although at a detailed level, water offers considerable landscape opportunities.

(v) Communications and Habitations

Fixtures of this type can place severe limits upon working and landscaping. The presence of habitations can impede the free disposition of mine facilities (due to noise and dust nuisance, blasting vibrations, etc.). Relocating public highways can be a most expensive and time-consuming process and very often these are left *in situ*. The unfortunate visual effect of a highway perched upon a spine, material having been excavated on each side, is well-known.

(vi) Boundaries

Ownership boundaries naturally limit the extent of the mining operation. Within large land holdings, ownership boundaries do not usually affect landscaping. However, if the site is confined, opportunities for aesthetic improvement may be circumscribed by such legal boundaries. Innumerable examples of this situation exist. Expansion of land holdings may be considered to overcome this problem or, if adjoining owners of minerals are also mineral operators, co-operative landscaping schemes can be devised. Not only does this course permit more satisfactory landscaping to be achieved, but it also avoids the sterilisation of mineral reserves along the boundary, thus proving of financial benefit.

(vii) Mining Plans

In conjunction with the information gained under (i)–(vi), details of the mine, as it exists and/or is planned, are required. The details required include: the mineral(s) to be worked (to enable any toxicity problems to be anticipated and revegetation possibilities assessed); method of working (quarry, strip mine, openpit, etc.); likelihood of serious subsidence occurring in underground mines; details of fixed plant (type, locations, noise output, etc.); extent of waste production and surface disposal facilities required; nature of waste; pit design at intervals throughout the life of the mine; transport facilities needed; services to be installed; and numerous other relevant aspects of the planned mine.

Virtually all of the information obtained can be stored upon topographic plans and sections. Depending upon the size of the developments, the scale may be 1/2500 or larger, usually up to 1/500. For complex proposals, or if the proposals are to be presented to the public, a scale model of the area is most valuable. Although costly, an accurate model not only aids public understanding of the scheme, but is of considerable assistance in designing many features.

The objective of the general plan is to discover the best ways of introducing a new development into the landscape with a minimum disruption to either visual amenity or the economic working of the mine. Although every facet of the information gathered may be drawn upon for this purpose, it is usually vital that critical viewpoints and the sight-lines into the mine development are explored at this stage.

Extensive on-site studies of the locality in which the mine is proposed are likely to disclose particular vantage points from which the mine will be highly visible—highways, laybys, public footpaths, etc.—to large numbers of people for short periods of time. Such locations are termed 'critical viewpoints' and the directions along which the observer would see the mine 'sight-lines'. Other viewpoints may be discovered such as dwellings, from which a continuous view might be obtained by the occupants only. Also there may be particular viewpoints from which certain features of the mine (*e.g.* fixed plant) would present an especially unfortunate visual effect and be a serious intrusion into the landscape.

Critical viewpoints are usually identified on site, but require verification by map or model. Vertical sections along the sight-lines can assist in determining whether structures of particular height will obtrude or not. In many climates it is important to take into account the time of year that the survey is being undertaken. A belt of mature deciduous trees will have a markedly reduced (if not negligible) screening effect in winter after leaf fall, and a summer survey can therefore be misleading. For this reason, it is sometimes helpful to predict the visual impact of the development on the basis of the terrain alone, taking into account vegetation as a secondary consideration. It is also necessary to incorporate the changes in the mine during its life, *e.g.* continued quarrying may cause intrusion upon new viewpoints many years after the implementation of a landscaping scheme.

3.4.2. Detailed Planning

The general survey often identifies matters of conflict between landscaping and other requirements. To resolve these, a more detailed study of certain aspects is needed.

Very broadly, visual impact problems may be lessened either by concealing the obtrusive feature, or improving its appearance, or a combination of both. It is, however, unrealistic to hope that most mineral operations can be totally concealed, and thus both approaches are usually required.

3.4.3. Concealment

Many aspects of the siting of mining and processing activities are sufficiently flexible to permit the selection of locations at which the maximum degree of natural screening is available. These are:

(i) Surface Excavation

This depends upon the mineral being worked. Many deposits of limestone, chalk, sand and gravel, etc., are areally of great extent and of relatively constant quality laterally. In such cases the precise location of the excavation is controlled to a large extent by ownership boundaries. It is frequently possible to orientate the working face to present the minimum visual impact from a critical viewpoint. Sometimes an area of winnable stone can be retained *in situ* as a screen, or the access cut can be curved to preclude a direct line of sight into it. Avoiding breaches in skylines or escarpments is usually a valuable contribution to the preservation of visual amenity (*see* Figs. 6–8).

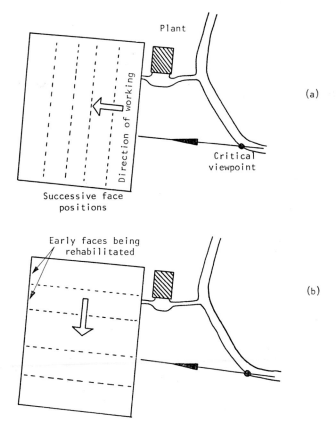

FIG. 6.   Orientating a quarry face to reduce visual impact. In (a) the quarry is worked along the sight-line and hence is visible throughout its life. In (b) the direction of working is across the sight-line so that disused faces can be rapidly rehabilitated.

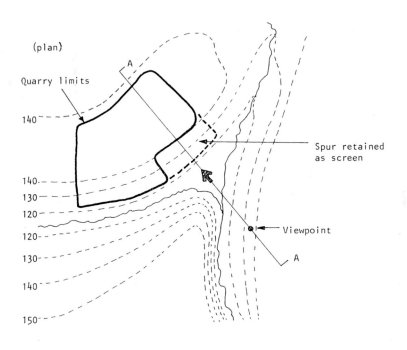

(plan)

Quarry limits

140

140.

130

120

120

130

140

150

Spur retained
as screen

Viewpoint

A

A

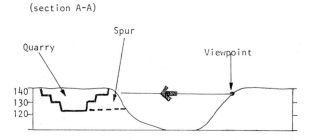

(section A-A)

Spur

Quarry

Viewpoint

140
130
120

FIG. 7.   Plan and section showing how the retention of a spur of ground can
provide screening.

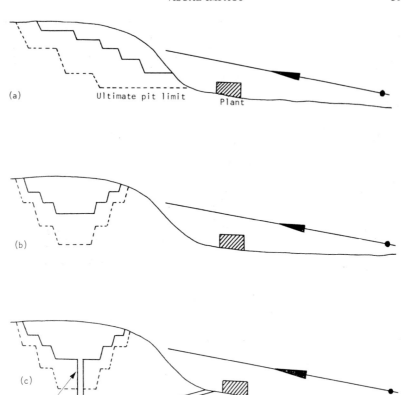

FIG. 8. Alternative methods of working a scarp site. In (a) the excavation is highly conspicuous. Preserving the scarp (b) conceals the excavation but haul roads would still be visible. In (c) the glory-hole method provides complete concealment, at increased cost.

If a deposit of this nature is massive in depth, it may be possible to limit visual impact problems by working in depth rather than laterally. Any aesthetic advantages during working may however be outweighed by hydrological factors and difficulties of rehabilitation. A further alternative, in the case of deposits in which lateral flexibility is limited by overburden depth, is to substitute underground mining for surface excavation. Many limestone mines in the U.S.A. and Britain have originated in such circumstances, but this would not be an economically feasible solution to a landscaping problem unless other, more important, factors were favourable.

A contrary situation arises in the case of minerals such as non-ferrous metals, or vein/seam deposits such as fluorspar and coal. Flexibility in locating the excavation is very limited, and almost any change from the mining optimum will result in wasted ore and serious detriment to the mine's economic feasibility. It is, however, sometimes possible to undertake dead work, such as access cuts, at visually acceptable points (Fig. 9).

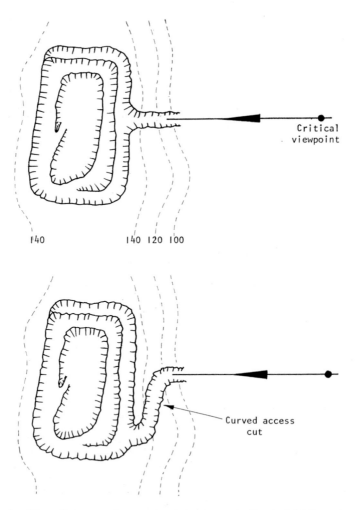

FIG. 9. Use of a curved access cut to prevent direct sight-lines into an excavation.

(ii) Tips

The location of surface waste tips can be chosen with a view to minimising visual impact but, except in fairly hilly or wooded terrain, there may well be inadequate natural screening.

(iii) Fixed Plant

Some surface plant can be relatively freely disposed around a site to meet visual requirements (e.g. repair shops, office blocks) whereas other equipment (processing plant, headframes, etc.) may be less flexible. The items least amenable to relocation from the technically optimum site are usually shafts and primary crushers; other sections of plant can be sited to take advantage of natural landscape features. Control of the height of the plant, so far as practicable, is of assistance.

It is therefore possible to adapt certain features of the mine design to benefit from whatever natural screening exists. It would be unusual for this to be wholly sufficient, and it is very common to construct artificial screening to conceal particular aspects of the operation. Screening can be achieved either by vegetation or banks or a combination of both.

(a) Vegetation Screens

The use of vegetation (usually trees) for screening purposes is common. If the screening function is to be fully performed it is important to realise the long term nature of this remedy.[3] Trees may be planted at any age, but are normally transplanted as young specimens less than about 3 years old, or else as semi-mature trees about 10–15 years old. Each method has its advantages. Young trees (1–2 years) can be purchased and planted at a cost per hectare of 2500 trees (1000/acre) of £400–£600, including site preparation. Maintenance (particularly weed control) will vary, often from £15 to £30 per hectare (£6–£12/acre) per annum for about five years from planting. After this time, the trees should be more or less maintenance free. Growth rates depend upon the species of tree planted and the site conditions but may be between 10 cm and 50 cm (4–20 in) per annum. A successful establishment rate of about 75–85% would be expected under the conditions typically found at mineral workings.

Even at an annual growth rate of 50 cm (20 in) height, it may well be a decade before the trees are high enough to screen anything. Consequently there can be advantages in using semi-mature trees specially grown and transplanted for the purpose. However, such trees can cost £50–£80 each. If planted at a density of about 1000 per hectare (400/acre), the total cost of such a scheme would be £50 000–£80 000/ha (£20 000–£33 000/acre) with maintenance commonly much higher than for young stock. Costs of this magnitude mean that semi-mature trees are rarely used at all at mineral workings—and never on any scale—being restricted to minor 'gardening' schemes at points where a rapid visual gain is desired. Moreover, semi-mature trees frequently 'go into check' upon transplanting, *i.e.* cease to grow. Check conditions may persist for several years and it is common for 1–2 year trees to overtake semi-mature trees thus rendering their high cost almost wasted. Details of methods of planting are given in Chapter 10.

Therefore, tree screens are normally composed of young stock. There are several disadvantages to be considered. These are:

(a) Choice of species is limited by soil conditions, climate, etc.
(b) Source of stock requires careful consideration; for example trees raised in lowland, sheltered nurseries are unlikely to thrive at upland, exposed mine sites.
(c) Deciduous trees may be the natural type found in the area, but evergreen conifers are, in most cases, the only trees to give all-year screening. The ability of deciduous trees to screen is much reduced in winter.
(d) Planting of 1–2 year stock needs to be undertaken at least a decade in advance of the need for screening. This implies that mine development also needs to be planned over the same timescale.
(e) Tree screening is very difficult if the mine is close to or above the natural tree-line altitude.
(f) Trees offer virtually no benefit in terms of noise reduction.

Nonetheless, in many circumstances a tree screen can be an ideal solution to visual impact problems. They have the benefit that, after initial establishment, they are almost maintenance-free. Unless extremely inappropriate species are chosen, trees are automatically accepted as 'natural' features of most landscapes, while they also offer such side benefits as the provision of wildlife refuges. Cases in which it is possible to augment and extend a nucleus of existing woodland have shown trees to be a very successful solution (see for example the case study in Section 3.5.1).

(b) Screening Banks

The construction of screening or banks of soil and overburden to conceal parts of mineral workings is an extremely common method of lessening visual impact. In Britain, a high proportion of large surface mines (notably hard-rock quarries, sand and gravel pits and open cast coal mines) construct amenity banks as a routine means of environmental improvement. The popularity of such banks is due to the relative ease of constructing them with labour and equipment already used at the mine, the fact that they form a convenient repository for waste material, and the rapidity with which the screening effect can be obtained—limited only by the earthmoving capacity which is applied to the task.

Amenity banks can perform a noise-baffling function in addition to aesthetic benefits if they are located appropriately (*see* Chapter 6). The full visual benefits are, however, only obtained if the bank is properly designed, particularly with respect to landform. A badly-contoured bank may be as intrusive as the feature which it is supposed to conceal. Because a bare earth bank would seldom be appropriate, it is normally vegetated, and the design of the bank and its plant cover are part of the same process.

The general survey should provide the basic data required for bank design. It should indicate:

(a) the location for the bank on the basis of the sight-lines which have been determined;

(b) the quantities (and rates of production) of overburden which will be available to build the bank;

(c) the land area available and other site characteristics;

(d) the landform which would be most appropriate to the particular setting.

Very frequently these factors are in conflict. Perhaps to match the bank contours to the natural topography would require an excessively large area of land, or a greater volume of waste than will be available, or slopes approaching or in excess of the natural angle of repose of the waste material. The ideal site to obtain the screening may be unsuited to bank construction because of, say, water-courses or marshland. In some cases, the waste material may be toxic and not amenable to revegetation. From a synthesis of all these matters, plus specific site features, the design of the landform of the bank can be obtained. Inevitably it is usually something of a compromise. Several aspects of the design process require emphasis:

(a) A choice may exist between a 'natural' or a 'geometric' landform. Subjective preference and site variations make generalisation difficult but a fairly widespread opinion is that it is preferable for large banks to be natural in form, *i.e.* replicating and imitating the existing contours in the area.

(b) To obtain sufficient height without exceeding the amount of over-burden, the space available, or the angle of repose, is often difficult. The problem can be overcome by building a bank which is not high enough (but otherwise meets the constraints imposed on it) and making up the height deficit with vegetation.

(c) An important choice which may exist is between locations near to the observer, or near to the feature being concealed. The former permits the same screening effect with a smaller bank, but usually results in *all* views being obscured, not merely the particular sight-line into the mine. The possibilities are shown in Fig. 10.

(d) If banks can be designed to act as noise baffles, this is advantageous. They are normally effective only when the noise source is close to the bank.

(e) The banks may be classed as 'temporary' or 'permanent' and short-life banks may be commensurately less well landscaped.

The nature of the available wastes has an important bearing upon the design of the bank. A stable structure is vital and this may cause slope angles substantially below those of natural repose to be selected, regardless of landscaping criteria. If toxic materials are present (coal shales, metal bearing wastes, etc.) these inhibit revegetation and should therefore be either avoided or buried. The surface layers of the bank should comprise those wastes most conducive to vegetation establishment.

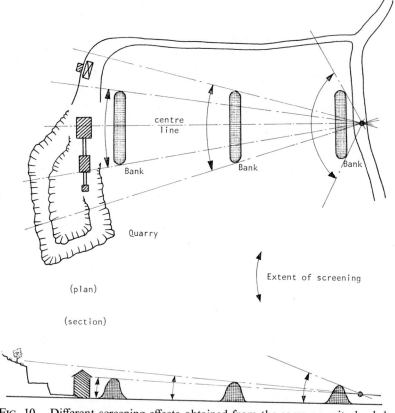

Fig. 10. Different screening effects obtained from the same amenity bank by varying its location.

Revegetation is considered in more detail in Chapter 10, but it should be noted here that rapid planting is important if soil erosion is to be prevented. Even if a shrub or tree cover is intended, preliminary seeding with grasses can be beneficial from this point of view. So far as possible, the plant cover selected should be appropriate to the locality, to achieve maximum visual integration of the structure.

3.4.4. Improving Existing Features

Modern mineral workings are on a scale that does not permit total concealment and thus it is also necessary to reduce the visual impact of features which will remain visible, by altering existing practices or designing for aesthetic appeal. The components of a mining operation which lend themselves to improvement are discussed below:

(i) Solid Waste Tips

Conventional tipping practice creates tips with the maximum visual impact, because the tip is built outwards to its final perimeter and therefore presents fresh waste surfaces throughout its life. In certain circumstances it is possible to reduce substantially the period during which the tip is devoid of vegetation by regrading and replanting as the tip progresses. Such methods have been employed in Britain by the National Coal Board and its Opencast Executive for tips at both open cast and underground mines. The main features of the method are shown in Fig. 11.

A technique which has considerable visual benefits is perimeter tipping. It is successfully operated in the West German brown coal mines and is illustrated in Fig. 12. It consists of the construction of an amenity bank around the perimeter of the tipping area, the outer face of the bank being rapidly revegetated. Within the screen so formed, conventional tipping can proceed. Further lifts can be added.[11] It is important that the volumes of overburden are known, in advance, with precision, because either a shortfall or a surplus of material would reduce the benefits of a perimeter tip.

(ii) Tailings Impoundments

Landscaping of tailings impoundments must be subordinated to the over-riding requirements of safety in design and operation of the facility. Nonetheless some scope for landform modelling exists (Fig. 13), but because many dams are progressively constructed during the life of the mine, landscaping is not always feasible until closure. Choice of site is probably the main way of controlling the visual impact of tailings dams.

(iii) Fixed Plant

The change over the last century in the external features of fixed plant at mines illustrates both the alteration of architectural taste and the influence of increased cost of traditional materials. In Britain embellishment of the most mundane buildings was commonplace—crenellated

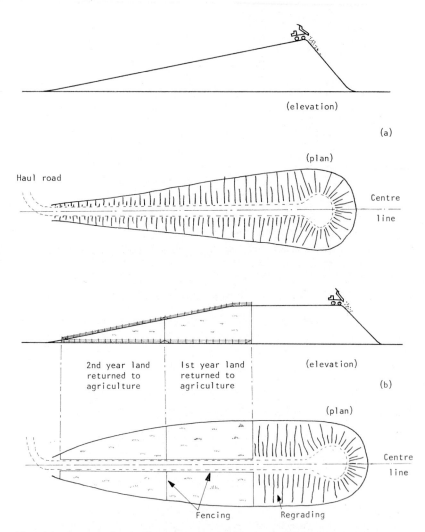

FIG. 11. Progressive landscaping of tips: (a) conventional tipping methods; (b) progressive regrading and seeding.

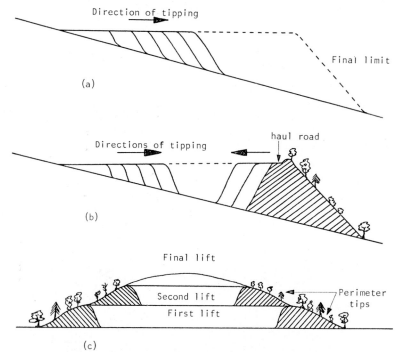

FIG. 12. The use of perimeter tipping to reduce visual intrusion: (a) conventional forward tipping; (b) perimeter tipping in the same situation; (c) method of perimeter tipping in flat landscapes.

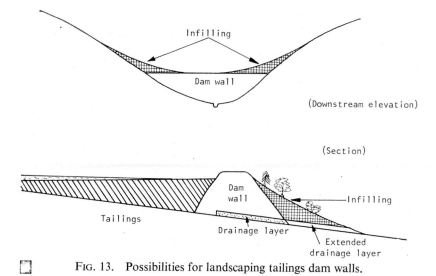

FIG. 13. Possibilities for landscaping tailings dam walls.

colliery boiler chimneys, Gothic adornments, elaborate brickwork and masonry—to the extent that many industrial buildings are now considered to be part of the country's architectural heritage. Such designs have given way to a preference (partly enforced by economic considerations) for severely functional structures. Buildings have to perform more functions than merely protecting the plant they house: provision of an acceptable working environment for employees, noise shielding and dust containment, as well as aesthetic requirements, have all to be considered during building design.

FIG. 14. This small, modern tin mine (Mount Wellington, Cornwall) is prominently situated, but its visual intrusion is reduced by the simple, uncluttered design of the surface facilities and the muted colours selected for them.

A fairly standard modern construction is a steel-framed structure with some type of lightweight external cladding, commonly either plastic, asbestos or steel sheet. To provide the additional properties required, an internal cladding or lining is often incorporated; for example, polyurethane foam with plaster board on the inside face, for thermal insulation.

The external texture of buildings is, except when viewed at close range, of less importance than the colour. The exact colours chosen depend upon the locality but it is usually found that muted pastel tones—greys, browns, olive green, sand, natural concrete, etc.—are most effective in rendering the plant inconspicuous (Fig. 14). Many cladding materials have a permanent colour incorporated in them. A wide range of colours is often obtrusive and limiting the colours used to only two or three—possibly associating them with the function of the structure (e.g. olive green mill buildings, dark brown conveyor housings)—is beneficial. In some cases it may be possible to use a colour which matches the colour of the dust found around the plant. This renders the dirt which accumulates upon buildings unobtrusive, although periodic cleaning is still required.

FIG. 15.   The surface buildings of a large modern colliery (Killoch, Ayrshire, Scotland) are inevitably a major intrusion in a rural landscape. Careful attention to architectural design has greatly improved this mine's appearance.

The disposition of plant around the site depends upon circumstances, but a regular grid layout can enable a unity of design to emerge. Compact layouts are usually more pleasing. An unfortunate effect is sometimes created by a multiplicity of conveyer belts connecting the various sections of the plant. It has been found that, if the slopes of conveyers are standard-ised as far as possible, and roofs of buildings are pitched to the same angle, the design has greater harmony when seen in silhouette. The 'scissors' effect where opposing conveyers cross is best avoided.

If the height of buildings can be reduced this usually reduces their impact. Excavation for foundations may permit buildings to be slightly sunk into the ground. In conjunction with external screening measures, lowering buildings may enable some to be concealed to a far greater degree. Some structures such as headframes, silos and chimneys are not usually amenable to height reduction and in such cases a clean, uncluttered design is the best remedy.

3.5. CASE STUDIES

The following case studies have been selected in order to illustrate some of the many aspects of landscaping at mineral workings.

3.5.1. Boulby Mine, U.K.

Cleveland Potash Ltd obtained planning permission in 1968 to construct a potash mine and processing plant within the North Yorkshire National Park. The environmental aspects of the mine planning, with particular reference to landscaping, have been described by Cleasby et al.[9]

The mine was designed to produce 1 500 000 t/a potash from a depth of 1070 m (3500 ft) and comprises two vertical shafts, ore storage silo, treatment plant and product-storage silo. The location of the mine within a National Park presented many environmental problems and site selection and landscaping were thus major considerations.

Site selection was influenced also by the need to obtain rail access, to dispose of tailings at sea, as well as by geological and mining factors. With the cost of shaft sinking and equipping some 25% of the total project cost of £40 million, any change from the technically optimum site could only be made at considerable cost.

Three possible sites were located, all within an undulating, farmland coastal belt between the moorland and the sea cliffs. For each site, visual surveys were made to ascertain critical viewpoints and the impact of the proposed development from each. Two sites were ruled out, primarily because the mine buildings would be silhouetted against the sea from many viewpoints. The third site, at the mouth of a wooded valley, avoided this problem, with the buildings being seen against a land background (Fig. 16).

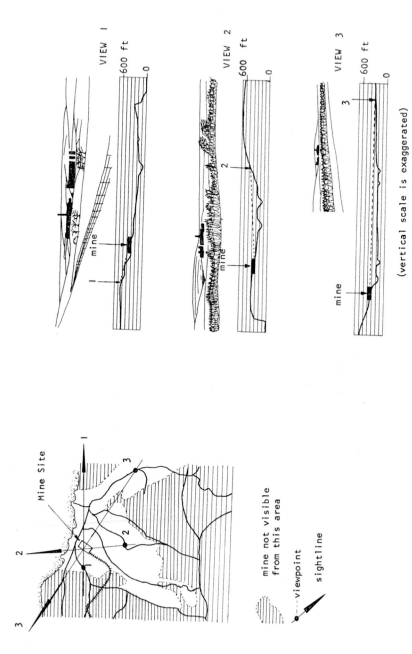

Fig. 16. Visual survey of the proposed site for Boulby mine (Cleasby et al.[9]).

(vertical scale is exaggerated)

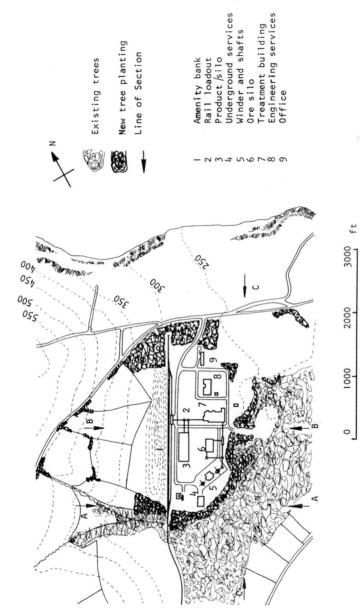

FIG. 17.   Final surface layout selected for Boulby mine (Cleasby *et al.*[9]). Section lines refer to Fig. 18.

Detailed site planning required close co-operation between management, technical staff, architects and landscape architects, and the result was inevitably a compromise between the unfettered requirements of each. The final layout achieved is illustrated in Figs. 17 and 18. Points of note include the very simple, rectangular layout, which is very compact (the area actually occupied is less than 35 ha [90 acres]), and the placing of the shafts at the extreme southern perimeter. This feature, though less desirable from a mining point of view, permitted the storage silos, treatment plant and interconnecting conveyers to be concentrated in one area. To avoid a confusing variety of angles, the slopes of conveyers and the pitch of roofs have been made the same ($15°$) wherever possible. The colours chosen for the mine buildings were greys, brown and buffs. Sight-lines down the valley slope to the mine have been interrupted by the construction of an amenity bank, while additional screening has been successfully incorporated by tree plantings as extensions to existing woodland. A small but important feature is the provision of a sinuous access road, thus preventing direct views into the mine site from the public highway.

3.5.2. Tara Mine, Ireland

The discovery and development of a major underground zinc/lead mine at Navan, Ireland, by Tara Mines Ltd is one of the most recent manifestations of Ireland's new importance as a non-ferrous mineral producer. Planning of mine development began in 1973 and it is anticipated that the mine, producing about 2 500 000 t/a ore, will commence production of concentrates in 1977. Although the mine is located in agricultural surroundings, it adjoins Navan town and a busy main road. Landscaping has thus been of great importance.[10, 12]

The company considered that, in dealing with visual impact problems, both the building complex and the immediate natural environment should be modified in order to obtain compatibility between these two elements, and with the landscape at large. Figure 19 shows the general layout adopted for the mine.

It is noteworthy that the layout of the buildings follows a strongly rectangular, grid pattern. Particular care has been taken to plan conveyer runs to avoid prominence. Buildings are to be clad in dark-brown insulated metal sheeting. Cylindrical bins are to be straw coloured, with blue for conveyers.

Facilities which are, normally, either unenclosed or only functionally enclosed within utility structures will be housed in custom-designed buildings—notably the coarse ore storage area. This provides both noise and dust control benefits as well as visual improvement.

Surface structures will be set within a new landscape composed predominantly of native species of trees and shrubs. Ornamental species and conifers will also be utilised for certain purposes. Earth amenity banks are

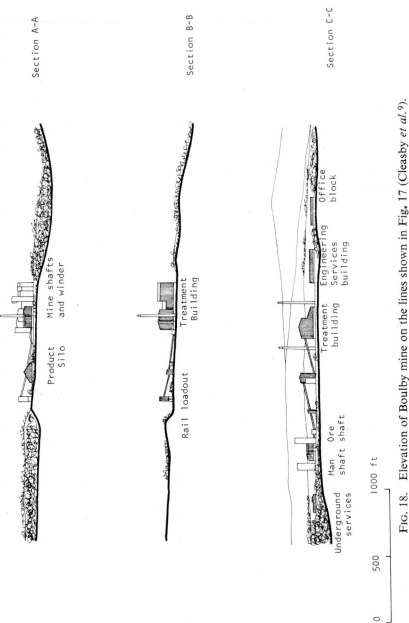

Fig. 18.   Elevation of Boulby mine on the lines shown in Fig. 17 (Cleasby *et al.*[9]).

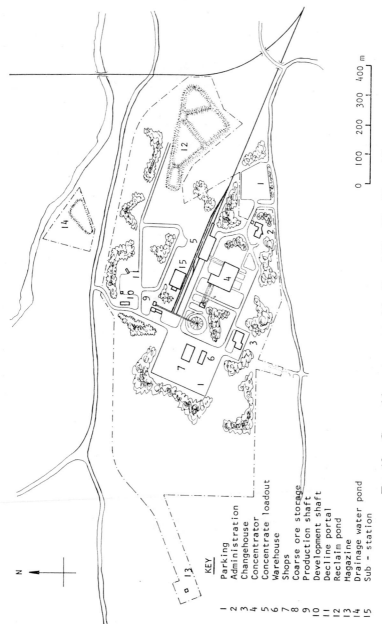

KEY

1 — Parking
2 — Administration
3 — Changehouse
4 — Concentrator
5 — Concentrate loadout
6 — Warehouse
7 — Shops
8 — Coarse ore storage
9 — Production shaft
10 — Development shaft
11 — Decline portal
12 — Reclaim pond
13 — Magazine
14 — Drainage water pond
15 — Sub - station

FIG. 19.   General layout proposed for Tara Mine, Eire (Frame[10]).

incorporated in the design, and these have been planned specifically to take account of noise attenuation as well as aesthetic benefits. To service the landscaping works, a tree/shrub nursery was established at an early stage: some 70 000 individuals of 70 species are held here, so that rapid progress will be possible with the landscaping as soon as mine development permits.

REFERENCES

1. Haywood, S. M. (1972). Landscape, *Quarry Managers' Journal*, **57**(3), 100–10.
2. Orr, J. G. (1970). Mineral extraction and the countryside, *Cement Lime & Gravel*, **45**(4), 83–8.
3. Clouston, J. B. (1971). The role of the landscape architect in quarrying, *Quarry Managers' Journal*, **55**(2), 39–44.
4. Jacobs, P. (1975). The landscape image, *Town Planning Review*, **46**(2), 127–50.
5. Casson, J. (1957). Landscape conservation in Lancashire, *Surveyor*, **116**(14), 953–9.
6. Fish, B. G. (1973). Quarrying and the landscape, *Cement Lime & Gravel*, **48**(10), 213–15.
7. Herbert, J. R. (1972). Landscaping in relation to disposal of waste materials, in *Aspects of Environmental Protection* (ed. S. H. Jenkins), I.P. Environmental, London, 433–42.
8. Downing, M. F. (1971). Landform design, in *Landscape Reclamation, Vol. 1* (University of Newcastle upon Tyne), IPC Science & Technology Press, Guildford, 32–42.
9. Cleasby, J. V., Sir Frederick Gibbard, Grimes, J. V. and Cuthbert, K. H. R. (1975). Environmental aspects of Boulby mine, Cleveland Potash Ltd, Yorkshire, in *Mining & the Environment* (ed. M. J. Jones), Institution of Mining & Metallurgy, London, 691–714.
10. Frame, C. H. (1974). Construction and operation of a mining complex in an urban area, in *21st Ontario Industrial Waste Conference*, Toronto, 35 pp.
11. Corner, J. T. (1968). Waste tip stabilization in the Ruhr, *Colliery Guardian*, **216**, 250–3.
12. *A Question of Environment* (1972). Tara Mines Ltd, Navan, 28 pp.
13. Haywood, S. M. (1974). *Quarries and the Landscape*, British Quarrying & Slag Federation publication INF10, London.

4

Air Pollution

4.1. INTRODUCTION

Man and most other organisms are entirely dependent upon air which is, moreover, the commodity required in the greatest quantity. About 14 kg (32 lb) of air per day are required by an adult, against 2 kg (5 lb) of water and 1 kg ($2\frac{1}{2}$ lb) of foodstuffs. A lack of air has an instantly deleterious effect, rather than the delayed results of deprivation of food and water.

Air is a mixture of about 13 gases, the three main components being nitrogen (78%), oxygen (21%) and argon (0·9%). Gaseous contaminants have usually been expressed as parts per million by volume, but now the concentrations of all contaminants are widely expressed as weight per unit volume, normally $\mu g/m^3$.

Air pollution caused by human activities has been a significant nuisance for nearly 700 years. The single most important reason for the increased seriousness of air pollution today is man's continued urbanisation. In 1950 it was estimated that half the world's population was rural, and half urban, while predictions suggest that the urban population will increase to some 80% of the total by the year 2000. As a result of the growth of urban communities there has been a marked concentration of aerial emission sources. The use of coal, originally for direct heating and today as an electricity source, results in the concentration of power generation in close proximity to cities. Likewise, the rise of the internal combustion engine as the motive force for almost all road and most rail transport causes exhaust emissions to be localised in cities. Thus, in the U.S.A., it has been estimated that half of the total emissions are released over less than $1\frac{1}{2}$% of the total land area.

The available statistics of air pollution are inadequate in scope and open to misinterpretation. Table 3 lists the main quantities of pollutants released by different industries in the U.S.A. It is clear that mining as such is too small to figure in these categories (other than coal cleaning), but secondary processing of mineral products by smelting and cement manufacture is a major source of industrial air pollution.

To place the industry figures in context, Table 4 gives estimates of the total aerial emissions from all sources in the U.S.A. In terms of overall tonnage, industry emits only 14% of the total pollutants. Thus, with coal

57

TABLE 3
QUANTITY OF AIR POLLUTANTS FROM MAIN U.S.A. INDUSTRIES
(derived from Ross[50])

| Industry | Total pollutants/annum | | (%) |
	$lb \times 10^9$	$(kg \times 10^9)$	
Petroleum refining	8·4	(3·8)	18
Non-ferrous smelters	8·3	(3·7)	17
Iron foundries	7·4	(3·4)	15
Kraft, pulp and paper mills	6·6	(3·0)	14
Coal cleaning and refuse	4·7	(2·1)	10
Coke (in steel manufacturing)	4·4	(2·0)	9
Iron and steel mills	3·6	(1·6)	7
Grain mills and grain handling	2·2	(1·0)	5
Cement manufacture	1·7	(0·8)	4
Phosphate fertiliser plants	0·6	(0·3)	1
Total	47·9	(21·7)	100

cleaning the sole categorised mining process, producing 10% of industrial pollutants, no more than 1·4% of the total U.S.A. air pollutant load is attributable to mining.

This is, however, a very misleading statistic. The practical importance of a pollutant depends upon its nature, concentration, and the point at which it is emitted. Thus sulphur oxides are very much more damaging than

TABLE 4
ESTIMATED AIR POLLUTION, BY TYPE AND SOURCE, IN U.S.A. IN
1970
(Council on Environmental Quality[57])

| Source | Pollutant (tonnes $\times 10^6$)/annum | | | | | |
	CO	Particulates	SO_x	HC	NO_x	Total
Transportation	111·0	0·7	1·0	19·5	11·7	143·9
Fuel combustion (in stationary sources)	0·8	6·8	26·5	0·6	10·0	44·7
Industrial processes	11·4	13·1	6·0	5·5	0·2	36·2
Solid waste disposal	7·2	1·4	0·1	2·0	0·4	11·1
Miscellaneous	16·8	3·4	0·3	7·1	0·4	28·0
Total	147·2	25·4	33·9	34·7	22·7	263·9

carbon monoxide, but either is more dangerous when emitted at ground level, where they are easily inhaled by man, than at more elevated positions. Comparison of total pollutant emissions, without qualification, is thus unhelpful.

There are five main atmospheric pollutants. Carbon monoxide (CO) is largely generated by the incomplete combustion of hydrocarbons in petrol engines. Hydrocarbons (HC) similarly arise from partial combustion of fossil fuels. The oxides of nitrogen and sulphur (NO_x, SO_x) are created by the burning of fossil fuels, particularly those containing sulphur. Particulates is an imprecise term including solid or liquid particles arising from combustion, abrasion or disturbance.

FIG. 20. Air pollution from a quarry processing plant; much of the visible emission is in fact steam from the coating plant.

Mining produces all these pollutants to some extent, but by far the most important and widespread are the particulates. Sulphur oxides (SO_2 in most cases) arise from smelting and are usually one of the most difficult environmental problems associated with that industry. Only in very rare cases have CO, NO_x or HC pollutants been known to cause external problems at mines, although CO and particulates can present very severe health hazards to mine employees at their places of work.

Full understanding of air pollution problems can only be obtained if the main features of the receiving atmosphere are comprehended. All emissions are made to the troposphere which extends from the surface of the earth to an altitude of about 12 km (7 miles). There is a fairly regular decrease in temperature with height (the lapse rate), from about 13°C at ground surface to −50°C at 10 km (6 miles). Above the troposphere is the stratosphere, to an altitude of about 50 km (30 miles), in which the temperature increases to about −10°C.

The nature of the lapse rate is extremely important in its effects upon pollutant dispersion and dilution (*see* Fig. 21). The adiabatic lapse rate (*i.e.* the decrease in temperature due to the expansion of the air with increasing altitude with no heat exchange) is about −1°C per 100 m (330 ft).

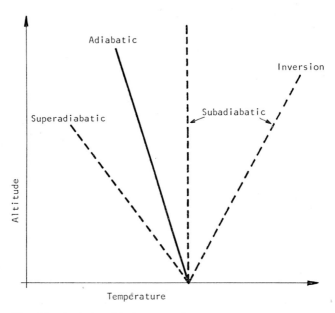

FIG. 21.   Relationship between air temperature and altitude.

If the actual decrease is faster than the adiabatic lapse rate, then the atmosphere is unstable and rapid mixing of the air occurs. If the decrease is slower than the adiabatic lapse rate, then stable conditions arise, culminating in the situation where there is an increase rather than a decrease in temperature with height. This is an inversion, and is of great significance in impeding the rapid dispersion of pollutants. The illustrations in Fig. 22 show how these conditions affect the behaviour of smoke plumes. As well as inversions, wind speed and direction, climate and topography all bear upon air pollution. These matters are discussed in more detail in Section 4.2.

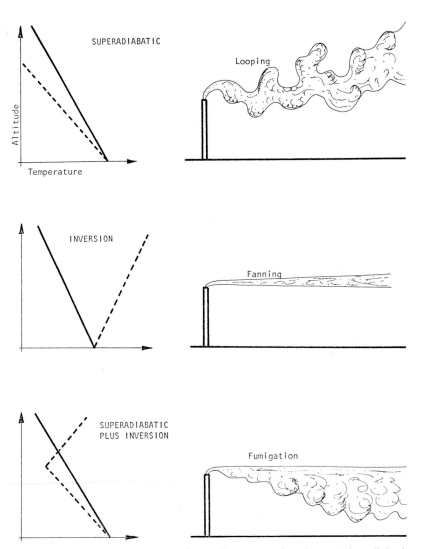

FIG. 22. The effect of atmospheric stability on smoke plumes. The adiabatic lapse rate is indicated by the solid line on the graphs, and the actual lapse rate by the broken line.

4.2. MINING POLLUTANTS

4.2.1. Nature and Effects of the Main Pollutants

Pollutants can be classed as gaseous or particulate, although in theory the dividing line is imprecise. Table 5 illustrates the range of particle diameters found in a variety of common substances; compared with gases such as CO_2 and SO_2 which have molecular diameters of about $0\cdot0005$ μm (5 Å), some particulate emissions are not dissimilar. For practical purposes however a division on the basis of particle diameter is valid, not least because control equipment differs on the same basis. Some types of air pollutant, *e.g.* smoke, contain both gaseous and particulate components.

TABLE 5
PARTICLE DIAMETER OF SOME COMMON
MATERIALS
(after Ross[50])

Substance	Diameter (μm)
Smog	<2
Clouds and fog	2–60
Mist	60–200
Drizzle	200–400
Rain	>400
Clay	<2
Silt	2–20
Fine sand	20–200
Coarse sand	200–2 000
Tobacco smoke	$0\cdot01$–1
Coal dust	1–100
Metallurgical dusts and fumes	$0\cdot001$–100
Ground limestone	10–1 000
Pulverised coal	3–500
Cement dust	3–100
Flotation ores	10–200

(a) Gaseous Pollutants

The gases most often of concern in mining are CO and SO_2.[28] Although CO_2 is also emitted by many processes it is seldom considered as a pollutant because its immediate effects seem negligible. Indeed, because CO_2 is a requirement for plant growth it has been argued that its release is beneficial. Opposing this view is the 'greenhouse' theory which suggests that increased CO_2 emission will cause an increase in temperature of the earth and melting of polar caps. Neither viewpoint has much observed data to back it and, because CO_2 control in mining is not an issue, it is not discussed.

CO is the world's most important gaseous pollutant. It is a colourless, odourless gas which, at high concentrations, causes death in humans by blocking the oxygen transport system of the blood. As such, the gas has been a major cause of accidental death among coal mining workers. At lower levels, sub-lethal effects occur: in urban areas, where 50 ppm CO is not an uncommon ambient level, impairment of the ability to choose, reduction of visual acuity, lassitude, etc., are among the effects. All are reversible by removing the person to air containing lower CO levels. There is little information on the effects of CO on other organisms and, as with CO_2, it can be doubted that CO has significant effects upon the external environment of the mine.

In contrast, the SO_x group of pollutants is highly destructive and common in several sectors of the mineral industry, particularly metal smelting and coal mining, but also brickworks and iron-ore sintering. The largest single source is the burning of high-sulphur fuels in power stations. The effects of SO_2 are felt by both man and other organisms at low concentrations of 0·3 ppm and respiratory irritation becomes marked at 5 ppm. Many plants are damaged by levels of 0·3–0·5 ppm, particularly conifers which are more sensitive than most other plants. Thus, at the smelters situated in the northern coniferous forests, a common feature is the destruction of trees around the smelters. One of the largest non-ferrous smelters in the world, at Trail, British Columbia, opened in 1896 and by 1925 was emitting 6300 tonnes of SO_2 per month, peaking at 9300 tonnes in 1930. When the area was studied in 1929–36, virtually no conifers survived nearer than 19 km (12 miles), while evidence of retarded growth of some species was found as far as 63 km (39 miles) south of the smelter. The effects on trees of SO_2 include injury to foliage and flowers, acidification of the soil, and eventual mortality of the tree.[26,45]

The position is complicated by the fact that, in the atmosphere, SO_2 can react with oxygen to form SO_3. This is strongly hygroscopic and combines with moisture to form sulphuric acid. The rates of these reactions are influenced by, among other things, the presence of particulates—such as ferrous iron and manganese-rich particles—which catalyse the formation of acid. The toxic effects of SO_2 are magnified 3–4 times in the presence of particulates.

When this H_2SO_4 is removed from the atmosphere, usually as an acid rain, damage may be caused remote from the initial source of the pollutant. Acid rain over Scandinavia derives in part from SO_2 emissions in the U.K. On a different scale, studies of remote lakes in Ontario suggest that acid fallout from the Sudbury nickel smelters, 65 km (40 miles) north-east of the study area, has acidified the lakes and destroyed fish populations.[48]

In addition to such biological effects, SO_x emissions also cause adverse chemical reactions in steel structures, power lines, some roofing materials, many natural stones, cement and paint.

There are many isolated instances of gaseous pollutants other than those above causing damage. One which deserves mention is fluorine compounds (gaseous and particulate) emitted from brickworks, often as hydrogen fluoride. This can cause damage to vegetation as well as fluorosis in livestock.[3]

(b) Particulates

Particulates result from the disintegration of solids and, if liable to remain in the atmosphere, may cause pollution. Liquid droplets, which behave similarly to solids, are also included in this category. Any process of combustion drying or abrasion, or any operation which allows particles to move in air or vice versa, has the potential to produce particulate pollutants. Mining, because it requires the excavation, comminution and transport of solids, has therefore an inherent tendency to cause this type of nuisance.

Particulates are classified by size as follows:

<0.1 μm diameter, usually resulting from combustion processes. Termed 'aerosols'. Undergo random (Brownian) motion and never settle; however, they coagulate by collision, and increase in size.
0.1–1.0 μm diameter, formed by condensation of vapours. Not affected by Brownian motion, but have settling time measured in months.
>1.0 μm diameter, and especially above 10 μm particles are formed by abrasion of solids, as well as by agglomeration. In these size ranges, particles have definite settling velocities.

Up to about 200 μm size, particles in suspension tend to fall out at uniform speed in calm air, because the acceleration due to gravity is counteracted by the friction of the air.

This constant speed is called the terminal velocity. For coarse particles above about 150–200 μm, this speed is found from Newton's law:

$$V_N = \left(\frac{8}{3} g \cdot D \frac{m_1}{m_2} \right)^{\frac{1}{2}}$$

where g = acceleration due to gravity, D = diameter of the particle, m_1 and m_2 = the specific masses of the dust and the carrier liquid and V_N = the terminal velocity.

For finer particles, Stokes' law applies:

$$V_S = \frac{D^2 g}{18\eta} (m_1 - m_2)$$

where η = viscosity of air and V_S = the terminal velocity. Examples of terminal velocities are given in Table 6.

The importance of particulate pollutants relates to several considerations. Especially in the case of human health, it is the actual concentration of airborne dusts within the respirable size range which is relevant. This range is taken to include all particles less than 5 μm diameter because these are likely to pass through the nasal tracts and enter the lungs. This is particularly so of particles smaller than 0·5 μm, while those larger than 5 μm are generally intercepted in the nose. Respirable particulates are of especial concern to employees. The other effects of dusts while in the atmosphere relate to visibility which can be seriously reduced.

TABLE 6
SETTLING VELOCITIES OF PARTICLES
$m = 1$ IN AIR

Particle size (μm)	Falling speed
200	1·2 m/s
100	0·3 m/s
50	70 mm/s
10	3 mm/s
5	0·7 mm/s
1	30 μm/s
0·5	7 μm/s

For other reasons, concern may be directed at those non-respirable particles which settle out. Fallout of particles causes discoloration of buildings, damage to paintwork, contamination of soils, vegetation and water, and mortality in livestock or other organisms.[27] The precise effects depend on the nature and concentration of the particles which are deposited.[2,29]

The effects of toxic dusts are easy to comprehend, but it is often forgotten that 'innocuous' dusts (limestone, for example) are also damaging. Recent work on limestone dusts illustrates that the diversity in forests is reduced as is the growth of some tree species. Soil changes may also occur.[56,58,59]

4.2.2. Sources of Pollution

Air pollutants from mining come from two types of source: easily defined sources such as chimneys, vents, exhausts, etc.; and dispersed sources such as dumps, stockpiles and haul roads. These are respectively point and non-point sources.

(a) Point Sources

Point sources of pollutants were, until recently, usually chimneys used to disperse combustion products plus residues of minerals being processed. Examples include ferrous and non-ferrous smelters, roasting, calcining, lime-burning, cement manufacture, asphalt coating, and drying, which produce particulates, F, SO_x, CO and organic vapours. In most of these cases, the concentration of emissions at chimneys for disposal has been an integral part of the whole process, for reasons of efficient draughting and combustion.

Today there is an increasing tendency for reasons of pollution control to convert hitherto dispersed sources to point sources. The pollutants which arise from these sources tend to be solely particulates, without combustion gases. Such sources include drilling, crushing and screening machinery, conveyers and transhipment points.

(b) Non-point Sources

These include sources which are laterally so extensive, or so unpredictable, that a specific point of emission cannot readily be identified. The major sources in this category are blasting (dust and combustion gases), haulage over internal and external roads, and general dust blow from working areas, tips, stockpiles, tailings dumps, etc. From all of these dust is the main pollutant. The only important gaseous pollution (SO_2 and CO) in the non-point source class arises from the spontaneous combustion of pyritic wastes mainly in colliery waste tips.

4.2.3. Dust Formation and Movement

Because of the importance of dust as a mining pollutant it is appropriate to examine in more detail the mechanisms of its formation and movement in the atmosphere.

Various attempts have been made to define the tendency to dust formation of particular minerals, and evolve a 'dustability index'. Three properties are concerned: brittleness, hardness and the applied force. Brittleness measures the inherent tendency to form dust, ductile substances being unable to store energy for the violent burst by which dust is produced. The stress required to cause deformation can be represented by hardness.

Thus coking coal is more prone to dust formation than lignite for, although they have the same hardness value, coking coal is more brittle. Likewise, when two substances of equal brittleness are compared, the one with the lower hardness will form more dust. Das[39] therefore defines a dustability index by the resultant effect of hardness and brittleness:

$$D = \frac{f(B)}{f(H)}$$

where D = dustability index, B = brittleness value and H = Vicker's hardness. However, no dust will be created unless a force is applied, and a dust forming index (DFI) can be defined, where:

$$DFI = \frac{f(B)}{f(H)} \cdot f(I)$$

and I = the induced factor causing dust production.

Once dust has been formed, or in the event of naturally fine-particle materials such as clays being present, the environmental nuisance from dust depends upon the opportunity it has to react with air flows of a velocity sufficient to carry it from the point of origin. Studies on the pick-up velocity (*i.e.* the air speed at which no settlement occurs) in pipelines have been made upon various coarse dusts, and examples are given in Table 7. If the dusts are wet, the air speeds are increased by 1 m/s (3 ft/s). These data correlate well with empirical experience at open cast minerals workings. In these cases, it is found that, at wind speeds of more than 5 m/s (16 ft/s) dust of above 100 μm will be raised from stationary dry surfaces and carried downwind for 250 m (800 ft). Wind speeds of 9 m/s (30 ft/s) can transport dust over 800 m (2600 ft).

TABLE 7

PICK-UP VELOCITIES OF DRY DUSTS
(Djamgouz and Ghonein[49])

Particle size (μm)	Air velocity, m/s (ft/s)		
	Granite	Silica	Coal
75–105	7 (23)	6 (20)	5 (16)
35–75	6 (20)	5 (16)	4 (13)
10–35	4 (13)	3 (10)	3 (10)

The above refers to the air speeds required to lift and carry particles, but in many mining situations the dust is raised by other means. On haul roads and at tip heads, for example, velocity is imparted to the dust by machinery and lower wind speeds suffice to transport the dust and prevent rapid settling-out. Tracer studies on asphalt roads demonstrate that, for newly deposited dust, as much as 1% can be resuspended per vehicle passage.[1] On internal mine roads there is often a continual potential for nuisance.

The three factors of inherent dustability, wind speed and imparted velocity mentioned so far are not the only ones. Others are the season and time of day, soil moisture and temperature, humidity, direction of wind and the relationship between wind direction and rainfall. Thus, in deciding whether any location may be at risk of dust nuisance, the proportion of time that *all* these factors are simultaneously unfavourable can be computed. To use a simple example, for a site worked 365 days a year, and a recipient 800 m (2600 ft) away: meteorological records show that winds in the correct direction occur on 38 days a year, but only on 16 days a year is their velocity in excess of 9 m/s (30 ft/s). Moreover, only during half the year is it likely that the site will be dry enough to permit any dust to be raised. Therefore, the recipient may be expected to be at risk on 8 days per annum. Such calculations are clearly capable of being made more valuable, given the requisite data on rainfall distribution, etc.

4.3. MEASUREMENT AND MONITORING

Measurement of air pollution levels is a complex and often expensive task, for which many methods have been devised.[28,38,51–53] This section reviews the procedures of greatest applicability to mining situations.

4.3.1. Ambient Measurement
(a) Smoke

Smoke has proved difficult to measure with quantitative accuracy, and existing methods ignore the gaseous component of the smoke in favour of particulates. The simplest method is to relate the shade of the smoke to a Ringelmann shade scale.[10] Such observations are merely visual comparisons and are too crude for all except preliminary surveys or similar work.

A more useful technique is to filter out solid components. If about 250 litres of smoke laden air are drawn through a 1 cm (0·4 in) diameter circle of white filter paper, a stain is obtained. This can be analysed by determining the proportions which are combustible, or soluble in carbon bisulphide, although such analyses are again not perfect. It is also feasible, with very precise equipment, to determine the weight of the filtrate. It is usually most satisfactory to make an optical comparison of the density of the stain with a standard scale by means of a photoelectric reflectometer.

For better accuracy, the filter should incorporate a flowmeter. Automatic operation can be introduced, whereby a roll of filter paper is mechanically advanced at, say, hourly intervals to obtain a more detailed record of diurnal variation. These methods estimate the largely non-settleable aerosol particulates. A standard apparatus and method is described by British Standard 1747: Part 2: 1969.

(b) Dust

The larger particles of dust emitted from mineral operations can be simply and fairly accurately estimated by the use of a dust deposit gauge.

British Standard 1747: Part 1: 1969 describes such a gauge, which is typical of those in use world-wide. It is illustrated in Fig. 23. In essence it consists of a collecting bowl at a standard height, 1·2 m (4 ft), from the ground surface; dust which settles into the bowl falls or is washed into a collecting bottle underneath. The collecting bottle is usually changed at monthly intervals, and the contents are analysed for: quantity of liquid collected; pH of the liquid; dry weight of undissolved solids; ash weight of undissolved solids; dry weight of dissolved matter. It is possible to analyse the contents for heavy metals or other constituents subject to allowing for contamination from the gauge and from the fungicide which is added to the collecting bottle on day 1 of the sampling period.

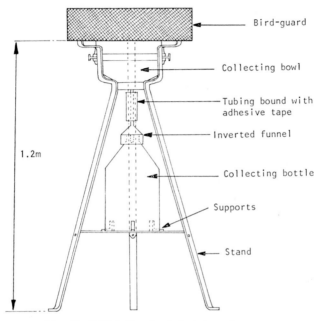

FIG. 23. British Standard dust deposit gauge.

The standard gauge provides an integrated record of total fallout over the period of exposure, known as time-average sampling. It can be refined into the C.E.R.L. directional deposit gauge (see British Standard 1747: Part 5: 1972). This does not measure vertical fallout, but horizontally-borne dust, and was developed because in high wind speeds the simple deposit gauge tends not only to be an inefficient collector but may indeed lose previously collected dust. The directional gauge is intended to avoid this, and provide data on the direction of the dust source. In effect, data from directional gauges measure the tendency of objects to become dirty. The dust collected is measured by optical obscuration, by the reduction in intensity of a previously calibrated light path.

It is clear that there is overlap between the functions of the deposit and directional types of gauge. There is also overlap between these gauges and smoke filters, because the latter will also collect some of the settleable dust in the smoke. The use of the smoke filter type of sampler, *i.e.* a high volume, short-term sampler, for dust measurement is often convenient. Used over 24 h—and often in automatic mode to collect hourly tape samples—such methods retain all particulates over 0·5 μm, in sufficient quantity for analytical determinations, and allow individual pollution events to be isolated in a way that monthly time-averaging cannot.

Other devices include impingers, in which particles are retained by blowing them on to a water surface, and impactors, in which retention is on a dry surface. These tend to be more applicable to heavily laden, enclosed atmospheres.

For very rapid dust sampling, simpler methods are available. Sticky paper, or greased glass sheets, can be suspended to trap particles for microscopic examination or, above a certain size, for counting. Quantitative measurements are not appropriate in this case. Another rapid and economical means of making preliminary measurements of fallout is by a petri dish, about 9 cm (4 in) diameter, with low vertical sides. At a time when weather forecasts predict neither wind nor rain for at least 24 h, dishes can be placed at suitable intervals. After exposure, the weight of material collected can be recorded.

(c) Heavy Metals

Aerial pollution by heavy metals can to some extent be detected by analysing dusts collected as described above. However, in recent years an effective method of biological measurement has been developed for this purpose. It depends upon the fact that certain mosses have excellent ion-exchange properties, presenting a very large surface area for exchange to occur. They have been found particularly suited to the collection and retention of airborne metals. The method consists of simply dispersing standard bags of moss about 4–5 cm (2 in) diameter over the area to be monitored, over a period of 3–4 weeks. After exposure, the moss is analysed for metals, correction being made for background levels. The method has yielded satisfactory results for Zn, Pb and Cd.[46,54]

(d) Sulphur Dioxide

Sulphur dioxide can be measured indirectly by the concentration of sulphates in the water collected in a dust deposit gauge, but this is not very satisfactory. Two methods are more commonly used for quantitative determination.

It is possible to incorporate in a smoke filter apparatus a bubbler between the filtration and pumping stages. The bubbler contains H_2O_2 at pH 4·5. H_2O_2 combines with SO_2 in the airstream to give H_2SO_4 and it is therefore possible to determine the amount of SO_2. In certain cases there can be interference from ammonia, hydrochloric acid, etc., so that the calculation may be taken as representing the net acidity of the air.

A common alternative is the lead dioxide method (British Standard 1747: Part 4: 1969) which is based upon the reaction

$$PbO_2 + SO_2 = PbSO_4$$

The method, which gives a monthly average result, consists of exposing a cylinder coated in a PbO_2 paste to the atmosphere in a louvred box. Sulphate is then determined by the barium chloride method.

4.3.2. Source Measurement

Measuring pollution emission at source, rather than by fallout, is complex but frequently essential to allow control measures to be properly designed. In most cases at-source measurement of pollutants is at stack emissions or similar discrete point sources.[42]

In stack sampling, the object is to perform isokinetic sampling. Seldom if ever is it possible to measure pollutant loads in a total stack emission, and therefore measurement has to be made upon a sample. An isokinetic sampling method is one in which the kinetic velocity in the sampling device is the same as in the chimney. Any deviation from 100% isokinetic will mean over- or under-estimation of small particles. The sampler inserted into the flue should be as small as possible if anomalous turbulence is to be avoided, and will normally measure temperature and flow rate, in addition to the pollutant. Where the flue gases are hot, as is usual, the sampling probe is electrically heated to prevent condensation in the sampling tube.

If isokinetic sampling is not possible, the % isokinetic can be estimated by the use of the formula

$$\% \text{ isokinetic} = 100 \, (M_a/M_c)$$

where M_a = mass emission rate of pollutant by area, and M_c = mass emission rate of pollutant by concentration. M_a and M_c are obtained by:

$$M_a = \frac{M}{t}\left(\frac{A_s}{A_n}\right)$$

$$M_c = M A_s \left(\frac{V_s}{V_n}\right)$$

where M = mass of pollutant collected in sampling time t, A_s = stack cross-sectional area, A_n = sample nozzle cross-sectional area, V_s = stack gas velocity and V_n = sample gas volume converted to stack conditions.

4.3.3. Monitoring Programme

Monitoring of air pollutants requires that the most appropriate measurement techniques be utilised over the most suitable intervals of time and space to achieve a properly defined objective. For example, if it is wished to detect isolated emission peaks, time-average samplers are useless. If it is desired to determine metal fallout as it affects livestock and foodstuffs then deposit gauges placed a metre above the ground are not sufficient. Monitoring thus requires that these matters be carefully specified. In addition, it is often important that the data obtained from a monitoring programme be correlated with meteorological factors, and with topography. In the case of cyclical operations at mining installations, precise details of these may enable correlations with observed pollution data to be obtained and thus the source of the pollutant may be isolated. Below are given case studies which illustrate various aspects of measurement and monitoring.

(a) Biological Monitoring

Little and Martin[54] describe the biological monitoring of heavy metal pollution (Pb, Zn, Cd) around a Zn/Pb smelter at Avonmouth near Bristol, U.K. Within an area of 250 km² (100 sq miles) around the smelter, 47 sites were selected for moss bag measurements. Each bag comprised 600 g (1 lb) of washed and air-dry *Sphagnum* moss in a fine-mesh nylon net, to give a loosely-packed near-spherical ball 4·5 cm (2 in) diameter. Sites were chosen such that bags could be hung freely, without chance of contamination by run-off from trees, etc., and more than 100 m (330 ft) from busy roads. Successive exposure periods ranged from 20 to 56 days. There was little vandalism (a problem with more conspicuous methods) and at least 83% of bags were recovered and analysed. Results were expressed in μg or ng of Pb, Zn or Cd per g of moss per day, after correcting for background.

Distribution maps were established by computer mapping to avoid subjective isolines. For each sampling period, a wind rose (a graphical presentation of wind direction and frequency) was constructed from hourly anemograph data, showing the length of time that wind was blowing from each of 12 compass points during the sampling period. An example of the type of result is given in Fig. 24 for cadmium dispersal, which illustrates the close visual coincidence between the wind rose and metal dispersion. When compared with standard deposit gauges, moss bag results show a strong positive correlation, although the results for moss bags tended to be higher. This is not unexpected in view of what has been said above.

(b) Geochemical Surveys

An apparent alternative to moss bag surveys is simply to analyse the metals accumulating in soils, or on leaves of grass and trees. In fact, although such geochemical analyses are very valuable from some points

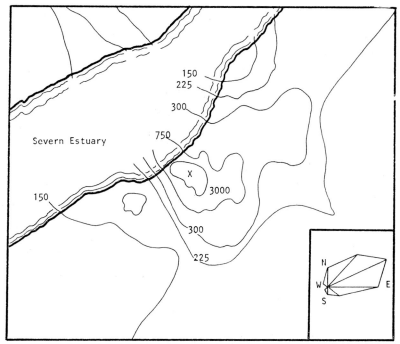

Fig. 24. Results of a moss bag survey of cadmium deposition around a lead/zinc smelter (at position X). Figures are ng cadmium/g moss/day. Wind frequency and direction during the survey period are indicated on the inset diagram (Little and Martin[54]).

of view, they are not of great use for examining overall emission and deposition. Each species of leaf has its own particular metal retention properties, pollutant accumulations on leaves reach a maximum because rain washes off much of the fallout, and there are grave difficulties in distinguishing in analysis between pollutants outside and inside leaves.

Geochemical analysis does however enable hazards to livestock and other organisms to be studied and in particular to follow the long term build up of pollutants. Its use is exemplified in the New Lead Belt studies in Missouri, U.S.A. (Wixson and Jennett[55]), in which accumulations of Pb, Zn, Cu and Cd in soils and leaf litter have been extensively monitored. An example of pollutant dispersal as a result of ore trucking is given in Fig. 25, which shows the accumulation of metals blown off road vehicles.

(c) Dust Deposit and Directional Gauges

Pluss and Strauss[31] describe a deposit gauge survey around an isolated cement works in Victoria, Australia. Twenty deposit gauges were located

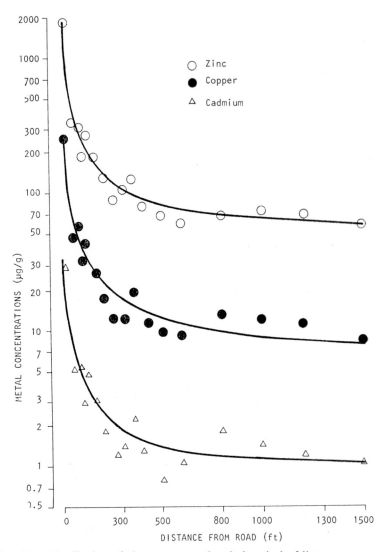

FIG. 25. Distribution of zinc, copper and cadmium in leaf litter near an ore-trucking route in Missouri (Wixson and Jennett[55]).

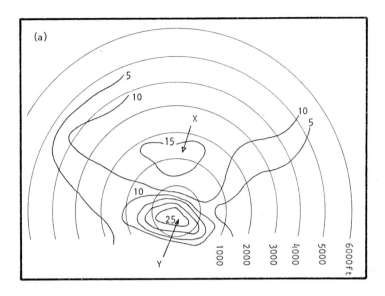

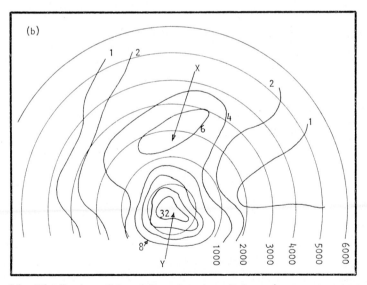

FIG. 26. Distribution of dust fallout (tons/sq mile/month) around an isolated cement works: (a) in first month; (b) in second month. Although the magnitude of deposition varies, both sets of results show a small peak around the quarry (X) and a larger peak around the cement works (Y). (Pluss and Strauss[31]).

at distances of approximately 150, 300, 600, 900, 1200, 1500 and 1800 m (500–6000 ft) from the works, concentrating on the leeward quarter. Wind speed and direction, temperature and humidity were recorded.

The unusual nature of the dust enabled it to be separated into cement and non-cement fractions, and an example of the results is shown in Fig. 26. Although the survey was rather crude it yielded much information, among which the rapid fall-off in dust deposition with distance is noticeable. Correlations with rainfall were obtained, while the survey also showed secondary maxima at the quarry and haul roads. The results have been utilised to design a 'green belt' around the works.

Another good example of the technique is work by Powell[47] on iron ore dust at Port Hedland, Western Australia. Here both deposit and C.E.R.L. directional gauges were used. A serious dust problem arose from the stockpiling for shipment of iron ore which contained a high proportion of <30 μm dust.

TABLE 8

IRON ORE DUST AT PORT HEDLAND,
WESTERN AUSTRALIA
(Powell[47])

Year	Mean total dirtiness	Mean % Fe_2O_3 in sample
1967	7·8	1·5
1968	8·0	16·6
1969	11·8	22·5
1970	17·6	37·1

The results obtained showed that, during strong cyclonic winds when dust conditions in Port Hedland were 'almost unbearable' the simple deposit gauge totally failed to reflect the situation, whereas the directional gauges recorded it well. The monthly total dirtiness figures ranged from 4·0 to 51·8 over the experimental period, compared with annual average figures in residential areas of 2·1–3·5 and a maximum monthly figure in these areas of 5·1. Table 8 gives a dramatic demonstration of the alteration in levels of dust deposition, and dust composition, which was disclosed by the survey. These results are closely related to the fact that iron ore shipment began in 1966 and by 1970 had built up to half a million tonnes per month.

4.4. AIR POLLUTION CONTROL

The selection of pollution control equipment requires consideration of the characteristics of the pollutant and its carrier gas, the available control equipment, and the emission levels which are to be achieved.

The pollutant may be particulate or gaseous and the airstream may be a mixture of both; it may be corrosive, and its concentration is important. The airstream characteristics are mainly temperature and flow rate, and the chosen equipment must be suited to these parameters, with due account of any variations and peaks in emission as a result of process variation. Emission standards are usually imposed by law.

To explain the importance of considering the concentration of the pollutant, suppose that the air to be cleaned contains a contaminant at 100 ppm. A treatment system of 95 % efficiency will give a 5 ppm effluent. If however the inlet concentration is reduced to 10 ppm, the treatment need be only 50 % efficient to achieve the same level of emission.

As has been explained, control methods to a great extent depend upon containing emissions and channelling them to defined points. Equipment for point source control can be sub-divided into that which removes particulates and that designed to purify gaseous emissions.

4.4.1. Control of Particulates: Point Sources

At its most basic, particulate control can be achieved by the total prevention of either release, or formation, of pollutants. Neither is easy to achieve.

An example of total enclosure exists at a Swedish quartzite plant of 800 tonnes/h capacity, opened in 1969.[11] The high silica content of the rock necessitated stringent dust control measures for health reasons, but there are associated environmental gains. A totally encapsulated plant was designed, with modified machinery where required so as to locate maintenance points (e.g. bearings) outside the capsule as far as possible. The only permanent openings to the processing machinery are at the primary crusher (input) end, and the final loadout of product. To prevent dust escape at these points, or via accidental leakages, the encapsulation has been put under a small negative pressure using a 30 000 m^3/h (18 000 ft^3/min) exhaust system to bag filters—an exhaust capacity only a tenth of the requirement for standard dust control measures. This saving on dust collection more than offset the costs of encapsulation.

Preventing dust formation requires that at crushing and handling points, and during such operations as stone cutting, the process material be sufficiently moist that dust will not be raised. Water is the usual medium, normally incorporating a wetting agent.[9,13,37] This reduces the interfacial tension between the dust and the water, and Table 9 illustrates the effectiveness of a non-ionic wetting agent in promoting the wetting of some common dusts 100 μm in size. Other wetting agents may be anionic or cationic. In a practical situation, the use of water plus wetting agent for sprays at a limestone primary crusher reduced dust fallout by nearly 60% compared with water alone. Tests on coal handling plants show similar results.

Another important matter is the moisture content of the mineral being processed. Water sprays are far more effective in preventing dust being raised than in 'knocking down' dust in the air and thus to prevent dust

TABLE 9

COMPARISON OF THE EFFECTIVENESS OF WATER AND
WATER + WETTING AGENT IN WETTING DUST
(Reilly[13])

Dust	Time to sink	
	Water	Water + wetting agent (1/750)
Crushed limestone	4 min bulk	Instantly
Pulverised fuel ash	45 s bulk	Instantly
Coke dust	30 min bulk	Instantly
Iron oxide	30 min bulk	$1\frac{1}{2}$ s
Bog iron ore	1·5 min	10 s
Limestone	0·5 s	Instantly
Talc	Would not wet	25 s
Lead dross	Would not wet	Instantly
Sinai ore	1 s	Instantly
Coal	2 h	10 s
Blue whinstone	1 s	Instantly

being raised the material must have an adequate water content. This varies considerably, according to the natural moisture content of the mineral when received for processing, the potential maximum moisture content, and the capacity of the mineral to retain water on its surface. Temperature is also an important consideration, for water sprays are not of great value if the temperature of the mineral exceeds about 60°C.

Therefore, an effective dust suppression system requires careful selection and positioning of spraying equipment to enable prewetting with water plus wetting agent. Typical locations for the use of sprays include crushers, tip heads, drawpoints, screens, etc. An alternative at some transfer points is to limit the height of fall of the material, or to enclose it during its fall, thus minimising the possibilities for interaction with winds.

Even a modest crushing and screening operation may require 20–40 sprays, each of which can throughput up to 1000 litres/h so that water requirements could be considerable. There are also cases where wetting the mineral is undesirable or impractical. Therefore many systems have been devised for dealing with dry dust which has already been raised. In essence they comprise a forced extraction system which draws dust-laden air from its point of origin into a separator, of which many types exist.

Some sources of particulates are already enclosed (*e.g.* driers, many crushers) but others are open and the dust has to be captured.[18-21] This is usually done by an exhaust hood and ducting to the separator. Operations such as crushing, conveying and tipping raise dust which requires capture velocities of 1–2·5 m/s (3–8 ft/s) at the source with, obviously, higher velocities in the ducting.

Once the dust has been captured, it must be removed from the airstream if atmospheric pollution is to be avoided. Five principles are employed for this purpose.

(i) Gravity

Settling chambers, effective for coarse dusts, use gravitational forces, by decreasing the air flow rate to less than the pick-up velocity.

(ii) Centrifugal Force

Used in both dry and wet collectors, centrifugation produces forces in excess of gravity and enables effective collection.

(iii) Inertia

The inertia of a moving subject will tend to keep it moving in a straight line, even though the gas flow may diverge. Thus the particle can be collected by impaction.

(iv) Interception (or filtration)

In principle a variation of inertial collection.

(v) Electrostatic Attraction

Widely used for dry collection.

Four different types of collector have been designed, using one or more of the above principles and their basic features are given in Table 10.[23,35,40] Table 11 illustrates the typical collecting efficiencies of each type, using a standard dust. Clearly this is not a wholly realistic test, but it demonstrates an important general point, namely that the efficiency of all types is high in the case of the large (10 μm) particles, but they differ markedly on small (1 μm) dusts. It is thus common for low-efficiency equipment to be used for pre-cleaning, removing the easily-captured large particles, to prepare the emission for more sophisticated treatment. This avoids overloading

TABLE 10
BASIC FEATURES OF PARTICULATE COLLECTORS
(modified from Ross[50])

Basic type	Specific type	Typical capacity (ft^3/min/unit of capacity)	Floor area (ft^2) for 100 000 ft^3/min capacity
Mechanical collectors	Settling chamber	20/ft^3 casing volume	2 600
	Baffle	1 200–3 600/ft^2 inlet area	300
	High-efficiency cyclones	3 000–3 600/ft^2 inlet area	125
Fabric filters	Manual cleaning	1–4/ft^2 fabric area	1 000
	Automatic shaker cleaning	1–4/ft^2 fabric area	1 000
	Automatic reverse-jet cleaning	3–8/ft^2 fabric area	600
Wet scrubbers	Impingement baffle	500–600/ft^2 baffle cross-sectional area	300
	Packed tower	500–700/ft^2 bed cross-sectional area	250
	Venturi	7 000–30 000/ft^2 throat area	100
Electrostatic precipitators	Single field	5/ft^2 collectrode area	270
	Multiple field	3/ft^2 collectrode area	500

the second stage cleaning plant, and enables a more economical overall installation. A common example is the use of multi-unit cyclones followed by electrostatic precipitation.

The main features of each type of collector are discussed below.

(a) Mechanical

Mechanical collectors are generally cheap to install and operate, but of low efficiency on the smaller particles which are of greatest concern.[12] The gravity settling chamber is no more than an enlarged length of ducting in which particles can settle out into hoppers. In the baffle type, the contaminated gas is introduced under a baffle of rods about 1·5 cm (0·6 in) apart. In order to reach the outlet the gas must make a sudden turn, while the dust particles, driven by inertia, are restrained below the baffle and settle out.

Cyclones, or centrifugal collectors, operate by creating a variety of contra-flow air movements.[14] The incoming air stream is forced into a descending vortex around the inner walls of an inverted cone. This throws the dust to the walls, whence it falls to a hopper; the cleaned gas ascends by an inner vortex to the exit. Different types of cyclone have been used,

TABLE 11

TYPICAL COMPARATIVE COLLECTING EFFICIENCIES
OF DUST COLLECTORS

(Squires[40])

Type of collector	Efficiency on a standard dust 80% 60 μm 30% 10 μm 10% 2 μm	Approximate efficiency at 10 μm	Approximate efficiency at 5 μm	Approximate efficiency at 1 μm
	%	%	%	%
High efficiency cyclones	84·2	85	67	10
Small multi-unit cyclones	93·8	96	89	20
Low-pressure drop cellular collectors	74·2	62	42	10
Spray tower	96·3	96	94	35
Self-induced spray collector	93·5	97	93	32
Wet impingement scrubber	97·9	99	97	88
Venturi high-pressure drop scrubber	99·7	99·8	99·6	94
Dry electrostatic precipitator	94·1	98	92	82
Irrigated electrostatic precipitator	99·0	99·0	98	92
Fabric filter	99·8	99·9	99·9	99·0
Fabric filter with pre-coat	99·9	99·9	99·9	99·9

but even the most advanced show a sudden fall-off in collection efficiency at smaller sizes.

(b) *Fabric Filters*

These are among the most varied and versatile of dust removal devices.[17] They are highly effective under a wide range of loadings but often expensive to install and operate. Commonly, woven-fabric filters have apertures of about 100 μm, but can collect particles down to 0·5 μm by virtue of the filtering effect of the layer of collected dust which rapidly builds up on the fabric. An alternative type of filter is the felt type with smaller interstices, which does not depend upon the formation of a dust filter-cake. Many different fabrics have been used for filters but none is usable at temperatures in excess of about 270°C.

Because the filter bags collect dust it is clear that there must be mechanisms for periodically removing it if the bags are to remain operative (3–5 y being the life of synthetic bag cloths). Therefore for intermittent processes, it is possible periodically to stop the air flow and either manually or, more usually, mechanically, shake the bag to dislodge the filter cake. Automatic control of shaking times is usual, for example by monitoring the draught loss due to build up of dust. Because the use of a single bag is usually inconvenient (while it is cleaned, no air treatment is possible) it is usual to employ several bags in parallel so that they can easily be taken out of service.

Another method of cleaning bags is by reversing the air flow using clean air to back-flush the bags and dislodge the dust. There are other methods, such as ring-jets or pulse jet cleaning which may also be applicable.

(c) Wet Scrubbers

The simplest form of wet scrubber is a gravity settling chamber with the addition of water sprays. Provision is made for recycling the water and removing the sludge which accumulates. To the chamber can be added wet impingement baffles so that, in addition to spray capture, particles impinge on the baffles and are carried down in a flowing water film. Again, by analogy with dry collection, wet cyclones are also employed.[15,44]

In packed tower scrubbers, an upward flow of gas to be cleaned moves against a downward flow of water in a tower packed with material such as glass or plastic spheres. Other types include the submerged orifice scrubber, in which the incoming gas is discharged below water, creating great turbulence and facilitating dust capture.

Such methods can be very effective, but the venturi scrubber is normally superior. The gas stream is passed through a venturi nozzle at velocities of 200–600 km/h (120–360 miles/h). Into the throat of the venturi is introduced the scrubbing water which is atomised by the gas flow. The gas then passes to a low-velocity chamber, where the wetted dust is cycloned off. Where very high pressure drops (75–250 cm [30–100 in] w.g.) are employed, the venturi scrubber is as efficient as a bag filter, but very much more compact. Several types have been evolved.

(d) Electrostatic Precipitators

These are the most efficient and expensive type of collector.[6–8,16,33] They usually have the lowest operating pressure drop (1–2 cm [0·5–1 in] w.g.) and thus low operating costs, but are sensitive to fluctuations in operating conditions and intolerant of conditions outside those designed for. The principle of operation is to apply a high voltage to a wire placed exactly between steel plates of different polarity. This charges the dust which precipitates on to the plates, from which the agglomerations can be cleaned mechanically or by washing. Precipitators are most widely used at power stations and cement kilns where their efficiency and tolerance of high temperature (up to about 540°C) make them essential.

4.4.2. Control of Gases: Point Sources

Some of the methods for particulate control, particularly wet scrubbing, may also effect reductions in gaseous pollutants. Purpose-built equipment is however usually required.

The oldest form of local air pollution control was by the use of chimneys to disperse the emission. Such methods are still very much used, but it is increasingly clear that they are *not* control measures, but merely reduce the incidence of serious local ground level pollution by introducing the emission into a larger volume of air. Prediction of dispersion of emissions from chimneys is most complex and inexact, and beyond the scope of this book.[10] However, in normal circumstances (*i.e.* level topography, no inversions, etc.) a rule of thumb estimate holds fairly well; the maximum downwind ground level concentrations of pollutants will occur at a distance of at least ten chimney heights from the chimney unless the chimney is less than $2\frac{1}{2}$ times the height of surrounding buildings when turbulence can interfere. The calculation is:

$$P_{max} = 1600E/uH^2$$

where $P_{max}(mg/m^3)$ = maximum ground level concentration, $E(lb/h)$ = rate of emission of pollutant from the chimney, $u(mi/h)$ = wind speed and $H(ft)$ = height of chimney.

True methods of gas control include scrubbers, adsorption and absorption, incineration and condensation. Absorption and scrubbing are usually the only methods applicable to the mineral industry. The equipment is essentially the same as that already described for dust control—packed columns and spray towers notably—but cannot function unless the liquid medium (usually water) is able to dissolve the gas. Dry absorption methods are being developed for large volume SO_2 emissions, namely dolomite injection, catalytic oxidation and others.

4.4.3. Control of Particulates: Non-point Sources

Attempts to control particulates arising from the numerous dispersed sources at mineral workings have only recently been made, and the results have so far been mixed. However, the only source which so far appears to defy control is dust from blasting. In this case the only practical control so far known is to blast only when climatic conditions preclude the spread of dust to particularly sensitive locations.

Another source of particulates is general dust-blow from surface excavations, dumps, etc. Normally these sources will comprise a very wide range of particle sizes and only produce dust as long as they are in active use. Dusting normally lessens by about a month after abandonment because the supply of fine particles is exhausted. An exception is fine-particle materials such as tailings, where dust can be a long-lived nuisance. For such situations revegetation or some other form of surface stabilisation is required and the methods are considered in Chapter 10. At active locations of this type, simple water sprays have been found adequate and can be automated.

Two main approaches to the collection of drilling dusts have been tried. Dry collection comprises a cyclone and filter which receive the dust via a flexible hose and collar around the drill hole.[4] The inherent vibration of most drill rigs obviates the need to provide filter-bag agitation. There have been problems in devising systems sufficiently rugged for this application, and also interference with operation of rigs, but these appear to have been overcome.

Wet drilling using foaming agents has also been attempted with success.[5,32,36] For the small diameter holes widely used in underground work, water alone is convenient to use, but for the larger holes common in surface excavations there can be problems in obtaining sufficient volumes of water. Control of the volume of water injected can be troublesome. Foaming agents of the organic detergent type retain some of these dis-advantages but synthetic foams have been successful. They are still at the experimental stage, but it has been shown that the advantages of dust suppression by foams can be offset by problems of foam preparation and poor bit cooling.

A major, intermittent dust source in many surface workings is haul roads, where dust arises from spillage from trucks and abrasion by their wheels. The most usual approach is to water all the haul roads by means of tanker vehicles, but this has the defect that the dust is not collected but remains on the road to cause further nuisance when it dries out. Conse-quently sweeping is also utilised. Control is facilitated upon properly compacted and graded haul roads, and especially if roads are coated, but such measures are usually impractical (for cost reasons) except on long-life haulage routes. Vehicle exhausts can be directed upwards to avoid creating excess turbulence at the road surface.

Transport by all methods (but especially road and rail) of the marketed product outside the mine property is a source of much dust nuisance. Tests on zinc concentrates transported in open rail cars in North America showed that, over a journey of 1700 km (1060 miles), each car lost 2·1% of its concentrate, equivalent to a loss of $330 per car.[24] Even if the product has a sufficient moisture content to prevent serious economic loss of this magnitude, small quantities can still be lost by local drying and blow-off. Other factors are leaking tailboards or hopper doors.

For dry materials such as cement, totally enclosed vehicles are employed, with pressurised unloading, but the dust from most mineral products can be controlled without the expense of purpose-built vehicles. Rigid fibreglass covers, tarpaulins and spraying with chemical binders to form resistant surface crusts, are all successfully employed. For metallic concentrates, wholly-sealed container transport has been mooted.

4.4.4. Control of Gases: Non-ponit Sources

The principal source in this category is spontaneous combustion of coal refuse tips.[34] Control is best directed at preventing ignition of the bank, which can be done during construction by tipping in successive layers, each layer being compacted. Once a bank is on fire, control is difficult. Watering and percolation often succeed but the fire may re-ignite when watering ceases, while the volumes of water needed are considerable. Blanketing with non-combustible materials to exclude air has also been widely used, although the covering is susceptible to erosion and failure. Excavation, saturation and controlled re-tipping is often the only solution.

4.4.5. Disposal of Collected Pollutants

Dust collected by the above methods requires careful disposal. Simply dumping dry dust enables it to cause another air pollution problem and the expense of collection may be wasted. It is therefore common to dispose of it as a sludge, even if collected dry, although attention to the possibility of water pollution is also required. A more intractable problem is the disposal of the sulphuric acid obtained by SO_2 removal. Reaction with lime to produce gypsum is widely practised, although this causes another disposal problem. There is sometimes a market for the acid.

REFERENCES

1. Sehmel, G. A. (1973). Particle resuspension from an asphalt road caused by car and truck traffic, *Atmospheric Environment*, **7**, 291–309.
2. James, P. W. (1973). The effects of air pollutants other than hydrogen fluoride and sulphur dioxide on lichens, in *Air Pollution and Lichens* (ed. B. W. Ferry, M. S. Baddeley and D. L. Hawksworth), Athlone Press, London, pp. 143–223.
3. Quellmalz, E. and Oelschläger, W. (1971). Fluorine contents of air and plants in the vicinity of a brick works, *Staub-Reinhalt. Luft*, **31**(5), 29–31.
4. Swedish drill dust-collectors on trial in U.K. (1973). *Quarry Managers' Journal*, **57**(9), 326.
5. Lewis, G. V. (September 1973). Foam drilling—settling the dust, *Canadian Mining Journal*, **94**(9), 42–8.
6. Gilliland, J. L. (1971). *Air pollution control in the portland cement industry*, American Institute of Mining, Metallurgical & Petroleum Engineers, Environmental Quality Conference, Washington D.C., paper EQC 56, 10 pp.

7. McKibbon, J. H. (September 1971). Selection of electrostatic precipitators to meet new pollution codes, *Canadian Mining & Metallurgical (CIM) Bulletin*, **64**(713), 82–4.
8. Estimating the performance of electrostatic precipitators (1973). *Process Technology International*, **18**(8/9), 339.
9. Cheng, L. (1973). Collection of air-borne dust by water sprays, *Ind. Eng. Chem. Process Des. Develop.*, **12**(3), 221–5.
10. Nonhebel, G. (1972). The regulation of air pollution in Britain, *J. Environmental Planning & Pollution Control*, **1**(2), 53–66.
11. Asklof, S. (1974). New system of quarry plant dust control, *Quarry Managers' Journal*, **58**(1), 24–7.
12. Merriman, A. D. (1964). The separation, collection and handling of grit and dust in industrial plant, *Quarry Managers' Journal*, **48**(9), 357–68.
13. Reilly, R. H. (1964). The use of water sprays for dust suppression in quarry plant, *Quarry Managers' Journal*, **48**(8), 313–22.
14. Koehle, H. (1970). Basic information on dry centrifugal collectors, *Canadian Mining Journal*, **91**(10), 85–8.
15. Malozzi, F. (1970). Wet collectors, *Canadian Mining Journal*, **91**(10), 80–5.
16. Norman, G. H. C. (1970). Electrostatic precipitation, *Canadian Mining Journal*, **91**(10), 69–72.
17. Dick, G. A. (1970). Fabric filters, *Canadian Mining Journal*, **91**(10), 72–9.
18. Hodgson, J. M. (1967). Dust control in quarries. 2-screening operations, *Quarry Managers' Journal*, **51**(6), 231–2.
19. Hodgson, J. M. (1967). Dust control in quarries. 3-impact crushers, *Quarry Managers' Journal*, **51**(7), 271–2.
20. Hodgson, J. M. (1967). Dust control in quarries. 4-rotary drying plants, *Quarry Managers' Journal*, **51**(8), 305–6.
21. Hodgson, J. M. (1967). Dust control in quarries. 5-batch heaters, *Quarry Managers' Journal*, **51**(9), 360–1.
22. Ireland, F. E. (1973). Dust, *Quarry Managers' Journal*, **57**(3), 85–91.
23. Squires, B. J. (1972). Removal of particulate matter from industrial airborne discharges, in *Aspects of Environmental Protection* (ed. S. H. Jenkins), Bostock, Hill & Rigby/IP Environmental, London, pp. 127–51.
24. Schwartz, P. L. (1974). Innovative rail transport systems cut concentrate losses, *World Mining*, **27**(1), 38–43.
25. Mitchell, E. R. (1971). Only people pollute, *Canadian Mining & Metallurgical (CIM) Bulletin*, **64**(712), 96–100.
26. Argenbright, L. P. (1971). Smelter pollution control, *Mining Congress Journal*, **57**(5), 24–8.
27. Schmitt, N., Devlin, E. L., Larsen, A. A., McCausland, E. D. and Saville, J. M. (1971). Lead poisoning in horses, *Arch. Environ. Health*, **23**, 185–95.
28. *Air Pollution Manual* (1960). American Industrial Hygiene Association, Detroit, Michigan, pp. 49–61.
29. Perry, H. and Field, J. H. (1967). Air pollution and the coal industry, *Trans. Soc. Min. Engineers*, **238**, 337–45.
30. Ireland, F. E. (1972). The scheduled processes, in *39th Annual Clean Air Conference*, National Society for Clean Air, Scarborough, 10 pp.

31. Pluss, D. H. and Strauss, W. (1972). Dust fall and concentration distribution measurements around an isolated cement works, in *International Clean Air Conference*, Melbourne, Australia, pp. 162–8.
32. Metzger, C. L. (1967). Dust suppression and drilling with foaming agents, *Pit & Quarry*, **60**(3), 132–8.
33. Singhal, R. K. and Chandan, J. S. (1966). Electrostatic precipitation, *Min. Mag.*, **114**(3), 170–6.
34. McNay, L. M. (1971). *Coal Refuse Fires, an Environmental Hazard*, U.S. Bureau of Mines IC8515.
35. Bowler, D. (1974). Controlling quarry dust, *Mine & Quarry*, **3**(4), 31–5.
36. Cole, H. W. (1975). The use of foam suppressants in the control of particulate emissions from grinding, crushing and transfer operations in the mining and rock crushing industries, in *5th Annual Industrial Air Pollution Control Conference*, University of Tennessee, Knoxville, Tennessee, 23 pp.
37. Pilz, K. W. (1972). Wet dust suppression brightens mineral processing picture, *Mining Engineering*, **24**(7), 81–3.
38. Measurement of dust emissions (1974). *Quarry Managers' Journal*, **58**(3), 93.
39. Das, B. (1973). Formation of dust, *Colliery Guardian*, **221**(4), 146–9.
40. Squires, B. J. (1973). Air pollution control in a modern industrial society, *Pollution Monitor*, No. 14, 15–20.
41. Hollowell, C. D. and McLaughlin, R. D. (1973). Instrumentation for air pollution monitoring, *Environ. Sci. & Techy.*, **7**(11), 1011–17.
42. Greenfield, M. S. (1973). Stack sampling keeps Government happy, *Canadian Mining Journal*, **94**(9), 56–8.
43. Code of practice for dust emissions (1974). *Quarry Managers' Journal*, **59**(5), 164.
44. Mallozzi, F. (1971). Wet collectors control dust and air pollution, *Canadian Mining & Metallurgical (CIM) Bulletin*, **64**(714), 77–84.
45. Slack, A. V., Falkenberry, H. L. and Harrington, R. E. (1972). Sulfur oxide removal from waste gases, *J. Air Pollution Control Association*, **22**(3), 159–66.
46. Roberts, T. M. (1972). Plants as monitors of airborne metal pollution, *J. Environmental Planning & Pollution Control*, **1**(1), 43–54.
47. Powell, R. A. (1972). The growth of an iron ore dust problem—Port Hedland, Western Australia, in *International Clean Air Conference*, Melbourne, Australia, pp. 146–51.
48. Beamish, R. J. (1974). Loss of fish populations from unexploited remote lakes in Ontario, Canada, as a consequence of atmospheric fallout of acid, *Water Research*, **8**, 85–95.
49. Djamgouz, O. T. and Ghonein, S. A. A. (1974). Determining the pick-up air velocity of mineral dusts, *Canadian Mining Journal*, **95**, 25–8.
50. Ross, R. D. (Ed.) (1972). *Air Pollution and Industry*, Van Nostrand Reinhold, New York and London.
51. Masters, G. M. (1974). *Introduction to Environmental Science and Technology*, John Wiley, New York and London.
52. Meetham, A. R. (1964). *Atmospheric Pollution, its Origins and Prevention*, Pergamon Press, Oxford.
53. Hesketh, H. E. (1974). *Understanding and Controlling Air Pollution*, 2nd ed., Ann Arbor, Michigan.

54. Little, P. and Martin, M. H. (1974). Biological monitoring of heavy metal pollution, *Environ. Pollut.*, **6**, 1–19.
55. Wixson, B. and Jennett, J. C. (1974). *Interim Progress Report* (New Lead Belt study), University of Missouri-Rolla.
56. Darley, E. F. (1966). Studies on the effects of cement-kiln dust on vegetation, *J. Air Pollution Control Assoc.*, **16**(3), 145–50.
57. Council on Environmental Quality (1972). *3rd Annual Report*, Washington D.C.
58. Brandt, C. J. and Rhoades, R. W. (1972). Effects of limestone dust accumulation on composition of a forest community, *Environ. Pollut.*, **3**(3), 217–25.
59. Brandt, C. J. and Rhoades, R. W. (1973). Effects of limestone dust accumulation on lateral growth of forest trees, *Environ. Pollut.*, **4**(3), 207–13.

5

Water Pollution

5.1. INTRODUCTION

Water is one of the earth's most abundant natural resources, and a resource which can be placed seriously at risk by the activities of the mineral industries. It is the medium of all life processes, which at a cellular level take place in aqueous solution, and the continued availability of pure water able to meet the demands of a growing human population is one of the most vital problems which the world faces.[1][2]

In nature, water passes through a cyclic process of evaporation and deposition, as shown in the hydrological cycle illustrated in Fig. 27. It is

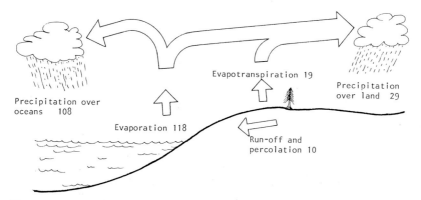

Evapotranspiration 19

Precipitation over land 29

Precipitation over oceans 108

Evaporation 118

Run-off and percolation 10

FIG. 27. Generalised picture of the world hydrological cycle. Figures are in 10^{15} gallons/a (Lvovitch[25]).

of particular importance that the overwhelming majority of the earth's water (some 99%) exists in forms which render it almost unusable directly for man's needs either because of salinity, or its physical nature (ice) or location in the ground.

A breakdown of water distribution is given in Table 12, from which it is apparent that usable fresh water in lakes and rivers, upon which we mostly rely, comprises no more than 0·0161% of the earth's total water. It is this

TABLE 12

DISTRIBUTION OF THE EARTH'S WATER RESOURCES
(Lvovitch[25])

Location	Water volume 10^{15} gal (10^{15} litres)		Percent of total
Oceans	362 000	(1 646 000)	94·2
Glaciers	6 350	(28 870)	1·65
Lakes	60·5	(275)	0·016
Soil moisture	21·6	(98)	0·006
Atmospheric vapour	3·7	(17)	0·001
River waters	0·32	(1·5)	0·000 1
Groundwater less than 0·8 km (0·5 mile) down	1 160	(5 270)	0·28
Total groundwater	15 800	(71 800)	4·13

small fraction of the water resources of the world which are most immediately prejudiced by mining and by many other human activities.

The hydrological cycle contains within it two self-purifying mechanisms. The more important is that due to evaporation from the earth's surface plus transpiration from vegetation (collectively termed evapotranspiration) which ensure that the water vapour returned to the atmosphere is purified. There are however some insecticides (e.g. DDT) which can bypass this purification process. The second mechanism is filtration, which occurs when water falling on the land surface percolates into the substrata and permeates through them. This, too, can be bypassed by molecules not susceptible to filtration. Additionally there are numerous biological purification processes which occur as a result of the activities of aquatic organisms.

Water pollution can be caused by two main types of action. These are:

(i) Introduction of substances (or certain forms of energy such as heat) into natural waters, causing physical and/or chemical changes.
(ii) Interception or diversion of all or part of a water resource.

The effects of these actions are as follows:

(i) The quality of the water may be adversely affected, rendering it less suitable (or wholly unsuited) for human consumption, or industrial use.
(ii) There may be ecological damage, altering the composition of (or eliminating) the natural biological communities inhabiting the water, and decreasing the diversity of organisms therein.

(iii) Water may cease to be available, in the required and accustomed quantities, at the points of use.

The mineral industries have only a small share of water use in quantitative terms. Table 13 shows the distribution of water withdrawal and consumption between the major classes of user in the U.S.A. in 1970. All industrial uses, excluding thermoelectric plants, accounted for 12% of water withdrawal (*i.e.* water abstracted from external sources) and only 6% of water consumed (*i.e.* water lost by evaporation, incorporation in product, etc.). Moreover, 88% of the water withdrawn was discharged again and, quantitatively, was available for reuse elsewhere.

TABLE 13
U.S.A. WATER WITHDRAWAL AND CONSUMPTION BY
MAIN USER
(derived from Murray and Reeves[89])

Use	Withdrawal (billion U.S. gallons/day)	(billion litres/day)	%	Consumption (billion U.S. gallons/day)	(billion litres/day)	%	Consumption as % of withdrawal
Irrigation	130	(492)	35	73	(276)	82	56
Public supplies	27	(102)	7	6	(23)	7	22
Rural-domestic	4·5	(17)	1	3·5	(13)	4	78
Industry	47	(178)	12	5·5	(21)	6	12
Thermo-electric	166	(629)	44	0·8	(3)	1	0·5
	374·5	(1 418)		88·8	(336)		

Of the 47 billion U.S. gallons (178 billion litres) per day withdrawn by U.S. industry as a whole, only 2% was taken by the mineral industries in 1962, a proportion which is likely to be almost the same today. This percentage includes usage in natural gas processing (accounting for 48% of the mineral industry's total water requirement) a sector of the industry which is outside the scope of this book.[27] In South Africa, the total percentage is somewhat higher as might be expected in view of the importance of the mining industry: mining takes 4·2% of all water used in the Republic.

The mineral industry's minute share of total water use is not indicative of a similar lack of importance in the overall water budget of the U.S.A. or anywhere else, for the quantitative picture is less crucial than the question of water quality. Although the industry consumes only 12 % of the water it abstracts, the 88 % balance is often seriously polluted and cannot be returned directly to the hydrological cycle without prior treatment.

The water actually used by the mineral industries, to which discussion has so far been confined, is only a portion of the total water influenced by mining activities. There is also a large volume of water which is casually affected by mining (for example, run-off, mine drainage, pumped mine water and groundwater flows) which does not enter into the mining processes directly but is nonetheless affected by them. Data are insufficient to enable quantification of volumes of water affected by causes of this type. In the U.S.A., the Bureau of Sport Fisheries and Wildlife has estimated that 18 400 km (11 500 miles) of stream channel have had their flow capacity reduced by a third or more as a result of sediments from mining, with a further 4000 km (2500 miles) affected to a lesser degree. Another U.S. survey, to 1st January 1967, indicated that 20 600 km (12 898 miles) of streams, and 449 natural or artificial lakes (representing in all over 113 000 ha [280 000 acres] of water) had been adversely affected by mining in terms of their suitability as fish habitats. It is not possible to apportion the damage among the 'process' and 'casual' categories, but the latter is probably the more important.

It is therefore apparent that in discussing water pollution, the volume of process water is an inadequate measure. Also required are the volume of other water influenced by mining, and the nature and extent of the decrease in water quality that mining may cause.

5.2. WATER IN THE MINERAL INDUSTRIES

Water is utilised in many stages of mining and mineral processing. These uses are briefly outlined below together with the main pollution potential of each.[27] A generalised picture of water in the mineral industries is given in Fig. 28.

(a) Mining

The use of water to remove mineral from the ground is a relatively uncommon technique and limited to a few industries. Hydraulic mining, performed by directing a high velocity jet of water at an unconsolidated deposit, is utilised mainly in gold and china clay mining. It is sometimes also found in the working of alluvial gem deposits, tin, wolfram, zircon, etc. The slurry which results contains the valuable mineral content and is therefore conserved so far as possible. However, once the water-borne sediment has been recovered, the silt-polluted water may contaminate surface or ground water, depending upon how it is discarded.

In dredging, water is used to float the dredge. The method is commonly

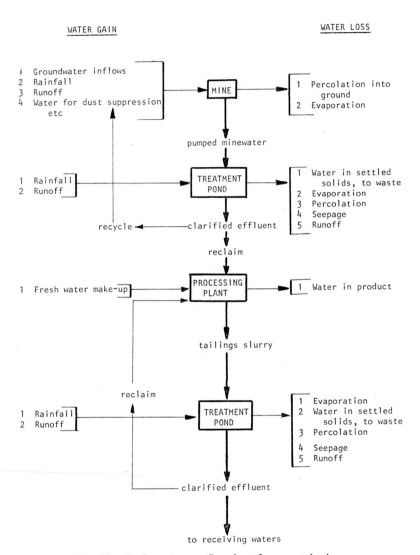

FIG. 28.   Basic water use flowsheet for a metal mine.

used to work alluvial deposits under fresh or sea water, and is particularly associated with tin, sand and gravel, and gold. The water pollution which is caused seems largely confined to silting. In 1962 in the U.S.A., 11·6% of total water usage in the mineral industries was for mining purposes, the major proportion being accounted for by hydraulic monitoring.

(b) Cooling Water

The main usage in mining is for cooling equipment such as crusher bearings, pumps and compressors. In these cases the water can be retained in a closed circuit and no pollution results; 4·6% of water use in U.S. mines is for this purpose. Slightly less common (3·8%) is the large-scale use of water in on-site electricity generating plants, where the main problem is that of thermal pollution of the waters receiving heated effluent.

(c) Process Water

This is defined as any water which, during mineral beneficiation processes, comes into direct contact with any raw material or product resulting from the process. Processing utilises by far the largest proportion of water, comprising 79% of all water used in U.S. mines. There are three sub-divisions of this category, the most important being the water used for wet screening, gravity separation processes (tables, jigs, etc.), flotation, heavy media separation, or leaching solutions. The other sub-divisions are washing water (for the removal of fines from metal ores, clays from stone, etc.) and the relatively small volumes of water used in wet scrubbers for air pollution control.

By virtue of its contact with the process feedstock, process water can become highly contaminated both physically and chemically. Commonly, process water contains residual quantities of un-recovered valuable mineral, large quantities of waste mineral matter, dissolved chemical constituents arising from these, and traces of any reagents which have been used.

(d) Transport Water

In some respects water used for transport can be considered as a type of process water. It is common to use water as slurrying medium to transport ores between different stages of beneficiation, to transport crude ore for treatment, or waste products for disposal. Indeed in certain instances, notably coal and chalk, the final product can be moved as a slurry from the mine to the consumer. Over short distances, a partial or total closed circuit, with no waste water discharge, can often be operated, but over long distances the expense of returning the transport water (which requires pipeline facilities to be duplicated) usually results in a total-loss system being used. The disposal of transport water has therefore to take into account its capacity for causing pollution.

(e) *Miscellaneous Water*

There are numerous uses of water which, although not usually very large in terms of volume (0·9% of total water use in U.S. mines) can frequently cause local problems. These water uses include dust control (sprays at crushers, conveyer heads, loading points, etc.), vehicle and other washing, drilling fluids, domestic and sanitary uses. Such waters may be contaminated by detergents, oils, suspended solids, organic matter, etc.

Apart from these uses of water, there are several routes by which mining can influence water without using it. These are as follows.

(i) *Pumped Mine Water*

Mine excavations usually have a water influx, either due to rainfall or to interception of groundwater flows. Mines using hydraulic backfilling have additional water from this source. This water is usually an unwanted feature of mineral working (though it can sometimes be used for processing, dust suppression, etc.) and may have to be pumped out. It can be contaminated by particulates, oils and grease, unburned explosives, ore chemicals, etc. U.S. copper mines pump between 0·008 and 1·769 m^3 (0·282 and 62·46 ft^3) water/tonne ore produced; there is no difference between open pits and underground mines in this respect.[88] Mine drainage often persists after mine closure and can present grave long-term pollution problems, especially in the coal and base metal sectors.[37,48,58,83]

(ii) *Run-off*

Run-off after rain or snow fall can give rise to serious pollution problems. The disturbed land caused by mining is usually very susceptible to erosion, and silting is thus a widespread result. A variety of other pollutants may also be transported into water courses by run-off. Studies in the Appalachian coalfield have shown suspended sediment loads of 10 000–50 000 mg/litre in streams during strip mining, compared with 150 mg/litre before mining began. The total loss by erosion from spoil banks can approach 800 tonnes/ha/a (300 tons/acre/a).[46]

(iii) *Percolation*

Polluted water stored in mine sumps or tailings impoundments can percolate into the ground and pollute aquifers. Percolation through the retaining walls of impoundments, with consequent pollution of surface waters, is very common.[64,80] The use of deep-well injection for water disposal can also have this effect.[61,68]

(iv) *Leaching*

Leaching of low-grade metalliferous ores (mainly copper) for metal production is merely an enhancement of a natural process which would

FIG. 29. Tailings disposal at a lead mine in the New Lead Belt, Missouri. Apart from the obvious land use and aesthetic problems, tailings disposal is usually a major source of water pollution.

often occur in any case. If pyritic waste heaps exist, some rainfall is likely to permeate into them, acidify and dissolve toxic ions before issuing from the heap and contaminating water courses.

(v) Groundwater Interception

An excavation is likely to intercept aquifers, which then become a source of mine water to be pumped away.[11] The consequence of interception can be diminution or elimination of borehole, spring or water course flows at points remote from the excavation.

FIG. 30. A relatively unusual form of water pollution is caused by this colliery in County Durham, England, where the mine waste is dumped at sea.

Predraining of highly permeable strata (*e.g.* chalk, sand and gravel) is sometimes practised. By lowering the water table, water supplies elsewhere may likewise be interrupted. Also, by creating an hydraulic gradient towards the predraining boreholes, it is possible to draw pollutants into originally pure groundwater. This has been one result of predraining some British chalk quarries located beside brackish estuaries. The problem is illustrated in Fig. 31.

(vi) Spillages

Spillages of oils, toxic reagents, etc., can often cause contamination of surface or groundwaters.

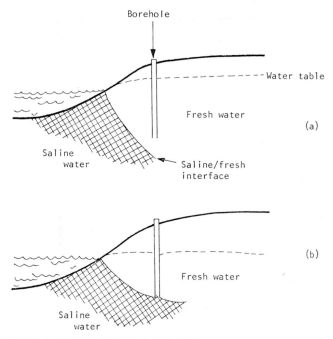

FIG. 31.   Effects of pumping from boreholes located near saline water (a); in (b) the saline water has been drawn into the fresh-water table.

(vii) Aerial Pollutants

Fallout of aerial pollutants is likely to affect surface waters.

The quantities of water used by various sectors of the mineral industry obviously vary considerably, according to the nature of the process employed, the availability of water, and the nature and quantities of material processed. A major survey carried out by the U.S. Bureau of Mines in 1962 produced detailed information on the water use in U.S. mines.[27] Water use was divided into the following categories.

(i)   New water: water brought onto the mine for the first time.
(ii) Recirculated water: water re-used to conserve new water, but excluding solutions recycled because of fixed metallurgical practices, *e.g.* sulphuric acid copper leaching solutions.
(iii) Total water: new water plus recirculated water.
(iv) Consumed water: water lost by evaporation or by incorporation into the product.
(v)   Discharge water: total water used, excluding recirculated and consumed waters.

TABLE 14

U.S. WATER USE BY MINERAL
(Kaufman and Nadler[27])

Mineral	Water use, 10^9 U.S. gallons/annum (10^9 litres/annum)				
	New	Recirculated	Total use	Discharged	Consumed
Anthracite	16·9 (64)	14·9 (56)	31·8 (121)	15·6 (59)	1·3 (5)
Bituminous coal	31·8 (121)	138·5 (525)	170·3 (645)	26·1 (99)	5·7 (22)
Clays	7·1 (27)	1·6 (6)	8·7 (33)	6·3 (24)	0·8 (3)
Copper ores	81·0 (307)	93·6 (355)	174·6 (662)	50·7 (192)	30·3 (115)
Gold	54·6 (207)	4·2 (16)	58·8 (229)	53·9 (204)	0·7 (3)
Iron ores	112·6 (427)	139·5 (529)	252·1 (955)	105·7 (401)	6·9 (26)
Lead and zinc ores	22·9 (87)	1·9 (7)	24·8 (94)	21·4 (81)	1·5 (6)
Phosphate rock	117·2 (444)	269·3 (1 021)	386·5 (1 465)	87·2 (330)	30·0 (114)
Sand and gravel	217·6 (825)	122·7 (465)	340·3 (1 290)	206·2 (781)	11·4 (43)
Crushed stone	50·4 (191)	16·0 (61)	66·4 (252)	47·0 (178)	3·4 (13)
Uranium ores	7·2 (27)	1·0 (4)	8·2 (31)	4·2 (16)	3·0 (11)

Table 14 illustrates the distribution between these categories of water used in the working and processing of the main minerals. There are clearly considerable differences between the different commodities: in the cases of clay, gold, lead/zinc and uranium mining, recirculated water was only a small percentage (less than 20%) of total water use. However, in the mining and processing of bituminous coal, copper and iron ores, and phosphate rock recirculated water constituted at least 50% of total water use, and as much as 81% in the case of bituminous coal. Table 15 shows the relative water use in terms of gallons/tonne ore processed. This is

TABLE 15

RELATIVE WATER USE IN U.S. MINES
(Kaufman and Nadler[27])

Mineral	Water use per tonne ore, U.S. gallons/tonne (litres/tonne)			
	Mining	Processing	Other	Total
Bituminous coal	10 (38)	524 (1 986)	13 (49)	547 (2 073)
Copper ores	18 (68)	861 (3 263)	166 (629)	1 045 (3 961)
Gold	1 124 (4 260)	72 (273)	109 (413)	1 305 (4 946)
Iron ores	5 (19)	1 428 (5 412)	173 (656)	1 606 (6 087)
Phosphate rock	958 (3 631)	4 819 (18 264)	262 (993)	6 039 (22 888)
Sand and gravel	55 (208)	452 (1 713)	1 (4)	508 (1 925)

likewise variable. Canadian studies show a range of $0.6–8.8$ m^3 ($21–311$ ft^3) water/tonne ore for processing base metal ores and gold. The usual requirements were $2–4$ m^3/tonne ($70–140$ ft^3/tonne). Many mineral processing plants do not require that new water be clean. Of the total new water used by the U.S. mineral industries, 53% was found by the survey to be fresh water (*i.e.* suitable for cooking and drinking), 44% contaminated (sewage effluent, mine drainage, etc.) and 3% saline (mainly sea water).

Water can be obtained from surface or subsurface sources, or from the sea. U.S. mines obtained 61% of their new water from surface sources, 36% from the ground and 3% from the sea. In arid areas, such as Nevada, the balance can be radically different:[77] in 1962, the mineral industries of this state obtained 75% of their new water from wells. 65% of water disposed of returned to surface watercourses and 16% was lost via evaporation or in the product, but only 16% was returned underground, thus causing a net depletion. This underground disposal consisted largely of uncontrolled seepage from tailings impoundments: the pollution this may cause has yet to be properly studied, but is potentially very serious.

5.3. NATURE AND EFFECTS OF WATER POLLUTANTS

There is no universally accepted definition of a pollutant, but it can be conveniently explained as 'a substance or form of energy which, if introduced into the environment, may or will adversely affect man or other organisms'. Although it is normally possible to measure, with a considerable degree of precision, the quantities of pollutants present in water, a definition such as that above still leaves scope for subjective assessment, notably in the meaning of 'adversely' and the importance which is attached to degrees of adversity.

Mining pollutants in water can affect man as well as other organisms, man being able to perceive the effects both in terms of nuisance and health hazard. Although rare, damage to human health has been recorded on occasions. The best-known example is that of cadmium-polluted water from a lead–zinc–cadmium mine in Toyama prefecture, Japan. The contaminated irrigation water was used in rice paddy fields. Since 1946 there have been numerous cases of 'itai-itai' (literally 'ouch-ouch') disease, a cadmium-induced degeneration of bones which can proceed to the extent that even coughing can cause multiple fractures of ribs. Up to 1965, nearly 100 deaths from itai-itai disease had been recorded.

Toxicity of this severity is rarely experienced by man or higher animals and, when it occurs, is largely the product of base metal mining. Lethal pollution levels are however very commonly experienced by aquatic organisms and there is still an abundance of surface water denuded of virtually all life.[53]

At a lower level, the pollutant may be either of a nature, or at a concentration, which causes some change in aquatic life, but by no means total destruction. Such effects are widespread throughout every sector of the industry.

The most common change is the reduction of diversity of the community, and this has been responsible for the development of a Biological Pollution Index to measure diversity:

$$d = \frac{s - 1}{\log_e N}$$

where d = the community diversity index, s = the number of species and N = the number of organisms. Commonly $d = 1\cdot0-5\cdot0$, the larger index figure implying greater diversity and consequently a more healthy body of water. When d tends towards $1\cdot0$, pollution damage should be suspected. An index of this type is not theoretically perfect since it tends to be weighted towards rather few organisms, but it proves useful in practice.

The lowest level of all is sometimes termed crypto-pollution. In this case organisms, including man, may possess detectable levels of a substance which has no discernible effect. The long term implications of exposure to very low levels of pollutants are as yet unknown.

Over the whole spectrum of mining, direct and immediate interference with human health is rare and the greatest impact is normally upon the aquatic ecosystem. It is therefore necessary to examine the broad features of this ecosystem before discussing the details of individual pollutants.

A generalised picture of the aquatic ecosystem is given in Fig. 32. It has two features. First, the food chain, which is the succession of consumption from phytoplankton (minute floating plants) to zooplankton (minute floating animals) to man or other consumer. Along a food chain of this type, initially negligible amounts of pollutants can become magnified or concentrated at each stage, so that the terminal consumer may ingest a damaging quantity. Second, the food chain is converted to a cyclical nutrient flow by the activities of scavenging organisms and micro-organisms which convert dead organic tissue back to inorganic nutrients which are once again able to enter the food chain. Such an ecosystem is balanced, and capable of automatically cleaning natural water which would otherwise become polluted with the dead remnants of the organisms within it. This self-cleansing capacity persists only as long as it is not overloaded, and as long as the cycle remains unbroken. The general ways in which pollutants act are either by overloading a particular conversion (*e.g.* excess organic material overloads the bacterial decay section) or by eliminating one or more stages (*e.g.* by killing fish). The natural waters in which such ecosystems occur can conveniently be classified into four types: standing fresh water (*e.g.* lakes); flowing fresh water (*e.g.* streams, rivers); sea water; and groundwater. There are connections and transitions

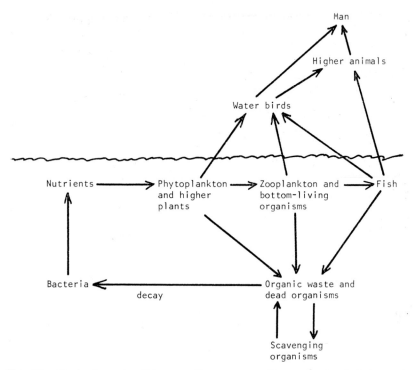

FIG. 32.   Basic elements of the aquatic ecosystem, illustrating food chains and
the recycling of organic matter.

between each, *e.g.* a lake empties into a river, which joins the sea, and thus
pollutants can be transferred from one to another. All natural waters are a
complex of chemical, physical and biological interactions, but it is possible
to distinguish these four types.

Standing water such as ponds and lakes is characterised by zonations of
flora and fauna around the perimeter, with the body of the water containing
plankton and fish communities. There is a tendency towards vertical
stratification of the major properties, as well as cyclic variations over 24 h
(diurnal) and 12 month (annual) periods. Pollutants introduced into lakes
tend to remain therein, other than losses via lake outflows, although their
forms and locations may be altered with time.

Flowing water is a particularly complex situation, for rivers alter not
only through cyclic, seasonal progressions, but also along their lengths,
according to depth, gradient, rate of flow, geology, salt concentration,
turbidity, etc. There is thus a multitude of biological habitats, but most
planktonic organisms adhere to the river bed sediments and rocks, or to

higher plants; there are considerable problems in maintaining plankton communities in flowing water. Pollutants introduced into rivers are naturally carried downstream but, after initial mixing, do not become further diluted except because of confluence with other water sources. Removal of the pollutant then depends on its biological degradation, or deposition. The general picture of introduction of a pollutant into a river is shown in Fig. 33.

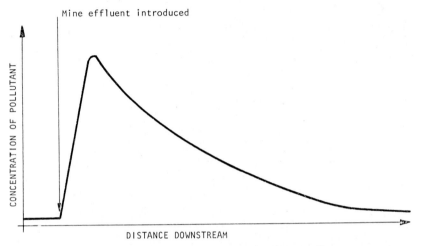

FIG. 33. Generalised picture of the introduction of a pollutant to a river. After a short period of mixing, by which the whole volume of water reaches peak pollutant concentration, the level of pollution gradually decreases.

In the marine environment, the range of habitats is restricted to the coastal (littoral) zone and the surface mass of water. The physical and chemical nature of the environment is different to that on land. Seasonal and vertical distributions likewise occur. Pollutants from mining seldom enter the seas directly, and rivers are usually the intermediaries. The sea's capacity for dilution, buffering and mixing is great, and it is difficult, other than in a few isolated cases, to discern the effects of mining pollutants among the much greater impact of man's other activities.[51,52]

Groundwater is very different from the other three types. It tends to have fairly constant physical and chemical properties. Its temperature is usually constant, at about 9°C for cold springs or 18–90°C in hot springs. Waters are often alkaline, with high levels of Ca^{2+} and HCO_3^- ions. Pollutants enter the groundwater by percolation, or because mining has penetrated the water table, or as a result of deliberate injection via waste wells.[61,64,68,80]

Natural filtration is very efficient for most pollutants, but there is the possibility of overloading these systems, or of bypassing them. Pollution of groundwater is one of the least-studied effects of mining, but the concern is usually with potable water, because groundwater supports few if any living organisms until it reaches the surface. Groundwater pollution is usually very long-lived whereas surface river pollution can rapidly be flushed out.

5.3.1. Individual Pollutants

Whether or not a particular substance is a pollutant depends on its nature, its concentration in the effluent, the total load discharged, the nature of the receiving water and its capacity to absorb the waste loading imposed upon it. Thus, in the appropriate circumstances, even distilled water can be a pollutant. Although the range of pollutants is very broad, it is possible to group them into relatively few categories, which are discussed below. It is first necessary to explain the term 'toxicity' and its measurement.

In sufficiently high concentrations, most chemicals can be toxic or lethal to aquatic life. Toxicity is defined in terms of the median tolerance limit (TL_m) for a particular organism over a particular time. The median tolerance limit is the range of concentrations of a pollutant in which the population of test organisms suffer 50% mortality within a specified time. The test organisms are usually fish (trout or goldfish are often used). Thus, toxicity measurement is a biological assay, and the range of concentrations must be qualified by the organism and time used (96 h is normal), as well as by the characteristics of the water used in the test, e.g. the TL_m96 of lead chloride for *Daphia magna* in Lake Erie water was 0·01–1·0 mg/litre. The TL_m96 is synonymous with the 96-LC_{50} (the 96 h median lethal concentration). From either expression can be derived the toxic unit (TU) for any pollutant.

$$TU = \frac{\text{Actual concentration}}{\text{Incipient lethal concentration (96-}LC_{50})}$$

A TU value of 0·1 for heavy metals is used in Canada as the upper limit for a 'clean' effluent. Values in the range 0·1–0·3 are marginal, and effects upon fish are often discernible in this region.

The units in which concentrations of pollutant are measured are in most cases, milligrammes of pollutant per litre of sample (mg/litre). This is virtually synonymous with the expression parts-per-million (ppm) which is being progressively abandoned. Exceptions to this are noted under the relevant heading.

(i) Organic Pollutants

One of the major non-mining categories (sewage), but a relatively small mining problem. Organic pollutants mainly comprise the proteins, fats, carbohydrates, etc., in sewage from mines. Coal is a peculiar, specific and widespread organic pollutant which is discussed separately, as are oils and certain organic reagents used in mineral processing. Mine effluents can sometimes have high organic levels caused by rotting vegetation in water storage lagoons, tailings ponds, etc. Coke ovens give rise to highly toxic tars containing phenols, cresols, naphthols, acridine, pyridine, etc., which are extremely dangerous to water life. Fish can be killed by less than 1 mg/litre of many of these chemicals.

Detergents from vehicle washing, etc., can be serious pollutants. Natural detergents (*e.g.* sodium oleate, sodium stearate) are not lethal in concentrations below 250 mg/litre in hard water (244 mg/litre $CaCO_3$), whereas synthetic detergents (*e.g.* sodium lauryl sulphate) are lethal above 6–7 mg/litre under the same conditions. Laboratory experiments on synthetic detergents have shown that 1 mg/litre can be sufficient to prevent growth of water plants.

As shown in Fig. 32 organic material (such as sewage) can be broken down to nutrients by micro-organisms; toxic substances such as phenols can also be degraded if at low concentration. With overloading, deoxygenation of water occurs (see below, Section 5.3.2).

(ii) Oils

Very commonly spilled in mines and quarries. On water, oil is aesthetically objectionable but, by forming a thin film over the water surface, can also interfere with re-oxygenation of water. Visible colour occurs with a film thickness of 0.038×10^{-4}–0.152×10^{-4} cm ($0.000\ 001\ 5$–$0.000\ 006\ 0$ in) thickness (*i.e.* 36–145 litres/km^2 [21–84 gal/sq. mile]). Film thicknesses of 10^{-4} cm (25.4×10^{-4} in) or greater interfere with re-oxygenation, and may coat the gills of fish. If oil contaminates boiler feed water, explosions may occur.

(iii) Cyanides

Common in gold mill waste waters at concentrations of 0.01–0.03 mg/litre. Cyanides are lethal to many fish at very low concentrations, as little as 0.04 mg/litre CN for trout. Over 5 years, 0.023 mg/litre seems lethal to trout, whereas smaller concentrations (0.009 mg/litre) can decrease the ability of trout to swim against water current.

(iv) Acids and Alkalis

Acidity and alkalinity can be conveniently expressed in terms of the hydrogen ion concentration. The exponent of the hydrogen ion concentration is termed the hydrogen potential or pH:

$$pH = -\log [H^+]$$

An acid is able to give off H^+ ions, the strength of the acid being determined by the extent of this release. An acid solution is one in which

$[H^+]$ is more than 10^{-7}, and alkaline or basic solution less than 10^{-7}. Thus 10^{-7} (pH 7) is taken as neutrality. Total acidity, including factors other than $[H^+]$, is expressed as mg/litre $CaCO_3$ equivalent.

Freshwater fish usually thrive in waters with pH values between 5·0 and 8·5. Sudden alterations in pH within this range can affect fish adversely, while if the range is extended beyond about 4·0–9·5 fish die. Acidic waters are particularly common in mining and the acid mine drainage problem is discussed below (Section 5.3.2); drainage of pH 2–6·5 is widespread.[53,69] Below pH 5, acid waters may also cause corrosion of metal or concrete structures, pumps, etc., and give rise to odour problems by liberating H_2S from river muds.[4,42] Highly alkaline effluents are less common, but are found in the cement industry and in the manufacture of concrete products. Over-neutralisation of acid waters can also cause problems of alkaline drainage.

(v) Base Metals

This category includes copper, lead, zinc, cadmium, nickel, chromium, arsenic, mercury, vanadium, beryllium and others. Many metals are necessary to health (Cu, Zn, Mn, Cr, etc.) in minute concentrations, but are highly toxic when present in excess at even low concentrations. The dividing line between healthy and toxic levels can be abrupt or, indeed overlap; algae may need 0·1 mg/litre Zn for growth, yet this level can be toxic to fish. Acidity aggravates the problem, which is predominantly found in the acid drainage of the base metal and coal mining sectors.[6] Examples of the toxic levels of some of these metals are given in Table 16.

Salts of heavy metals apparently act on fish by coagulating mucus on the gills, resulting in asphyxia.

(vi) Fluorides

Mainly found in the waste waters from fluorspar mining, toxicity is seldom a problem because such effluents are usually hard-water which precipitates F as Ca/Mg salts. In soft water, F can be quite toxic, with a TL_m range of 2·6–6·0 mg/litre F in rainbow trout.

(vii) Dissolved Solids (Soluble Salts)

It is common for mine effluents to have high levels of dissolved solids such as chlorides, nitrates, phosphates or sulphates, of sodium, calcium, magnesium, iron and manganese. The main source is dissolution from contact with the rock, but nitrate also arises from unburned ANFO explosives.

The effects of these compounds vary greatly according to concentration. At low levels nitrates and phosphates are nutrients and their release by mining may cause rapid growth of algae with subsequent deoxygenation (see Section 5.3.2). At higher levels, the character of the water may be altered, with deleterious effects upon fish. Fresh water contains no more

TABLE 16
TOXIC LEVELS OF HEAVY METALS IN FRESH WATER

Metal	Lethal concentration (mg/litre)	Organism
Cu	0·01–0·02	'Fish'
Cu (as $CuSO_4$)	0·14	Trout
(as $CuSO_4$)	0·75	Perch
(as $CuSO_4$)	2·1	Black bass
(as $CuSO_4$)	0·1	Blue-green algae
Zn	0·15–0·7	'Fish'
Be	0·2 (TL_m96 in soft water)	Fathead minnows
Cr (as $NaCrO_4$)	0·1	Water fleas
(as $NaCrO_4$)	20·0	'Fish'
Pb	(0·1–)0·5–1·0	Bacteria
Pb	0·01–10·0	Water fleas
Pb	3·3–10·0	Tadpoles
Pb (as $PbCl_2$)	3·0–60·0	Green algae
Pb	2·4 (TL_m96 in soft water)	Fathead minnows
Pb	>75·0 (TL_m96 in hard water)	Fathead minnows

than 100 mg/litre salts as Cl and excessive release of soluble salts may convert it to brackish (100–1000 mg/litre) or even salt water (more than 1000 mg/litre). Note that sea water contains about 20 000 mg/litre Cl. Increases in Cl above 1000 mg/litre can kill many fish and water weeds, although some fish (*e.g.* roach) can tolerate 6000 mg/litre Cl.

Bicarbonates, sulphates and chlorides of calcium and magnesium cause hardness of water, and may make it unsuitable for further industrial use without softening. Soluble salts of iron and aluminium can react in bicarbonate alkaline conditions to give precipitates of insoluble hydroxides, a particularly common problem in many mine drainage streams.

(*viii*) *Organic Reagents*
The number of reagents used in mineral processing runs into many hundreds. They can be classed broadly as follows:[57]

Frothers: these are to toughen the bubble wall during flotation. Frothers usually consist of molecules with two constituents of opposite (polar and non-polar) properties, *e.g.* cresol $CH_3C_6H_4OH$. Often these are volatile compounds and not all of them appear in the waste water.

Collectors: xanthates or other chemicals, which collect non-floating minerals and attach them to gas bubbles formed during flotation, are included in this category. Xanthates are dithiocarbonates, such as sodium isopropyl xanthate $(CH_3)_2CHO—C\!\!=\!\!S—SNa$. They are water soluble.

Depressants: these are used for depressing one mineral while another is being floated. They include cyanides, acids, etc.

Others: include pH modifiers, activating agents, flocculants, coagulants, dispersants. There is also a great variety of inorganic chemicals.

Many organic reagents are highly toxic (*see* Table 17). However, it is unusual to detect in the effluent from mine tailings areas concentrations of more than 5 mg/litre of any commercial reagents, and typically the figure is 2 mg/litre or less. This is well below the threshold of toxicity for most reagents. Reagents may also contribute to oxygen demand in the water, but generally they seem to cause few problems.

TABLE 17

TOXICITY OF SELECTED MILLING REAGENTS

Reagent	Lethal concentration (mg/litre)	Organism
Cresylic acid	1·0	Goldfish
Ortho cresol	11·2 TL_m96	Catfish
Ortho cresol	29·5 TL_m24	Carp
Cresols	6–50	Many plankton organisms
Potassium amyl xanthate	0·1–1	Water fleas
Aerofloat 238 promoter	210·0 TL_m96	Trout
(sodium disecondary butyl	152·0 TL_m96	Salmon
dithiophosphate)	7·58 TL_m96	Oysters

(*ix*) *Colour*

Discoloration of water, by dissolved or suspended matter, is not uncommon. Discoloration is not, of itself, harmful (*e.g.* many peaty waters support good fish populations) but can often be an aesthetic problem. However, in some cases, colour can be both unsightly and damaging: the reddish deposits of precipitated ferric hydroxide in some mine streams are an example. If deoxygenation of such water occurs, black ferrous sulphide may be formed. Colour is measured in standard platinum–cobalt units.

(*x*) *Suspended Solids*

The transport by flowing water of solid particles in suspension is a natural phenomenon, aggravated in the case of mining by the creation of disturbed areas of land susceptible to erosion.[5,8,9] A wide variety of inert inorganic materials arises from mining, together with a relatively minor contribution from organic solids. If, in the latter category, coal is included, the organic fraction becomes of great importance. Coal is a particular discolourant, and its phenolic content can also be damaging.

Suspended solids interfere with self-purification of water by diminishing light penetration (by up to 50%) and hence photosynthetic activity, and damage fisheries by silting-over of food organisms. Discharge of china clay wastes which are neither toxic nor deoxygenating, produced such silted conditions in Cornish rivers that no algal life could survive. There is also aesthetic nuisance. If the solids are at all abrasive (as are many mineral wastes), damage to fish, plants and pumps, etc., can be caused, while silting may cause flooding or interference with navigation. It is difficult to specify harmful levels of suspended solids, but it appears that, below 75 mg/litre, little injury to fish occurs. Nonetheless, some countries employ standards of 10 or 20 mg/litre for domestic and industrial effluents.

(xi) Turbidity

Turbidity arises from the non-settleable fraction of the suspended solid load.[20] It may either be a true colloidal dispersion, or very fine suspensions of coarser particles. There is no good distinction between these fractions, but colloids are often taken as having a particle size of 1–500 nm. The effects of turbidity on fish life appear relatively small. Turbidity is normally measured by a standard candle turbidimeter, and expressed in Jackson or Formazin turbidity units (JTU or FTU) or silica (SiO_2), all as mg/litre.

(xii) Thermal

Heated effluents are normally found at power stations or other establishments using cooling water. Mine and mill waters are often higher in temperature than the receiving water, particularly pumped minewater from deep mines. Even mill discharges can be 5–7°C higher than ambient. Heated effluents are extremely damaging: the dissolved oxygen content of the water decreases (oxygen solubility bears an inverse relationship to temperature) while the oxygen-demanding biochemical reactions are speeded up. Deoxygenation can therefore result. Lethal temperatures for some fish are quite low: 25°C for trout.

Other effects of a temperature rise are: increased toxicity of any pollutants present; increased growth of undesirable weeds; and interference with hatching of fish eggs.

(xiii) Radioactivity

There have always been a number of radioactive elements in the earth and most natural waters have appreciable background radioactivity.[16,17] Radium, the decay product of uranium, is the most potent cumulative poison known. Uranium milling processes preferentially dissolve the uranium, and leave in the tailings pond radium concentrations of 1–2100 pCi/litre (picocuries per litre, 1 pCi $\equiv 10^{-9}$ g/litre). The International Commission on Radiological Protection specifies the maximum concentration of dissolved radium in water for human consumption as 3 pCi/litre. Thorium, another uranium decay product, is found in tailings pond solutions at up to 100 000 pCi/litre; the maximum recommended concentration is 2000 pCi/litre.[74]

Aquatic flora and fauna absorb and concentrate radioactivity discharged in effluents. The effects on man of low doses are principally genetic, while no effects upon aquatic organisms appear to have been recorded in respect of mining radioactivity.

5.3.2. Specific Pollution Problems

Water pollution problems in mining are seldom, if ever, attributable to any one specific pollutant. It is general for several pollutants to be found in any single waste water stream. The possible combinations of pollutant, even under the thirteen broad groupings above, are numerous. There are, however, four major problems—acid mine drainage, eutrophication, deoxygenation and heavy metal pollution—which are the result of combinations of factors, and which are recognised as the most serious water pollution situations to be found in mining.

(i) Acid Mine Drainage (AMD)

AMD problems are predominantly chemical, and of great complexity. Originally considered to be a problem associated only with coal mining, and particularly with abandoned mines, AMD is now known to occur as a result of working many other minerals. Any deposit containing sulphide minerals, and particularly pyrite, is a potential source of AMD.

AMD is produced when a sulphide reacts with air and water to form sulphuric acid. The basic process occurs in three stages:[1,10,14,18,19,21,69,73,81,84]

(1) The oxidation of the sulphide, usually FeS_2. If the reaction takes place in dry surroundings, water-soluble ferrous sulphate and sulphur dioxide are formed:

$$FeS_2 + 3O_2 = FeSO_4 + SO_2$$

More commonly, the reaction occurs in the presence of water, with the direct formation of sulphuric acid:

$$2FeS_2 + 2H_2O + 7O_2 = 2FeSO_4 + 2H_2SO_4$$

(2) Ferrous sulphate, in the presence of sulphuric acid and oxygen, can oxidise to produce ferric sulphate (water-soluble). This transformation is not controlled by the presence of water, but it appears that a bacterium (*Thiobacillus ferro-oxidans*) is an essential mediator and,

if not actually responsible for the oxidation, at least greatly accelerates it. The reaction is:

$$4FeSO_4 + 2H_2SO_4 + O_2 = 2Fe_2(SO_4)_3 + 2H_2O$$

(3) The ferric iron so produced combines with the hydroxyl $(OH)^-$ ion of water to form ferric hydroxide. This is insoluble in acid, and precipitates:

$$Fe_2(SO_4)_3 + 6H_2O = 2Fe(OH)_3 + 3H_2SO_4$$

An alternative to this reaction occurs because the ferric iron may also enter into reaction with iron sulphide and 'backtrigger' further oxidation, thus accelerating acid formation:

$$Fe_2(SO_4)_3 + FeS_2 = 3FeSO_4 + 2S^0$$
$$S^0 + 3O + H_2O = H_2SO_4$$

The sulphuric acid generated tends not to be found as high concentrations of free acid, due to further reactions with other minerals. It must be emphasised that the reactions above are very far from being the whole picture: many side reactions take place while, in some conditions, partially oxidised sulphur compounds (sulphites, thiosulphates and polythionates) are intermediates during acid formation. These thio-salts are found in alkaline oxidation conditions, at pH 7 or above.

TABLE 18
AMD FROM VARIOUS SOURCES

Parameter (mg/litre)	Source						
	1	2	3	4	5	6	7
pH value	3·2	3·4	3·0	2·3	2·6	2·6	2·0
Acidity (CaCO₃)	647	230	1 560	3 180	1 600	3 800	14 600
COD	338	—	—	50	30	110	270
TS	20 220	—	4 110	6 170	4 180	—	—
TSS	trace	—	—	15	355	—	—
TDS	—	1 200	—	6 155	3 825	—	—
Ferrous iron	2 305	68	445 ⎫	280	960	1 310	⎰1 750
Ferric iron	238	12	208 ⎭				⎱1 450
SO	—	800	—	2 800	2 280	4 050	7 440

1. Klein.[69] Colliery AMD in Lancashire.
2. Wilmoth et al.[41] Colliery ferrous-iron discharge in Pennsylvania.
3. Davis et al.[90] Colliery AMD; U.S.A.
4–7. Hawley.[76] 4: underground mine water at an Ontario uranium mine. 5: Ontario iron mine waters. 6: Ontario abandoned FeS tailings decant. 7: Ontario uranium mine, tailings seepage.

The bacterium *T. ferro-oxidans* thrives in highly acid environments (pH 1·5–3·0) and can function as low as pH 0·9. The organism, which requires oxygen for growth, can also transform nickel, copper, zinc, molybdenum and other metallic sulphides.[30]

The nature of the drainage, and hence its effects, vary greatly, but some examples of AMD from various sources (for AMD can arise from processing as well as mining) are given in Table 18. A comparison of a stream above and below the point of discharge of mine water is given in Table 19.

TABLE 19

EFFECTS OF MINE EFFLUENTS UPON RECEIVING WATER QUALITY

Parameter (mg/litre)	Before AMD	After discharge[a]	*Pb/Zn effluent*[b]		
			Control[b]	Mill effluent into tailings pond	Stream below tailings discharge[b]
pH value	7·4	3·8	7·5	8·0	6·9
CL	50	56	—	—	—
COD	4·4	4·6	—	—	—
BOD$_5$	3·4	0·5	—	—	—
TSS	45	252	0	40 582	8
Ferric	8	100	—	—	—
Total Fe	—	—	<0·1	0·3	<0·1
Cu	—	—	<0·05	1·1	<0·05
Na	—	—	0·7	13·2	4·2
K	—	—	0·3	21·9	4·1
Ca	—	—	5·4	46·0	29·2
Mg	—	—	1·6	13·8	11·0
Zn	—	—	<0·1	<0·1	<0·1
Cd	—	—	<0·02	<0·02	<0·02
Mn	—	—	<0·1	0·7	0·2
Pb	—	—	<0·1	<0·1	<0·1
Sb	—	—	<5·0	<5·0	<5·0

[a] Klein.[69]
[b] Williams *et al.*[71]

These tables illustrate some of the numerous objectionable features of AMD: low pH, which restricts or eliminates most organisms in the water, high total acidity and sulphate levels, high iron, and high total solids (TS). There is usually a fairly low level of suspended solids (TSS), the majority being dissolved (TDS). High ferrous iron levels are undesirable because the demand for oxygen created by their transformation to the ferric form tends to seriously deplete oxygen levels. The ochre colour of precipitated ferric hydroxide—which can be yellow to reddish/brown depending upon conditions—is aesthetically objectionable. In the U.S.A., coal mine drainage has been divided into the four classes shown in Table 20.

A secondary effect of AMD is that it renders heavy metals soluble and hence greatly increases toxicity problems. Even in cases where the AMD is derived from marcasite, pyrite or pyrrhotite associated with coal, sufficient quantities of base metals exist to create pollution, while the working of sulphide metalliferous deposits obviously increases the chances of metal pollution.

TABLE 20
MINE DRAINAGE CLASSIFICATION
(U.S. E.P.A.[91])

Parameter	Class 1 Acid discharge	Class 2 Partially oxidised and/or neutralised	Class 3 Oxidised and neutralised and/or alkaline	Class 4 Neutralised and not oxidised
pH	2–4·5	3·5–6·6	6·5–8·5	6·5–8·5
Acidity (mg/litre $CaCO_3$)	1 000–15 000	0–1 000	0	0
Ferrous iron (mg/litre)	500–10 000	0–500	0	50–1 000
Ferric iron (mg/litre)	0	0–1 000	0	0
Aluminium (mg/litre)	0–2 000	0–20	0	0
Sulphate (mg/litre)	1 000–20 000	500–10 000	500–10 000	500–10 000

In conclusion, an AMD problem is likely to develop at any mine if

(a) the mineral contains pyrite;
(b) the mineral does not contain sufficient carbonate or other alkali to neutralise the acid;
(c) pyrite is rejected and discarded into an impoundment not designed for its retention.

(ii) Heavy Metal Pollution

The heavy or base metals are those with a density greater than 5, and comprise some 38 elements in total. Not all of these are of significance in mining situations, where the elements most commonly of concern are Zn, Cu, Pb, Cd and Hg. At even extremely low concentrations, some metals are lethal if regularly ingested; they are therefore among the most important mining pollutants.

The sources of heavy metals vary, but are largely restricted to discharges and drainage from coal and metal mines. Some examples of metal concentrations in mine drainages are given in Table 21 where a comparison is provided with the World Health Organisation standards for drinking

TABLE 21

HEAVY METAL CONCENTRATIONS IN MINE EFFLUENTS AND RECEIVING WATER

Metal (mg/litre)	Source						
	1	2	3	4	5	6	7
pH value	2·0	6·4	2·6	7·7	7·7	11·0	6·5–9·2
Zn	11·4	2·15	0·97	0·28	0·01	3·40	15·0
Ni	3·2	0·0	0·39	—	—	—	—
Co	3·8	—	0·47	—	—	—	—
Cu	3·6	0·22	0·96	0·12	0·01	0·18	1·5
Mn	5·6	0·58	83·0	0·98	0·28	0·84	0·5
Pb	0·67	0·0	0·0	0·1	0·1	0·05	0·1
Cd	0·05	—	0·02	0·01	0·01	—	0·01
V	20·0	—	4·3	—	—	—	—
Mo	—	—	<0·5	—	—	—	—
Cr	—	—	0·15	—	—	—	—
As	0·74	—	0·38	—	—	0·05	0·05
Mg	106·0	16·0	12·0	—	—	—	150·0
Ca	416·0	—	101·0	—	—	—	200·0
K	69·5	—	5·1	—	—	—	—
Na	920·0	—	7·0	—	—	—	—
P	5·0	—	0·2	0·1	0·1	—	—

1–3. Hawley.[76] 1: Seepage from uranium tailings; 2: Cu/Pb/Zn tailings decant; 3: iron mines, pumped water.

4–5. Gale et al.[87] Missouri New Lead Belt. 4: Effluent after dilution in stream; 5: Control stream.

6. Purves.[60] Tin mine tailings decant, U.K.

7. World Health Organisation.[70] Maximum levels for drinking water.

water. By far the most polluted water is the seepage from uranium tailings (1) and the pumped water from an iron mine (3). The more alkaline samples (2, 4, 5 and 6) carry lower metal burdens. On theoretical grounds, a pH of 9 or greater must exist before many metals will precipitate, but pH 10–12 is required for some metals. Details are given in Table 22 but it should be noted that theoretical solubilities vary in practice.

TABLE 22
THEORETICAL RELATIONSHIP BETWEEN
pH AND METAL SOLUBILITY

Metal ion	pH	Minimum solubility achieved (mg/litre)
Fe^{3+}	7·6	0·000 1
Cu^{2+}	8·9	0·000 5
Zn	9·2	0·08
Pb	9·3	6·0
Ni	10·2	0·000 9
Fe^{2+}	10·6	0·001
Cd	11·1	0·002

A valuable study of the effects of metallic pollutants has been undertaken at the lead mines in Cardiganshire, Wales, closed by 1921. Examination of rivers draining the mining areas, which in the 1919–22 period had dissolved lead concentrations of 0·2–0·5 mg/litre, showed only 14 species of invertebrate animals. By 1922/3, when mining had ceased, the lead concentration stabilised at 0·1 mg/litre (even at flood times) and 29 species were found, as well as a greater number of individuals of these species and a regrowth of algae and higher plants. Ten years later (1931/2) lead levels had declined to 0·02 mg/litre, and the number of species had dramatically increased, to 103. More recent surveys, and comparisons with unpolluted rivers, nonetheless show that the mine-influenced rivers are seriously depleted in species variety.[54–56,63] Studies on colliery-influenced streams show a similar picture.[43,81] Very broadly, the more advanced (in an evolutionary sense) the organism, the greater its susceptibility to metal pollution.

The toxicity of metallic ions is influenced by factors other than pH. In natural waters the toxic ions sodium, calcium, potassium and magnesium are present in proportions such that their toxicity is mutually counteracted; additionally they antagonise other ions (Pb, Zn, Cu) and reduce their toxicity also. The converse effect is synergism where pairs of metals (Ni and Zn, Cu and Zn, Cu and Cd, for example) are together more toxic than their cumulative individual toxicities would suggest.[53]

(iii) Eutrophication

Natural fresh water lakes undergo an ageing process which theoretically turns them into dry land. 'Young' lakes—formed after glaciations, for example—have low nutrient levels and low productivity of the organisms inhabiting them, and are termed oligotrophic ('few foods'). As the lake ages, it becomes more nutrient-rich, due to accumulation of sediments and organic debris, and is then termed eutrophic. The term eutrophication refers to the process which, although a natural one, can be drastically speeded up by human activities such as the discharge of mine effluents. This is termed cultural eutrophication.

Mining effluents may be toxic, but they can also be nutrient-rich: for example, in phosphates, nitrates, silica, etc. In this case, the natural nutrient/consumer cycle is upset (Fig. 34). Normally the population of an

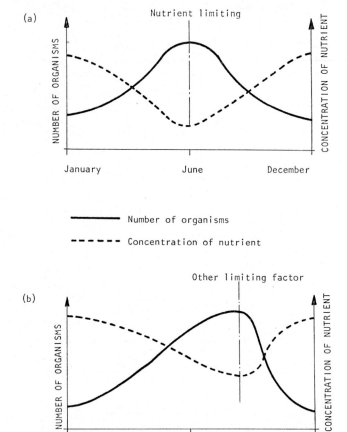

FIG. 34.   Natural cycles occurring in lakes are illustrated in (a) where a population of organisms increases until a factor (such as a nutrient) becomes limiting. If a nutritious pollutant is added (b) the population increases to a higher level and peaks later in the season, dying rapidly when some other factor becomes limiting.

organism increases until one of the essential factors for growth is depleted, *i.e.* growth is limited by the nutrient which is least available in relation to requirements. When this point is reached the population declines again. The artificial addition of a nutrient, however, prevents that nutrient becoming limiting and the population then increases beyond its natural level until another factor becomes limiting—commonly oxygen. The result is that an algal 'bloom' occurs, sometimes choking watercourses, with the subsequent oxygen depletion destroying fish and other populations. The large quantities of dead algae can then produce noxious decay products which aggravate the pollution.[62]

(iv) Deoxygenation

Most water living organisms survive only in aerobic conditions, *i.e.* require atmospheric oxygen, dissolved in the water, for respiration. This dissolved oxygen (DO) is consumed by respiration and/or chemical transformations, and is replenished by the process of photosynthesis in green organisms (which produce oxygen as a waste product) and by re-aeration through the surface of the water. Several instances of mining resulting in deoxygenation have already been quoted and it is one of the major consequences of pollution.

The minimum quantities of DO required in water depend upon the individual organisms. Most fish require at least 5 mg/litre DO for at least 16 h per day, and never less than 3 mg/litre for 8 h. Most natural waters contain 8–10 mg/litre. The DO content is reduced, at a relatively steady rate over a 24 h period, by respiration. During daylight hours, it is replenished by photosynthesis, and it is continuously transferred from the atmosphere across the air–water interface. At peak photosynthesis times, there can be DO excess, such that O_2 diffuses out of the water.

There are two other routes of DO depletion which are frequently aggravated by mining. Micro-organisms such as bacteria utilise organic wastes as foods, converting them to CO_2 and H_2O in aerobic conditions. During aerobic decomposition, DO is consumed. Therefore, an excess of organic material can reduce DO levels to zero because the rate of DO consumption is far greater than its rate of replenishment. All aerobic aquatic life is thus killed. There remain certain organisms capable of decomposing such wastes anaerobically (without DO); in these cases, the end products are not only innocuous CO_2 and H_2O, but also ammonia, methane and hydrogen sulphide.

To quantify this process, the important measure of biochemical (or biological) oxygen demand, BOD, has been developed. The BOD of an initial introduction of organic waste (BOD_L) which will demand oxygen during decomposition declines exponentially with time but it is normal to express BOD in terms of oxygen utilised over 5 days (Fig. 35). Complete oxidation may take 21 days, hence BOD_{21}. When the test is performed

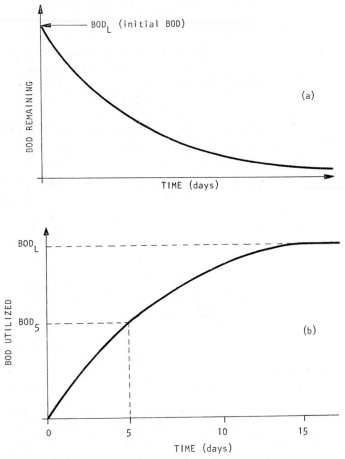

FIG. 35. Relationship between time and (a) BOD remaining and (b) BOD utilised.

under standard conditions the results are termed the BOD_5 and expressed in mg O_2/litre. The relationship between total BOD (BOD_L) and BOD_5 is:

$$BOD_L = \frac{BOD_5}{1 - \exp(-5k)}$$

where k = reaction rate constant.

Domestic sewage has a BOD_5 of around 200 mg/litre; compare this with the 8–10 mg/litre saturation value of O_2 in most waters, and the potential for creating anaerobic conditions is obvious.

The other DO-consuming processes are those of chemical oxidation, for example of sulphides and low valency metal salts. This is particularly relevant to mining, as has already been discussed, and this class of DO consumption is measured by the chemical oxygen demand, COD. COD includes everything likely to demand oxygen, whether or not also bio-degradable, and is also expressed as mg/litre. Therefore if the whole of the pollution is caused by biodegradable organic compounds, $COD = BOD_L$. However, if non-biodegradable oxygen-demanding pollutants are present, $COD > BOD$. COD values of many hundreds can occur in mine effluents, and COD estimations are generally of greater applicability than BOD to mine effluents.

Apart from these direct effects upon DO, temperature increases reduce the solubility of oxygen (from 15 mg/litre at 0°C to 7 mg/litre at 35°C). Demand for oxygen is increased as the number and activity of organisms increases, *i.e.* during the summer, during daylight, and as a result of nutrient additions.

5.4. CONTROL OF WATER POLLUTION

The fundamental requirement for the successful control of water pollution is a knowledge of the quantities and qualities of all waters which may in any way be affected by mining, the quantities of water required during mining and processing, and the quality of these process waters after use. Unless a properly designed monitoring programme has been carried out, any control measures which are taken are unlikely to result in the desired result at the optimum cost.[13,82]

5.4.1. Monitoring
A monitoring programme has to contain the following features.

(*i*) *Definition of Objectives*
Monitoring is usually constrained by climate, accessibility, budget and time, so that unless the objective is clearly understood it may not be achieved. Possible objectives include: baseline studies prior to mining; assessment of current damage; prediction of effects of mining; water-use possibilities, etc.

(*ii*) *Selection of Parameters to be Measured*
The foregoing text gives some idea of the range of pollutants which can occur. The parameters chosen should, where possible, err on the side of superfluity, particularly if baseline studies are being undertaken. It can be useful to analyse preliminary samples for every conceivable contaminant— known as a 'Scan Analysis'. Some of the more important factors to be measured are summarised in Table 23. These vary from mine to mine.

TABLE 23
MAIN COMPONENTS OF A
MONITORING PROGRAMME

Physical	
	Temperature
	Turbidity
	Water flows
Chemical	
	Conductivity
	Alkalinity
	pH
	Hardness
	Colour
	DO
	COD/BOD
	Nitrogen
	Phosphorus
	Metals
	TS/TDS/TSS
Biological	
	Phytoplankton
	Zooplankton
	Benthic organisms
	Fish
	Water fowl

In all cases, seasonal variations must be elucidated, and perhaps diurnal cycles as well.

(iii) Selection of Sampling Locations
   Correct selection is vital if valid samples are to be obtained. Locations should enable a representative sample to be obtained, and be easily accessible for routine sampling. There should be sufficient locations to enable 'freak' results to be noticed and to allow all important positions at which the mine may affect water quality to be monitored. To measure the extent of interference, control stations must also be selected. It is wise to have about 10% of the sample stations as controls. Figure 36 indicates a typical sampling layout.

(iv) Sampling Procedures
   Sampling is not as simple as it at first appears, and trained personnel are often required, especially for biological sampling. For lakes, an appropriate sampler is one able to take a volume of water from any desired depth and bring it unchanged to the surface. Stream sampling can be done by hand, from well-mixed sections of the water. In order to detect annual variations,

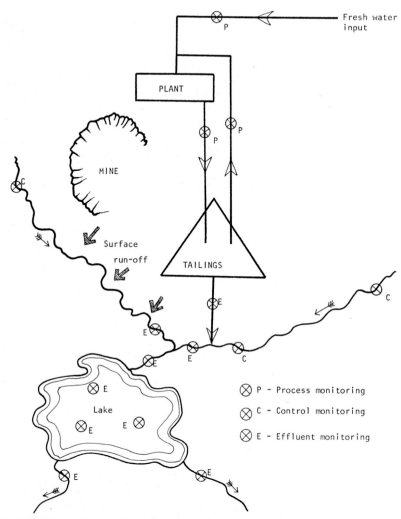

FIG. 36.   Basic components of a water quality monitoring scheme.

samples should be taken at least once per month. Frequency is partly dictated by variations in effluent quality.[14] A process discharging a continuous effluent stream of regular quality may only require daily sampling; a plant operating intermittently, with a highly variable effluent, may be sampled hourly or even continuously. A monitoring programme for baseline purposes should last at least 2 years. All samples should be fully labelled.

(v) Analysis

Pre-preparation of the samples, such as fixing of phytoplankton on the spot, should be done where necessary prior to transmission to the laboratory. Standardised analytical methods should be selected, and rigorously followed throughout. In the event of a method being altered during a monitoring programme samples should be analysed by old and new methods until the relationship between the two is ascertained. Different methods of analysis frequently give different results.[91]

In some cases analyses can be carried out at the sampling stations, either by fixed automatic or portable manual equipment.[49] Automatic equipment is routine for in-plant or other secure stations requiring frequent sampling; DO, pH, temperature, conductivity, TSS, Si, F, NH_3, Na, Cl and many other parameters can be monitored on this basis.

Remote sensing is occasionally used for broad surveys of pollution, using normal colour, infra-red or other types of aerial photography. Satellite-borne multi-spectral scanning has been experimented with.

The detection and monitoring of groundwater flows is expensive. However, any mine which is expected to work below the water table, in permeable rock, should undertake investigations.

5.4.2. Water Control

The physical control of volumes and routes of water at a mine is frequently a major task. It is required to enable the volume of water used in mining to be minimised, to prevent contamination of unpolluted water, and to intercept polluted water and divert it to the appropriate treatment facilities (Section 5.4.3). Control is complicated by the fact that so many sources of water pollution are not point sources and hence are difficult to intercept.[2] The following sections discuss the main water control techniques.[65]

(i) Preplanning

The baseline survey should have identified all waters which may be at risk from a proposed mine. All the aspects of the mine which may cause pollution require examination, so that every phase of the operation can be designed to avoid contamination. It is invariably more satisfactory to avoid pollution rather than subsequently treat water.

(ii) Controlled Mining Techniques

In surface mining there are numerous ways in which water pollution can be lessened by the choice of an appropriate method of mining. Water almost inevitably enters the mine, and diversion ditches may be required to channel it to a sump. Here, settling can occur and (perhaps) oils can be removed, before pumping out for any further treatment. It may be possible to predrain the mine area; the water which is pumped from the boreholes thus remains uncontaminated.

A major source of surface-mine pollution in the U.S.A. is contour mining for coal. Here, the mining method should avoid steep slopes which are susceptible to erosion; should bury pyritic wastes promptly to avoid oxidation; and should maintain a barrier at the low wall to prevent uncontrolled run-off of water. Wherever possible, prompt backfilling can reduce the area of disturbed land and hence lessen polluted run-off.[46] These techniques are illustrated in Fig. 37.

In underground mines there is also some scope for prevention of pollution formation. In coal mining, a large part of water drainage is via roof joints opened as a result of caving. In shallow mines, subsidence may allow water ingress from surface sources. Consequently any mining method which reduces the fracture of overlying strata can be advantageous; most

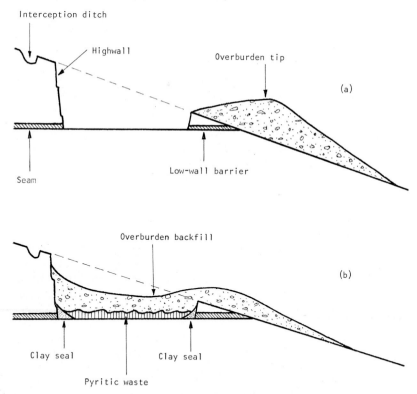

FIG. 37. Design of contour mining method to minimise pollution: (a) by leaving a low-wall barrier to contain run-off; (b) in reclamation, the exposed coal seam is sealed with clay, pyritic wastes are backfilled, and overburden is graded back.

methods, however, decrease the percentage mineral extraction, or otherwise increase costs. In coal mining, working of the seam downdip rather than updip such that it can be flooded at the cessation of mining reduces acid formation due to the exclusion of oxygen (Fig. 38). The design of seals or barriers must take account of the maximum water pressure that will be experienced.[38]

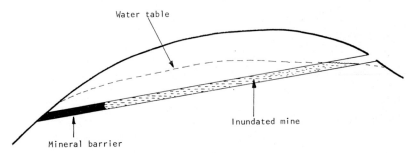

Fig. 38.   Pre-planned flooding of a mine worked down-dip to prevent pyrite oxidation.

(iii) Erosion and Infiltration Control

Infiltration into wastes can result from subsurface water movements, 'leakage' from abandoned mines, or downward percolation of surface waters and rainfall. The outflows of waters which have percolated through contaminated material may be highly polluted. Control of infiltration therefore requires either that the wastes be isolated from the water supply, or that their permeability is decreased. Erosion occurs as a result of rapid water flow over susceptible wastes, and regrading, compaction, diversion and revegetation are the usual techniques.

Permeability can be decreased by compacting the waste, or isolating it with concrete, asphalt, clay or other impervious seal. Clay is the cheapest material and the most practical. An impermeable layer of material can be used as a seal over adit entrances or auger holes, prior to backfilling, to prevent water egress and infiltration into the waste (Fig. 37).

An alternative is to divert surface waters before they reach the waste, and convey them around the area. This reduces pollution and erosion, as well as reducing the volume of contaminated water subsequently produced. Examples are given in Figs. 39, 40 and 41. The installation of underdrains can ease tip drainage and improve stability, and lessen the residence time of water in the tip.

In underground mining there is only limited opportunity to reduce infiltration. The main methods are to seal and grout old bore holes and fissures, and to reduce water penetration through shaft linings.

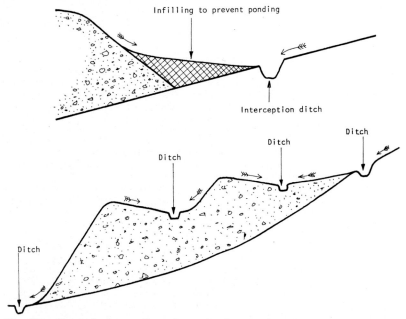

FIG. 39. Control of run-off on tips, using interception ditches and grading to prevent ponding and percolation through tip.

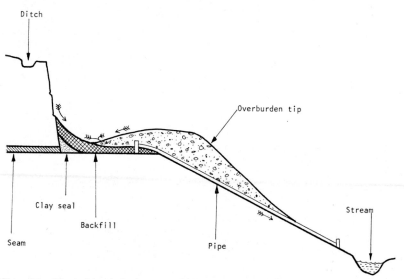

FIG. 40. Use of buried pipe to achieve water control at a reclaimed contour strip mine. Note clay seal over coal seam and grading of backfill to channel run-off to pipe.

FIG. 41.　Elaborate drainage control, including concrete-lined culverts, installed on the Aberfan tips after the disaster.

(iv) Handling Polluted Water

If the formation of polluted water cannot be prevented, techniques must be devised to handle the contaminated flows.

The single most important control technique is water re-use. If a closed-circuit system can be approached or attained, then discharge of effluent can be reduced or eliminated. The principal components of closed recycle systems are treatment ponds for mine water and mill effluent, with associated pumps. Re-use of effluent is mainly practised in drier areas, for in high-rainfall conditions discharge of surplus water may be inevitable. Provision is required for storm-water storage or diversion. Some mines can achieve 100% recycling during summer but discharge water in winter.[50,59,75]

Re-use is also complicated by the quality of water required for mining and processing. In particular, multi-stage milling circuits, using floating and depressing agents in sequence, can suffer from reagent build-up interfering with the flotation process. Lagoons and impoundments function as settling areas to remove suspended solids, but chemical treatments are often required as well.

A technique restricted to arid regions is the use of evaporation ponds to reduce waste water volumes. Lagoons are deliberately constructed with large surface area and shallow depth; if there is an excess of water (*e.g.* pumped minewater) it is therefore possible to evaporate much of it and lessen the volume to be treated by other means.

(v) Regrading

The regrading of mined lands is an important pollution control technique, which can bury pollution-forming materials, reduce erosion and landslides, and eliminate ponding. It is however a part of the wider topic of rehabilitation, the primary objective of regrading usually being land re-use rather than pollution control. The landforms which are created by regrading are controlled to a large degree by the desired after-use but, from the water pollution viewpoint, regrading should achieve gentle slopes which are neither susceptible to serious erosion, nor ponding of water, but which provide adequate conditions for revegetation (see below).

(vi) Revegetation

The procedures for revegetating mined land are discussed in Chapter 10. However, a vegetation cover can often be very effective in reducing water pollution. An herbaceous ground cover stabilises disturbed surfaces, reduces the velocity of run-off and can more-or-less eliminate erosion. Vegetation removes large quantities of water from the soil by the process of transpiration; however, it does not always decrease the infiltration of water into spoil. In particular, trees are relatively ineffective, at least during the decade

FIG. 42. A restored open cast coal site, awaiting the spreading of topsoil. Because the site is very susceptible to erosion by water at this stage, a temporary settling pond has been installed (right) to prevent pollution of nearby streams with suspended solids.

or so of initial growth. Revegetation is usually the cheapest and most satisfactory stabilising method, but should be applied as part of an overall rehabilitation scheme if maximum benefits are to be achieved.

(vii) Mine Sealing

Sealing of abandoned mine entrances, drainage levels, etc., is an important way of preventing water pollution, but can be difficult to apply.[38] The usual objective is to prevent the outflow of polluted water, inundate the workings, and thus prevent oxidation of pyritic materials.[14] Mine seals can be designed to withstand any likely head of water, but the seal is only a small part of the whole containment system. The major part is the perimeter of the mined-out area, the condition of which is often difficult to ascertain. Sealing failures therefore usually occur because the pressure of the impounded water causes it to break out at weak points such as outcrops, fractures, subsided areas, etc. Many seals leak around their edges, due to problems of anchoring them to the strata.

There are considerable dangers in mine sealing, for the ultimate head may not be predictable or controllable. Sudden failures can (and have) cause major pollution destruction and indeed loss of life and property. It is usual therefore to install some form of pressure reduction system, to prevent the maximum designed head from being exceeded. Other methods of preventing oxidation include coating the pyrite with a chemical barrier to air and water, and anti-bacterial agents.[22,32,33] Inundation is still the only practical method.

FIG. 43. To intercept seepage through the tailings dam wall, some metal mines have a small reclaim dam with a pump (right) to return seepage to the main dam.

5.4.3. Water Treatment

The complexities of surface and sub-surface hydrology are such that even the most careful application of the measures described above is unlikely to prevent entirely the formation of polluted water, which requires treatment before discharge. Mine drainage is most commonly treated to remove those pollutants which present a threat to aquatic life, but in some cases the treated effluent may have to form part of a public water supply and be treated to potable water standards.

Waste water for treatment is normally collected in impoundments. These may range from small ponds taking a few tonnes/day to major tailings dams covering several square kilometres and receiving tens of thousands of tonnes/day of slurry wastes. Settling ponds are often used in sequence, sometimes in conjunction with clarifiers or thickeners. The main object of all such ponds is to settle-out and store the large proportion of the solids in the effluent. The relatively clear decanted liquid can then be passed to secondary facilities for any chemical treatment required. However, the larger tailings ponds also allow chemical changes such as oxidation to occur. Obviously the effectiveness of any pond varies according to the retention time of the effluent, which can range from as little as 4 h to several months. Commonly a minimum retention time of 30 days is employed, plus the capacity to hold run-off from the predicted 20-year storm event in order to reduce uncontrolled discharges. The advantages and disadvantages of large tailings ponds are summarised in Table 24. Below are discussed the major treatment processes normally applied to mine effluents.[88]

TABLE 24

TAILINGS PONDS AS TREATMENT SYSTEMS[67]

Advantages	Disadvantages
1. Performs large number of processes, especially TSS reduction	Lacks responsive means of control; hard to optimise the processes performed
2. Often high treatment efficiency	Large land area needed; major influence on hydrology
3. Often the only way of storing solids long-term	Severe rehabilitation problems; long-term safety hazard
4. Evens-out effluent flows	Difficult to isolate from surface run-off
5. Little operating expertise required	Major design expertise required
6. Common and familiar method	High installation costs

(i) Neutralisation

Acidic effluents can be neutralised with any alkaline materials.[23,24,28,29,41,44,45,72,73] By proper alkali selection, neutralisation can also effect precipitation of metals as hydroxides, as well as anions such as fluoride, phosphate and sulphate. The choice between the common alkalis (Table 25) is determined by cost, reactivity, availability, convenience of handling, volume of sludge produced and desired effluent quality. The most commonly used alkalis are lime and hydrated lime.

TABLE 25
COST COMPARISON OF NEUTRALISING AGENTS

Alkali	Basicity factor[a]	Cost ($/tonne)	Cost ($/tonne basicity)
Quick lime (calcium oxide)	1·786	25·35	14·19
Hydrated lime (calcium hydroxide)	1·351	27·56	20·40
Crushed limestone (calcium carbonate)	1·000	8·82	8·82
Dolomite (calcium magnesium carbonate)	0·543	25·90	47·70
Magnesite (magnesium carbonate)	1·186	27·56	23·24
50% Sodium hydroxide	1·250	83·77	67·02
50% Sodium carbonate	0·943	39·68	42·08
Ammonium hydroxide	1·429	71·65	50·14

[a] Equivalent neutralising ability, grammes of $CaCO_3$—equivalent per gramme of alkali.

Many metals precipitate out as insoluble hydroxides (Table 26) at particular pH levels. However, some metals, such as Zn and Al, will redissolve in very alkaline solutions, which can create difficulties if the effluent contains more than one metal. Precipitation of metals generally reduces their level in the effluent to 1 mg/litre or less. Removal of iron is hindered by the fact that, in fresh drainage, the ferrous ion predominates, which precipitates out at pH 9·5. If the iron can be oxidised to the ferric form before neutralisation, the pH can be kept much lower.

Calcium carbonate as limestone is the cheapest source of neutralising capacity, but is not the most effective. Until recently, its use was hampered by 'blinding' and slow reaction rates. It is effective mainly for ferric drainage, and a small particle size is advantageous. Limestone treatment is therefore still at the pilot plant stage.

TABLE 26
MINIMUM pH VALUE FOR
COMPLETE PRECIPITATION OF
METALS AS HYDROXIDES

Metal ion	Minimum pH
Sn^{2+}	4·2
Fe^{3+}	4·3
Al^{3+}	5·2
Pb^{2+}	6·3
Cu^{2+}	7·2
Zn^{2+}	8·4
Ni^{2+}	9·3
Fe^{2+}	9·5
Cd^{2+}	9·7
Mn^{2+}	10·6

Lime neutralisation is a routine process very widely employed. It comprises four steps. First the effluent is neutralised with lime (in slurry form usually) and mixed for 1–2 min. Aeration is then undertaken for 15–30 min, to oxidise iron to the ferric form. The drainage is then settled and classified, and the precipitated sludge disposed of. Plants have been constructed with flow rates of 681–37 850 m³/day (2400–1 336 000 ft³/day), at capital costs of \$43–\$382/m³ capacity (1973 costs) and operating costs in the range 3–25 cents/1000 m³/mg/litre acidity. Recent laboratory studies indicate that if lime treatment is combined with ozonation, better metal removal is achieved, and at lower pH.

There are two important points in connection with neutralisation. First, unless all the acid-forming capacity is removed from the effluent, re-acidification may occur at distances remote from the point of discharge.[16,17] The addition of excess alkali can obviate this, but the discharge of highly alkaline water can itself be damaging. Discharges should usually be in the range pH 6–9. This is illustrated in Table 27 for a discharge initially pH 5–6.

Second, neutralisation results in substantial quantities of precipitated sludge, often of no more than 1–5% solids by weight, which can present a difficult dewatering and disposal problem. Simple lagoons are the usual method currently used, sometimes alternately to allow air drying and subsequent tipping.

(ii) *Flocculation*

Some reagents, such as lime, ferric compounds and aluminium sulphate, are added to waters to promote settling of suspended solids. Flocculation

TABLE 27

DOWNSTREAM ACIDIFICATION OF TAILINGS EFFLUENT
(Schmidt and Conn[92])

Parameter (mg/litre)	Miles (km) downstream of discharge						
	0·2 (0·3)	1·15 (1·85)	2·01 (3·23)	4·65 (7·48)	5·55 (8·93)	6·96 (11·20)	13·09 (21·06)
pH value	3·5	5·0	5·6	2·6	2·5	2·6	3·0
SO$_4$	1 574	1 235	1 230	1 515	1 498	1 520	1 245
Cu	0·010	0·036	0·01	<0·01	0·005	0·005	<0·01
Zn	0·38	2·0	1·8	2·5	1·4	1·4	1·8
Pb	1·0	1·1	3·0	1·2	0·65	0·15	0·70
Fe total	1·5	7·8	16·6	32·9	27·5	36·0	32·5
Ferrous	0·1	6·0	3·7	22·0	1·7	—	6·0
Ferric	1·4	1·8	12·9	10·9	25·8	—	26·5
COD	492	406	324	135	25·9	26·9	63·9

is applied after easily-settled solids have been removed, and is particularly useful for colloidal clays, etc., which settle naturally with great difficulty. Phosphate slimes in Florida are one waste type for which no successful treatment yet exists.

(iii) Precipitation

Although lime and limestone are the most common precipitating agents, as discussed above, other chemicals are also used. Sulphides are extremely effective in reducing metal concentrations—of mercury, for example—but limited to alkaline waters to avoid the generation of poisonous hydrogen sulphide.

Co-precipitation involves the removal of materials from solution by incorporating them within the particles of another precipitate.

The standard method of removing radium, which is not easily precipitated, is co-precipitation with barium chloride in the presence of excess sulphate. Almost all radium is removed as the sulphate, to a level of 1 pCi/litre.

(iv) Reduction

Only applied in mining to a limited extent at present, in the cementation of copper leachates. Possibly this method may also be applied to hexavalent chromium in waste waters, as it is in other industrial processes.

(v) Oxidation

Aeration and oxidation are used for promoting the ferrous-ferric transformation, as already noted. Other applications are in cyanide removal,

CN being oxidised to cyanates (CNO) and then to carbon dioxide and nitrogen. Excess chlorine and a pH of 10–11 are required. Aeration is useful in removing a variety of other COD-producing pollutants from waste waters.

(vi) Biological

Biological treatment of effluent has been applied at one lead mine in Missouri.[93] Eutrophic conditions are utilised to encourage algal blooms, the algae trapping and assimilating suspended and dissolved metals. Dead algae are collected in a final polishing pond before effluent is discharged. Such an ingenious system can only function in a climate which allows adequate algal growth throughout the year, which precludes its application at many mines.

(vii) Untried Methods

There is a great variety of water purification methods which are used to obtain domestic water but which have not been applied to mining. They include adsorption on activated carbon, ion exchange, desalination, ultra-filtration, reverse osmosis, solvent extraction, evaporation, distillation, electrodialysis and freezing.[34–36,39,40,88] Some of these are no more than laboratory trials, while most suffer from technical or economic limitations. However, considerable research is being devoted to their applications in mining and improvements may be anticipated.

5.5. WATER QUALITY STANDARDS

The success or otherwise of pollution prevention and treatment measures can only be judged by reference to a standard, and such standards are widely applied by law in many countries. To define standards is an immensely complex task, requiring consideration of the following matters.

(i) Waters Included

Existing standards commonly cover only surface water. Groundwater has received little attention.

(ii) Basis of Standard

Many effluents are discharged through pipes, culverts, channels, etc., and are termed point sources. Run-off is a non-point source. Point source effluents are usually the only discharges to which effluent standards apply.

(iii) Location of Measurement

Waste water can be sampled before discharge, or after dilution in the receiving water. Both systems are used.

(iv) Receiving Water Variations

The significance of discharge varies according to the nature of the receiving water. It is usually more practicable to specify universal standards, rather than apply different standards to each mine. Both approaches are used.

(v) Parameters

Selection of parameters to be used in the standards tends to cover those which appear to be the most important, and those which can be easily and routinely measured. The parameters vary according to the type of effluent.

(vi) Composition

Two matters are of concern: first, the concentration of pollutant in a discharge; and second, the total quantities involved. For example, an effluent may contain 0·5 mg/litre Cu (0·5 g/m^3) but clearly the effects may vary according to the volume of effluent; 1000 m^3 (35 000 ft^3)/day represents 0·5 kg (1 lb) Cu but 10 000 m^3 (350 000 ft^3)/day is 5 kg (11 lb) Cu discharged, which may be beyond the capacity of the receiving water. Thus both concentration and total waste load can be specified in standards. Because of variability of waste composition, standards often include both 24 h maximum and 30-day average values.

Standards now often take account of the synergistic effects, by specifying that the sum of the ratios of actual:toxic levels for each metal should be less than one. Thus, if the toxic levels for coarse fish for Pb and Zn are 1·0 mg/litre and Cu 0·1 mg/litre, and the actual concentrations 0·5, 0·5 and 0·1 respectively, then

$$\frac{0·5}{1·0} + \frac{0·5}{1·0} + \frac{0·1}{0·1} = 2·0$$

and the effluent is too polluted.

(vii) Costs

Achievement of standards requires that the appropriate technology be available at a price that mines can afford to pay. The technology is available for most water treatment requirements, but many national standards are qualified by the phrase 'economically achievable'.

The extent to which standards are applied varies from country to country and from mineral to mineral. Most non-metal/non-coal mines, for example, frequently have no more than suspended solids/turbidity, pH and oils specified. Where groundwater may be particularly at risk, there may be a prohibition on working below the water table. The main sources of water pollution, metals and coal, usually have to comply with more detailed standards.

The actual levels of pollutants in effluents specified in standards are in a state of flux, and it is fruitless to discuss in detail the situation which exists at the moment. However, some examples are given in Table 28. The major feature of these data is the stringency of the standards which are already being applied; in many respects, effluents have to be equal to, or better than, WHO drinking water standards. The goal in the U.S.A. is, in fact, no discharge of pollutants by 1985 and the probability is that many countries will follow suit.

TABLE 28

MINE EFFLUENT STANDARDS

Parameter (mg/litre)	See footnote							
	1	2	3	4	5	6	7	8
pH units	6–9	6–9	7–9	6–9	6–9	6·5–8·5	5·5–9·5	6·5–9·2
TSS	10	20	30	30	30	—	25	—
Oil and grease	1	10	—	—	—	—	2·5	0·3
CN	0·01	—	—	—	0·1	—	0·5	0·05
Cd	0·005	0·05	0·01	—	—	0·01	—	0·01
Cu	0·05	0·05	0·1	—	—	—	1·0	1·5
Fe	0·1	1·0	—	4·0	—	0·3	—	1·0
Hg	0·001	0·005	—	—	—	—	—	0·001
Pb	0·05	0·2	0·05	—	—	0·05	1·0	0·1
Zn	0·05	0·5	0·5	—	—	5·0	5·0	15·0
COD	—	50	—	—	—	—	75	—
Cr	—	0·2	0·05	—	—	—	0·5	0·05[a]
Mn	—	0·5	—	—	—	0·05	—	0·5
BOD$_5$	—	—	20	—	20	—	—	—
As	—	—	0·01	—	—	0·05	0·5	0·05
Total toxic metals	—	—	—	—	0·5	—	—	—

[a] U.S. drinking water standard.
1. Draft U.S. standards for copper mills in wet areas, 30-day average.[88]
2. Draft U.S. standards for Pb/Zn mine water, 30-day average.[88]
3. Irish standards for Tara Pb/Zn mine; post dilution measurements. Pre-dilution levels specified for Pb (0·02), CN (0·1) and Fe (1·0).
4. U.S. effluent limitations for coal mines, 1973.[91]
5. Generalised British standards.
6. Petracco.[86] Australian standards.
7. Stander et al.[85] South African general standards.
8. World Health Organisation, maximum permissible concentrations in drinking water.[70]

REFERENCES

1. Klingensmith, R. S. (1962). Technical aspects of control of drainage from active mines, in *Proc. 1st Symposium on Coal Mine Drainage*, Bureau of Environmental Health, Pittsburgh, Pa., pp. 35–9.
2. Lorell, H. L. (1962). *ibid.*, pp. 41–52.
3. Wheeler, W. H. (1962). Reclamation of strip mined land in Pennsylvania, *ibid.*, pp. 71–3.
4. Klashman, L. M. (1962). Improvement of acid polluted streams by water resources development, *ibid.*, pp. 75–7.
5. Clausen, H. T. (1973). Ecological aspects of slimes dams construction. *J. Sth African Inst. Min. Metall.*, **74**(5), 178–83.
6. Ruthven, J. A. and Cairns, J. (1973). Response of freshwater protozoan artificial communities to metals, *J. Protozool.*, **20**(1), 127–35.
7. Prosser, M. J. (1974). *Plant Discharge to the River Boyne*, British Hydromechanics Research Association, report of research 1238, BHRA, Bedford.
8. Stall, J. B. (1972). Effects of sediments on water quality, *J. Environ. Quality*, **1**(4), 353–9.
9. Oschwald, W. R. (1972). Sediment-water interactions, *ibid.*, 360–5.
10. Rudd, R. T. (1973). The impact of slimes-dam formation on water quality and pollution, *J. Sth African Inst. Min. Metall.*, **74**(5), 184–92.
11. How Moons Hill quarry faces are worked (1973). *Mine and Quarry*, **2**(2), 9–15.
12. Prince, A. T. (1971). The water environment, *Canadian Mining & Metallurgy (CIM) Bulletin*, **64**(712), 51–5.
13. Frame, C. H. (1974). Construction and operation of a mining complex in an urban area, in *21st Ontario Industrial Waste Conference*, Toronto, 35 pp.
14. Cairney, T. and Frost, R. C. (1975). A case study of mine water quality deterioration, Mainsforth Colliery, County Durham, *J. Hydrology*, **25**, 275–93.
15. Wixson, B. G. and Jennett, J. C. (1974). Interim Report to NSF/RANN (New Lead Belt study). University of Missouri-Rolla.
16. Ontario Water Resources Commission (1971). *Water Pollution from the Uranium Mining Industry in the Elliot Lake and Bancroft areas*, Vol. 1, OWRC, Ontario.
17. Ontario Water Resources Commission (1972). *Water Pollution from the Uranium Mining Industry in the Elliot Lake and Bancroft areas*, Vol. 2, OWRC, Ontario.
18. Smith, E. E., Shumate, K. S. and Svanks, K. (1968). Sulfide to sulfate reaction studies, in *2nd Symposium on Coal Mine Drainage Research*, Coal Industry Advisory Committee, Pittsburgh, Pa., pp. 1–11.
19. Singer, P. C. and Stumm, W. (1968). Kinetics of the oxidation of ferrous iron, *ibid.*, pp. 12–34.
20. Deul, M. (1968). Turbidity measurements as an indicator of solids content of neutralized mine water, *ibid.*, pp. 35–8.
21. Kim, A. G. (1968). An experimental study of ferrous iron oxidation in acid mine water, *ibid.*, pp. 40–5.

22. Shearer, R. A., Everson, W. A. and Mausteller, J. W. (1968). Reduction of acid production in coal mines with use of viable anti-bacterial agents, *ibid.*, pp. 98–106.
23. Mihok, E. A. and Chamberlain, C. E. (1968). Factors in neutralizing acid mine waters with limestone, *ibid.*, pp. 265–73.
24. Calhoun, J. A. (1968). Treatment of mine drainage with limestone, *ibid.*, pp. 386–91.
25. Lvovitch, M. I. (1972). World water balance, in *Proc. Reading Symposium*, Vol. 2, UNESCO, Geneva.
26. Masters, G. M. (1974). *Introduction to Environmental Science and Technology*, John Wiley, New York and London.
27. Kaufman, A. and Nadler, M. (1966). *Water Use in the Mineral Industry*, U.S. Bureau of Mines, I.C. 8285, Washington, D.C.
28. Ford, C. T. (1970). Selection of limestones as neutralizing agents for coal mine water, in *3rd Symposium on Coal Mine Drainage Research*, Coal Industry Advisory Committee, Pittsburgh, Pa., pp. 27–51.
29. Scott, R. B. and Wilmoth R. C. (1970). Neutralization of high ferric iron acid mine drainage, *ibid.*, pp. 66–93.
30. Lau, C. M., Shumate K. S. and Smith E. E. (1970). The role of bacteria in the pyrite oxidation kinetics, *ibid.*, pp. 114–22.
31. Ramsey, J. P. (1970). Control of acid mine drainage from refuse piles and slurry lagoons, *ibid.*, pp. 138–44.
32. Wallitt, A. L., Jasinski, R. and Keilin, B. (1970). Prevention of acid mine drainage: silicate treatment of coal mine refuse piles, *ibid.*, p. 180.
33. Shearer, R. E., Mausteller, J. W., Everson, W. A. and Zimmerer, R. P. (1970). Characteristics of viable anti-bacterial agents used to inhibit acid-producing bacteria in mine wastes, *ibid.*, pp. 188–99.
34. Mason, D. G. (1970). Treatment of mine drainage by reverse osmosis, *ibid.*, pp. 227–40.
35. Kremen, S. S., Nusbaum, I. N. and Riedinger, A. B. (1970). The reclamation of acid mine water by reverse osmosis, *ibid.*, pp. 241–66.
36. Rose, J. L. (1970). Treatment of acid mine drainage by ion exchange processes, *ibid.*, pp. 267–78.
37. Riley, R. V. and Rinier, J. A. (1972). Reclamation and mine tip drainage in Europe, in *4th Symposium on Coal Mine Drainage Research*, Coal Industry Advisory Committee, Pittsburgh, Pa., pp. 1–14.
38. Foreman, J. W. (1972). Evaluation of mine sealing in Butler County, Pennsylvania, *ibid.*, pp. 83–95.
39. Holmes, J. and Schmidt, K. (1972). Ion exchange treatment of acid mine drainage, *ibid.*, pp. 179–200.
40. Wilmoth, R. C., Mason, D. G. and Gupton, M. (1972). Treatment of ferrous ion acid mine drainage by reverse osmosis, *ibid.*, pp. 115–56.
41. Wilmoth, R. C., Scott, R. B. and Hill, R. D. (1972). Combination limestone-lime treatment of acid mine drainage, *ibid.*, pp. 244–65.
42. West, E. C. (1972). Corrosion of navigation facilities, *ibid.*, pp. 344–6.
43. Herricks, E. E. and Cairns, J. (1972). The recovery of stream macrobenthic communities from the effects of acid mine drainage, *ibid.*, pp. 370–98.

44. Ford, C. T. (1974). Use of limestone in AMD treatment, in *5th Symposium on Coal Mine Drainage Research*, National Coal Association, Louisville, Kentucky, pp. 205–228.

45. Grandt, A. F., McDonald, D. G. and Yocum, H. (1974). Studies of lime-limestone treatment of acid mine drainage, *ibid.*, pp. 229–45.

46. McCarthy, R. E. (1973). Preventing the sedimentation of streams in a Pacific Northwest coal surface mine, in *Research & Applied Technology Symposium on Mined-Land Reclamation*, National Coal Association, Pittsburgh, Pa., pp. 277–86.

47. Perhac, R. M. (1972). Distribution of Cd, Co, Cu, Fe, Mn, Ni, Pb and Zn in dissolved and particulate solids from two streams in Tennessee, *J. Hydrology*, **15**, 177–86.

48. Hawley, J. R. (1973). *Mining Effluents and their Control in the Province of Ontario, Canada*, presented at Annual Meeting of SME/AIME, Chicago, Illinois, 26 pp.

49. Stedman, D. O. (1972). Automated methods of specific chemical analysis in water pollution control, in *Aspects of Environmental Protection* (ed. S. H. Jenkins), IP Environmental, London, pp. 89–125.

50. Browne, R. N. and Butler, H. R. (1970). *Water Reuse at INCO's Sudbury Mills*, presented at Annual Meeting of Canadian Mineral Processors, Ottawa, Ontario, 14 pp.

51. Salazar, R. C. and Gonzales, R. I. (1973). Design, construction and operation of the tailings pipelines and underwater tailings disposal system of Atlas Consolidated Mining & Development Corporation in the Philippines, in *Tailing Disposal Today* (ed. C. L. Aplin and G. O. Argall), Miller Freeman, San Francisco, pp. 477–511.

52. Evans, J. B., Ellis, D. V. and Pelletier, C. A. (1973). The establishment and implementation of a monitoring program for underwater tailing disposal in Rupert Inlet, Vancouver Island, British Columbia, *ibid.*, pp. 512–52.

53. Jones, J. R. E. (1964). *Fish and River Pollution*, Butterworths, London.

54. Carpenter, K. E. (1924). A study of the fauna of rivers polluted by lead mining in the Aberystwyth district of Cardiganshire, *Ann. appl. Biol.*, **11**, 1–23.

55. Carpenter, K. E. (1925). On the biological factors involved in the destruction of river-fisheries by pollution due to lead mining, *Ann. appl. Biol.*, **12**, 1–13.

56. Carpenter, K. E. (1926). The lead mine as an active agent in river pollution, *Ann. appl. Biol.*, **13**, 395–401.

57. Hawley, J. R. (1972). *Use, Characteristics and Toxicity of Mine-Mill Reagents in Ontario*, Ministry of the Environment, Ontario.

58. Williams, R. E. (1975). *Waste Production and Disposal in Mining, Milling and Metallurgical Industries*, Miller Freeman, San Francisco.

59. Joe, E. G. and Pickett, D. E. (1975). Water re-use in Canadian ore-concentration plants—present status, problems and progress, in *Minerals and the Environment* (ed. M. J. Jones), Institution of Mining & Metallurgy, London, pp. 133–48.

60. Purves, J. B. (1975). Aspects of mining and pollution control in southwest England, *ibid.*, pp. 159–79.

61. Tofflemire, T. J. and Van Alstyne, F. E. (1973). Deep-well injection, *J. Water Poll. Control. Fed.*, **45**(6), 1103–8.
62. Foehrenbach, J. (1973). Eutrophication, *J. Water Poll. Control Fed.*, **45**(6), 1237–46.
63. Abdullah, M. I. and Royle, L. G. (1972). Heavy metal content of some rivers and lakes in Wales, *Nature*, **238**, 329–30.
64. *Ground water pollution from subsurface excavations* (1973). U.S. Environmental Protection Agency, Washington, D.C.
65. *Processes, procedures and methods to control pollution from mining activities* (1973). U.S. Environmental Protection Agency, Washington, D.C.
66. *Effluent guidelines and receiving water quality objectives for the mining industry in Ontario* (1973). Ministry of the Environment, Ontario.
67. Bell, A. V. (1974). The tailings pond as a waste treatment system, *Canadian Mining & Metallurgical (CIM) Bulletin*, **77**, 73–8.
68. Stefanko, R. (1970). Subsurface disposal of mine water, *Trans. Society of Mining Engineers, AIME*, **247**, 54–60.
69. Klein, L. (1962). *River Pollution Vol. 2: Causes and Effects*, Butterworths, London, pp. 119–22.
70. *International Drinking Water Standards* (1971). World Health Organisation, Geneva.
71. Williams, R. E., Wallace, A. T. and Mink, L. L. (1971). Impact of a well managed tailings pond system on a stream, *Mining Congress Journal*, **57**(10), 48–56.
72. Deul, M. (1969). Limestone in mine drainage treatment, *Mining Congress Journal*, **55**(10), 88–91.
73. Rivett, L. S. and Oko, U. M. (1971). Tailings disposal, generation of acidity from pyrrhotite and limestone neutralization of wastewater at Falconbridge's Onaping mines, *Canadian Mining & Metallurgical (CIM) Bulletin*, **74**, 186–91.
74. Beverly, R. G. (1968). Unique disposal methods are required for uranium mill waste, *Mining Engineering*, **20**(6), 52–6.
75. Bailey, R. P. (1970). Case for reclamation of mineral processing water, *Canadian Mining Journal*, **91**(6), 87–92.
76. Hawley, J. R. (1972). *The Problem of Acid Mine Drainage in Ontario*, Ministry of the Environment, Ontario.
77. Holmes, G. H. (1966). *Water Requirements and Uses in Nevada Mineral Industries*, U.S. Bureau of Mines, I.C. 8288, Washington, D.C.
78. Roe, L. A. (1966). Basic water management concepts for mineral processing projects, *Engineering & Mining Journal*, **167**(9), 189–94.
79. Cederstrom, D. J. (1971). Hydrologic effects of strip mining west of Appalachia, *Mining Congress Journal*, **57**(3), 46–50.
80. Emrich, G. H. and Merritt, G. L. (1969). Effects of Mine Drainage on Groundwater, *Ground Water*, **1**, 27–32.
81. Campbell, R. S. and Lind, O. T. (1969). Water quality and aging of strip-mine lakes, *J. Water Poll. Control Fed.*, **41**(11), 1943–55.
82. Oko, U. M. (1971). Monitoring waste water quality at Falconbridge's Sudbury operations, *Canadian Mining Journal*, **92**(6), 72–5.
83. Schmidt, J. W. and Conn, K. (1971). Abatement of water pollution in the base metal mining industry, *Canadian Mining Journal*, **92**(5), 49–52.

84. Hawley, J. R. and Shikaze, K. H. (1971). The problem of acid mine drainage in Ontario, *Canadian Mining Journal*, **92**(6), 82–93.
85. Stander, G. J., Henzen, M. R. and Funke, J. W. (1970). The disposal of polluted effluents from mining, metallurgical and metal finishing industries, their effects on receiving water and remedial measures, *J. Sth African Inst. Min. Metall.*, **71**(5), 95–103.
86. Petracco, F. F. (1973). Pollution control in the base metal mining industry, *Australian Mining*, **65**(2), 50–2.
87. Gale, N. W., Hardie, M. G. and Jennett, J. C. (1972). Transport of trace pollutants in lead mining waste waters, in *Trace Substances in Environmental Health*, **6**, 95–106.
88. *Draft development document for effluent limitations guidelines and standards of performance for the ore mining and dressing industry, point source category* (1975). U.S. Environmental Protection Agency, Washington, D.C.
89. Murray, C. R. and Reeves, E. B. (1972). *Estimated Use of Water in the United States in* 1970, U.S. Geological Survey, Circular 676, Washington, D.C.
90. Davis, D. W., Brown, T. S. and Long, B. W. (1972). Dewatering sludge by using rotary vacuum pre-coating filtration, in 4*th Symposium on Coal Mine Drainage Research*, Coal Industry Advisory Committee, Pittsburgh, Pa., pp. 201–33.
91. *Methods for identifying and evaluating the nature and extent of non-point sources of pollutants* (1973). U.S. Environmental Protection Agency, Washington, D.C.
92. Schmidt, J. W. and Conn, K. (1969). Abatement of pollution from mine waste waters, *Canadian Mining Journal*, **90**(6), 54–60.
93. Gale, N. L., Hardie, M. G., Whitfield, J. and Marcellus, P. (1974). The impact of lead mine and mill effluent on aquatic life, presented to *University of Minnesota Mining Symposium*, Duluth, 20 pp.

6

Noise (Excluding Air Blast)

6.1. INTRODUCTION

Noise is normally defined as objectionable or unwanted sound. Sound is produced by a source causing vibrations in the medium surrounding it. These vibrations are propagated as waves in the form of pressure variations and are termed sound waves if they fall within the range capable of exciting the sense of hearing. The most common propagating medium is air, in which sound travels with a velocity of 344 m/s (1130 ft/s) at 20°C and normal atmospheric pressure. The two most important aural characteristics of a sound wave are frequency and amplitude.

The range of audible frequencies varies from one individual to another but generally lies between 20 and 20 000 Hz although the ability to hear the higher frequencies reduces with age. The human ear cannot distinguish all frequencies within the audible range with equal facility. This is illustrated in Fig. 44 and it can be seen that sounds between 1000 and 4000 Hz are normally most easily heard.

The amplitude of the sound wave corresponds to the change in pressure caused by the wave and thus is related to the intensity of the sound. This change in pressure is usually called the sound pressure and the range to which the ear responds is approximately 0·000 02–20 Pa. The faintest sound that can be heard is termed the threshold of audibility and the sound pressure at which physical discomfort becomes apparent is the threshold of feeling or audible pain. The ratio of these two thresholds is about 10 million to one and it has been found inconvenient to use a linear unit to cover such wide limits. Consequently a logarithmic unit, the decibel (dB), is normally used. The decibel is ten times the logarithmic ratio of two powers. Sound power is proportional to the square of sound pressure and hence the number of decibels between two pressures p_1 and p_2:

$$dB = 10 \log_{10} \left(\frac{p_2}{p_1}\right)^2 = 20 \log_{10} \left(\frac{p_2}{p_1}\right)$$

As stated above, the decibel expresses a ratio between two sound pressures and it is necessary to use a reference pressure in order to standardise decibel values. By international agreement the reference level

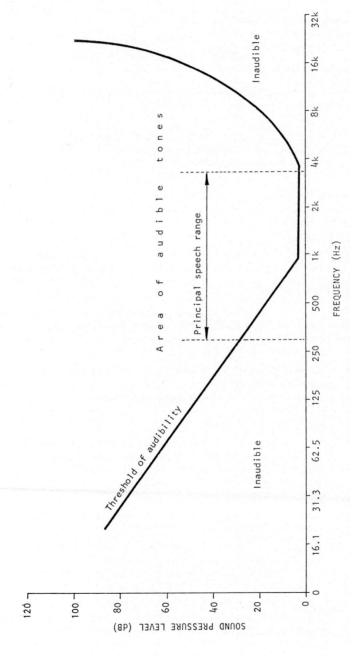

FIG. 44. The auditory field (after Bell[1]).

is 0·000 02 Pa which is approximately the threshold of audibility of a 1000 Hz pure tone. Thus, for a sound pressure of p, in pascals:

$$\text{sound pressure level} = 20 \log_{10} \left(\frac{p}{0 \cdot 000\ 02} \right) \text{dB}$$

Because the scale is logarithmic, sound pressure levels expressed in decibels cannot be added arithmetically. For example, 60 dB + 60 dB = 63 dB and not 120 dB. The addition of 3 dB represents a doubling of sound pressure level. This is an important point in discussing noise problems. The use of the decibel scale permits large differences in sound pressure to be expressed in numbers of convenient size, but it can lead to intuitive errors in understanding the increase or reduction in noise levels. If the nature of the decibel is not understood, a reduction from 63 to 60 dB can seem insignificantly small whereas it does in fact represent a halving of the sound pressure level.

Because the human ear does not respond with equal sensitivity at all frequencies within the audible range, sound pressure level by itself does not accurately express a listener's subjective impression of the loudness or intensity of the sound. The unit of loudness level is the phon, which is derived by a subjective comparison with a 1000 Hz reference tone. When the reference tone and the sound under consideration are judged to be equally loud, the loudness level of the sound in phons is numerically equal to the sound pressure level of the reference tone. Figure 45 shows equal loudness contours and it can be seen, for example, that a sound of 60 dB at 1000 Hz gives the same subjective impression of loudness as a sound of 68 dB at 100 Hz. Because loudness level is a subjective phenomenon, there may be some disagreement between listeners, but differences seldom exceed 6 phons.

Noise is measured with a sound level meter which consists of a microphone, amplifier, weighting networks and an indicating meter. As is apparent from the discussion above, it is sometimes desirable that the meter should give a direct reading of loudness rather than sound pressure level and it is for this reason that the weighting networks are incorporated. Historically three networks, A, B and C, have been used with frequency response characteristics simulating the 40, 70 and 100 dB equal loudness contours respectively. However, extensive experience has shown that the A weighting correlates well with subjective noise ratings even at high levels and consequently it is in general use whenever a measurement corresponding to the subjective reaction to a noise is required. The sound level meter is an internationally standardised piece of equipment, of value because it provides a simple means of ranking noises of similar character, but it should not be assumed that the reading necessarily corresponds precisely with the loudness level in phons.

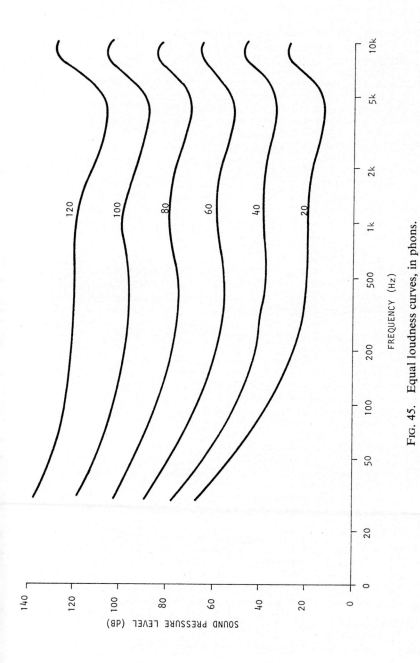

FIG. 45. Equal loudness curves, in phons.

When recording sound level meter readings it is important to state which, if any, weighting network was used, *e.g.* 65 dB(A). Table 29 gives the dB(A) values of a number of typical noises.

Noises may be characterised not only by loudness, but also by tonal range and duration. As will be discussed later, all three of these factors can be important in assessing the problems which a particular noise can cause.

TABLE 29

TYPICAL NOISE LEVELS

Noise level (dB(A))	Typical examples	Relative sound energy	Subjective assessment
120	Threshold of pain	1 000 000 000 000	
110	Pop music concert	100 000 000 000	
	Compressed air powered rock drill at 1 m (3 ft)		Deafening
100	Steel riveter at 5 m (16 ft)	10 000 000 000	
90	Heavy diesel vehicle at 8 m (26 ft)	1 000 000 000	
	Muffled pavement breaker at 7 m (23 ft)		Very loud
80	Ringing alarm clock at 1 m (3 ft)	100 000 000	
	Small air compressor at 7 m (23 ft)		
70	Inside small car	10 000 000	
60	Normal conversation	1 000 000	Loud
50	Quiet office	100 000	
40	Rural House	10 000	Moderate
30	Whisper	1 000	
20	Quiet Room	100	Faint
10	Rustle of a leaf	10	
0	Threshold of audibility	1	Very faint

Where it is desirable to investigate the tonal characteristics of a noise, the sound level meter can be used with octave band or one-third octave band pass filters for frequency analysis. By international agreement the bands are centred on a series of preferred frequencies, an example of which is given in Table 30. Figure 46 shows a typical frequency analysis.

A point source in free air emits sound equally in all directions and thus at any instant the sound is distributed evenly over the surface of a sphere whose centre is the source. The intensity or sound power per unit area decreases as the square of the distance from the source following an inverse square law. A doubling of distance from the source therefore yields a theoretical attenuation of 6 dB and consequently when measuring

TABLE 30
STANDARD OCTAVE FREQUENCY BANDS

Band (Hz)	Cut-off frequencies (Hz)
Low pass	31·5–90
125	90–180
250	180–355
500	355–710
1000	710–1400
2000	1400–2800
4000	2800–5600
High pass	5600–8000

noise levels from a point source it is essential to record the distance between the source and the point of measurement.

In practice the operation of the inverse square law is modified by atmospheric and site conditions. There is some reduction in airborne sound due to air absorption. This is insignificant for low frequencies, which closely follow the theoretical attenuation, but is detectable at higher frequencies, e.g. at 4000 Hz the attenuation due to air absorption is approximately 1 dB per 65 m (200 ft). An increase in relative humidity

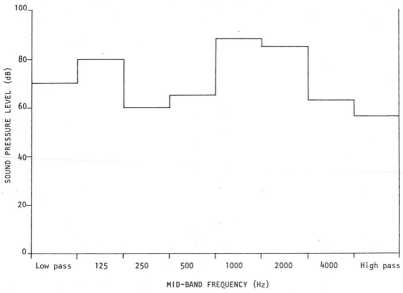

FIG. 46.   Example of an octave band analysis.

causes slightly greater attenuation and this effect is most marked at higher frequencies.

The velocity of sound decreases as temperature decreases and variations in atmospheric temperature can thus affect the propagation of sound, causing the sound rays to follow curved paths rather than straight lines. If temperature decreases with altitude, as is normally the case during daytime, sound rays are diffracted upwards and noise problems are reduced. This is illustrated in Fig. 47 which shows the effects of a temperature increase with altitude and of a change in temperature gradient from

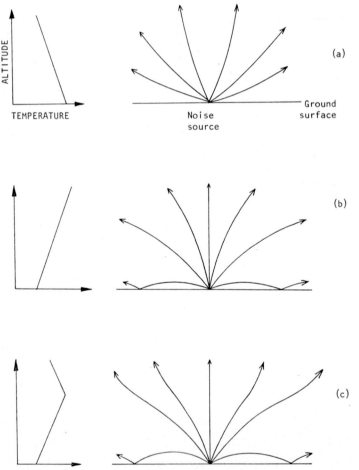

FIG. 47.   Effect of temperature gradients upon sound path: (a) negative gradient; (b) positive gradient; (c) inversion.

negative to positive, commonly called a temperature inversion. In both cases some sound is refracted back towards the surface of the earth and increased noise levels are experienced where a focus occurs. Positive temperature gradients and temperature inversions are common at night and partly explain the better audibility of sound at night which is frequently reported. The lack of background noises is also partly responsible for this.

Variations of wind velocity with altitude similarly cause diffraction of sound rays, although usually the effect is less critical than temperature. Under normal atmospheric conditions wind speed increases with altitude, and sound rays are diffracted downward in the wind direction and upward against the wind, thus causing increased noise levels downwind and improved attenuation upwind. Occasionally wind velocity decreases with altitude causing the unusual situation that attenuation is increased downwind and noise levels increase upwind.

When measuring noise levels for comparison purposes it is desirable to record the atmospheric conditions prevailing in addition to the noise reading and the distance from the source.

Noise is reflected by the ground and the surface of walls, buildings and other structures and is diffracted around boundaries causing sound energy to penetrate into geometrical shadows. To minimise inconsistencies, which could amount to several decibels, British Standard (BS) 4142: 1967 recommends that environmental noise measurements be made at a height of 1·2 m (4 ft) and at least 3·6 m (12 ft) away from any reflecting structure.

6.2. THE PROBLEMS OF NOISE

It is usual to segregate noise problems into those which are physiological and those which are psychological. However, the World Health Organisation defines health as 'a state of complete physical, mental and social well-being, and not merely an absence of disease and infirmity'. If such a comprehensive definition is accepted, the problems caused by noise could all be described as health problems.

It has long been recognised that exposure to high noise levels can result in temporary or permanent impairment of hearing ability. A temporary threshold shift is a reduction in the threshold sensitivity of hearing which gradually passes off when exposure to the noise ceases, and a permanent threshold shift occurs when the ear does not recover. It is not easy to identify positively a permanent loss of hearing occasioned by noise since, except in extreme cases, the effect is only apparent over a number of years and there is in any case a natural loss of hearing acuity with increasing age. The problem has been extensively investigated in a number of countries and a variety of standards, codes of practice and legislation has resulted. Although there are minor differences between the accepted standards in various countries, the concept of noise exposure receives universal acceptance. This concept relates a particular noise level to a maximum exposure time. Table 31 shows the Code of Practice issued by the U.K. Department of Employment. There is little difference between this and the International Organisation for Standardisation (I.S.O.) recommendations, the Office of Safety and Health Administration (O.S.H.A.) standards used in the U.S.A. and the criteria adopted in other countries.

TABLE 31
U.K. DEPT. OF EMPLOYMENT CODE OF PRACTICE

Sound level (dB (A))	Maximum exposure (h/day)
90	8
93	4
96	2
99	1
102	0·5
105	0·25
108	0·125

The U.K. code recommends that no one with unprotected ears shall be exposed to a sound level above 135 dB or to an impact noise exceeding 150 dB and describes a method for calculating an equivalent continuous sound level when the level of sound varies throughout the day.

Hearing damage is a serious potential hazard for employees within the mining industry. However, it is most unlikely that any member of the general public would be subjected to a noise level exceeding 90 dB(A) by any aspect of mining except blasting or heavy vehicle movements and then only for limited periods of time. Consequently it may be concluded that the noise problems created by mining in the external environment are psychological rather than physiological.

Nuisance or annoyance is a subjective phenomenon and hence is not open to precise specification. In Britain, nuisance is nowhere defined by statute but is a concept well established in common law and represents 'unlawful interference with a person's use or enjoyment of land, or of some right over or in connection with it' (Reed v. Lyons and Co., 1945). A famous and much quoted judgement (Walker v. Selfe, 1851) described the nature and extent of interference which constitutes a nuisance, with particular reference to noise:

'every person is entitled as against his neighbour to the comfortable and healthy enjoyment of the premises occupied by him and in deciding whether, in any particular case, his right has been interfered with and a nuisance thereby caused, it is necessary to determine whether the act complained of is an inconvenience materially interfering with the ordinary physical comfort of human existence not merely according to elegant or dainty modes and habits of living but according to plain and sober and simple notions obtaining among English people'.

Many states afford their citizens some form of protection against nuisance and, whatever the form of legislation, face essentially the same problem of translating definitions such as the one given above into a well-defined, practical, enforceable code which permits objective assessment of a subjective reaction.

Complaints of noise nuisance generally arise when a noise interferes with work, communication, recreation or sleep. The reaction to a particular noise depends upon many factors including the age, occupation and susceptibilities of the listener; the location (urban, rural, residential, industrial); the time of day or night; and the characteristics of the noise.

A noise can only intrude if it differs in character from the normal ambient noise. The most important factors in determining the intrusiveness of a noise are sound pressure level, frequency characteristics and duration. Most objective attempts to assess nuisance adopt the technique of comparing the noise with actual ambient noise levels or with some derived criterion.

Table 32 shows the approximate reaction of the majority of people to changes in sound pressure level. It may be concluded that any continuous noise which increases the ambient noise level by 10 dB or more is likely to give rise to a substantial number of complaints. This is the basis of British Standard 4142: 1967, which provides a means of assessing the nuisance potential of industrial noise. If a fresh noise source raises the ambient noise level by 10 dB(A) or more (after making corrections for time of day, type of locality and tonal characteristics) then complaints will arise. This approach has the virtue of being relatively simple and has received general acceptance for continuous noises which do not fluctuate markedly and are not predominantly within a small frequency range. The importance of

TABLE 32

HUMAN REACTION TO INCREASES IN SOUND PRESSURE LEVEL

Increase in sound pressure level (dB)	Human reaction
Under 5	Unnoticed to tolerable
5–10	Intrusive
10–15	Very noticeable
15–20	Objectionable
Over 25	Very objectionable to intolerable

location is illustrated in Table 33, which shows the approximate maximum background noise levels likely to be acceptable in a variety of situations.

A number of methods for determining nuisance from fluctuating or intermittent sources are in use. The simplest and most easily enforceable is to adopt a maximum sound pressure level which must not be exceeded. A more sophisticated technique is to base a standard upon the calculated

TABLE 33

MAXIMUM ACCEPTABLE BACKGROUND NOISE LEVELS

Location	Maximum background noise level ($dB(A)$)
Theatres and concert halls	30
Schools	35
Houses at night	35
Private Offices	40
Houses in daytime	45
Restaurants	45
Public Offices	50
Typing Offices	55
Shops	55
Workshops (depending on use)	50–75

equivalent continuous noise level (L_{eq}) or upon the L_{10}, L_{50} or L_{90} values, which are the noise levels exceeded for 10%, 50% and 90% of the time during which measurement takes place. The Wilson Report suggested that the L_{10} standard for the living rooms and bedrooms of houses should be as shown in Table 34, but this suggestion was never implemented because of its stringency. Figure 48 illustrates the use of equivalent continuous noise level standards by the U.S. Dept. of Housing and Urban Development.

TABLE 34

WILSON REPORT SUGGESTED L_{10}
RESIDENTIAL STANDARDS[4]

Situation	L_{10} standard ($dB(A)$)	
	Day	Night
Country areas	40	30
Suburban areas, away from main traffic routes	45	35
Busy urban areas	50	35

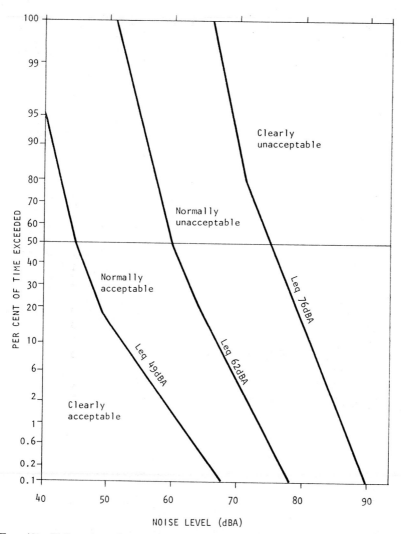

FIG. 48. U.S. noise criteria for residential areas, using calculated equivalent noise levels (after Beranek [5]).

The problem of using standards based on L_{eq}, L_{10}, L_{50} or L_{90} is that monitoring can be very expensive and tedious and thus can render the standard difficult to enforce.

Noises with their energy concentrated in a narrow band of the frequency spectrum can be especially intrusive, particularly if the frequency is high. Typical examples are the hum or whine of a fan or generator. It is customary to make allowance for such tonal characteristics by arbitrarily increasing the measured sound pressure level but this method is not entirely satisfactory since complaints of tonal noises are received at sound pressure levels indistinguishable from the ambient noise level.

In conclusion, there are reasonable satisfactory techniques for assessing nuisance from steady, continuous, atonal sources but noises which fluctuate, are intermittent or markedly tonal in character present problems which have not been entirely resolved. Some important noise sources in the mining industry fall into the latter category.

6.3. NOISE SOURCES AND LEVELS

The three major categories of noise source in mining are fixed plant, mobile plant used internally, and external transport movements.

Fixed plant comprises a very wide range of equipment including crushers and grinders; screens; conveyers and storage bins; ore processing machinery; fans and air conditioners; electrical gear; workshops and loading facilities. Fixed plant is normally located in one or more central areas of the mine and is frequently enclosed to protect men and equipment from the elements and to improve security. It would be meaningless to quote 'typical' overall noise levels emanating from fixed plant sites since the range is very wide depending upon the size of the operation, the degree of screening and the quantity and type of equipment used. Table 35 indicates the range of noise levels commonly associated with various fixed plant installations.

Mobile plant used on site is associated with drilling, blasting, loading, haulage or service operations. A great variety of type and size is found throughout the mining industry and some examples of noise levels are given in Table 36.

Some external transport movements are associated with all mining operations, the purposes normally being the supply of materials, the despatch of valuable product and waste and the movement of personnel. Although almost every method of transport is used by the mineral industry, the preponderance of traffic, measured in tonne-kilometres, travels by road and it is this method which causes the greatest noise problems. Chapter 9 deals fully with the general environmental problems of external transport and discussion here is confined to the question of noise.

TABLE 35
NOISE LEVELS FROM FIXED PLANT INSTALLATIONS

Equipment	Noise level (dB(A))	Measurement location
Electrical ventilation fans	90–100	At 5 m (16 ft)
Compressed air fans	up to 110	At 5 m (16 ft)
Jaw crusher	90–100	Operator position
Cone crusher	92–98	Operator position
Compressed air hammer	104–112	Operator position
Drill sharpeners	102–122	Operator position
Ball mill	up to 100	Operator position
Conveyors	82–113	Operator position
Pumps	89–100	Operator position
Flotation	63–91	Inside flotation building
Furnaces	73–104	Inside furnace building
Quarry plant area	88–102	Various external sites in general plant area
3000 ft³/min (85 m³/min) Compressor house with corrugated enclosure	52	300 m (1000 ft)
7500 tonnes/day Concentrator	70	100 m (330 ft)

TABLE 36
NOISE LEVELS FROM MOBILE PLANT

Equipment	Noise level (dB(A))	Measurement location
Compressed air rock drill	110–115	At 1 m (3 ft)
	98	At 15 m (50 ft)[a]
Large portable compressor	80	At 7 m (23 ft)
	81	At 15 m (50 ft)[a]
7 m³ (10 yd³) dragline	90–92	Operator's cab
Diesel trucks	74–109	Driver's cab
	88	At 15 m (50 ft)[a]
Electric shovels	78–101	Operator's cab
Graders	76–104	Operator position
Dozers	84–107	Operator position
	87	At 15 m (50 ft)[a]
Locomotives	75–95	Driver position
Rotary drills	72–100	Operator position
Front end loaders	83–101	Operator position
Scrapers	92–104	Operator position
	88	At 15 m (50 ft)[a]

[a] Figures used by Environmental Protection Agency, U.S.A.

In general, once on the public highway it is impracticable to segregate mineral traffic from other road users. The maximum noise levels generated by road traffic are controlled by law in most countries and, although there is some variation, kerbside values of 88–92 dB(A) are usually the maximum permitted for heavy vehicles.

The actual levels reaching the general public from a mine site depend upon the noise generated by fixed and mobile plant, the distance between the noise sources and the listener and the degree of attenuation occurring along the noise path. Continuous noise levels exceeding 75 dB(A) are unusual (and over 80 dB(A) very unusual) outside the mine perimeter, although mobile plant operation may occasion higher values for limited periods of time. Presently available evidence suggests that exposure to noise levels of this magnitude, even for 24 hours per day, cannot cause hearing damage. However, dependent upon location, noise levels of 75–80 dB(A) have a serious nuisance potential. The Wilson Report recommends a maximum of 75 dB(A) in urban areas near main roads and heavy industrial areas, and 70 dB(A) elsewhere. Furthermore these standards apply from 7 a.m. to 7 p.m. and more stringent, but unspecified, levels are suggested for night working which is a feature of many mines.

It is therefore apparent that a proportion of mining operations can cause nuisance during the daytime and that night working is likely to be a contentious issue for any mine located close to residential or inhabited rural areas.

6.4. REMEDIAL MEASURES

The proportion of energy transmitted as noise is very small for all mining equipment. For example the compressed-air rock drill, a relatively noisy machine, converts less than 0·1 % of the energy delivered to it into noise. The problems of noise reduction are compounded when it is realised that a diminution of 10 dB requires the dissipation of 90 % of the noise energy and 20 dB requires 99 % dissipation.

There are three fundamental control techniques available: to reduce the noise energy generated at source, to isolate the source, or to increase the attenuation or absorption between the source and the listener. In effecting noise control, the source is the logical place to begin. Reduction of noise energy at source is usually simpler and cheaper than attempting isolation, attenuation or absorption.

Most noise originates from mechanical vibrations but some can also arise from electromagnetic, aerodynamic or hydrodynamic forces and other causes such as sympathetic resonance. Noise at source may be reduced by changing to another machine, altering the design, construction or installation, or improving maintenance.

The substitution of an inherently quieter type of equipment may be practicable. For instance compressed air powered machines, which have a particular problem with exhaust noise, are often amenable to this remedy; replacing a compressed air fan with an electric fan of the same work capability can decrease the noise level by up to 20 dB. Where economically practicable it can sometimes be advantageous to replace a machine by a larger, slower running unit of the same type, thus reducing the overall noise level.

Noise control in design, construction and installation must take account of the inherent sources of vibration and the way in which they are coupled to acoustic radiations. In mechanically vibrating machines the coupling may be direct, *i.e.* sound waves are created directly by the vibrating component, but often vibrations are transmitted to other parts of the machine and its mounting and some of these may prove more effective acoustic radiators than the original source. Detailed design techniques are beyond the scope of this book but some of the important factors for consideration are changing the mass or stiffness to avoid resonance, the use of rigid construction and casings with isolation mountings, designing balanced machines, and lining or treating surfaces acoustically.

Noise of aerodynamic origin may be reduced by the use of reactance or absorption type silencers. Reactance mufflers are applicable where the noise is caused by repeated pulses of gas, as in the exhaust of diesel and compressed air powered machines. The muffler acts as a shock absorber, cushioning the pulses and smoothing the flow of gas. Absorption mufflers are used where there is a continuous noise such as a blower or pump. The muffler sub-divides the air stream and passes it through noise-absorbent material. Both types of silencer have the disadvantage of constricting the exhaust, thus creating a back pressure and reducing the efficiency of the machine and this constitutes a practical limit on the degree of noise dissipation attainable.

Regular maintenance of machinery can effect significant reductions in noise levels. Most equipment tends to become noisier with age because parts become worn, clearances increase and components have greater freedom to vibrate. Correct lubrication and prompt replacement of worn parts slows down this process and thus helps to reduce noise levels. Regular tightening of bolts and checking for rattling or other new sources of acoustic radiation further assists noise reduction.

Table 37 gives some practical examples of noise reduction at source.

If the inherent vibrations cannot themselves be reduced in power to an acceptable level, fixed plant noise sources may be isolated by enclosure.

The machine casing itself may act as an isolating enclosure if it has a minimum of openings and is free from resonance. Alternatively the noise source may be located in an acoustically designed building, room or other structure. Attention must be paid to noise transmitted through the

TABLE 37

EXAMPLES OF NOISE REDUCTION AT SOURCE

Noise source	Remedial measure	Noise reduction (dB(A))
Road breaker moil point	Elastomeric damping collar	7
Electric ventilation fan	Absorption silencer	15
Compressed air fan	Complete overhaul	12
Pneumatic rock drill	Exhaust hose + reactance silencer	9
Pneumatic loader	Reactance silencers	8
Rock loading chute	Lined with conveyor belting	7

structure as well as airborne vibrations if insulation is to be effective. Typical reductions achieved are shown in Table 38. These are average values and, in general, high frequencies undergo greater attenuation than low frequencies. Possible disadvantages of enclosure are that the view of the machine is impeded, access to it is limited, and any operator obliged to work within the enclosure is subjected to increased noise levels.

The listener may be insulated from the noise source by the use of ear protectors and this is an important method of preventing hearing damage in employees. However, it is of little relevance in reducing the noise problem external to the mine.

As discussed previously, attenuation and absorption occur naturally at about 6 dB for each doubling of distance as noise travels from the source to the listener. Noise nuisance may be diminished or eliminated by taking advantage of this fact in siting noisy activities. Fixed plant may, where practicable, be located as far from areas of potential nuisance as possible and both internal site roads and external vehicle routes may be selected

TABLE 38

EFFECT OF SOUND INSULATION

Insulation	Attenuation dB (average of all frequencies)
Absorptive treatment of sound reflecting surfaces	3–5
Complete enclosure of machine and lining with fibre glass	8–10
Single glazed window	25
Double glazed window	35
22 cm (9 in) Brick wall	45–50

having regard to the possibility of noise nuisance. In practice, the requirements of cheap, efficient operation are likely to limit flexibility in siting both fixed plant and vehicle routes. Also mining operations (especially surface mining) are extremely complex and it is difficult to predict the noise levels at the mine boundary. Excavations can screen noise and the effect of topography and the location of the noise source are important. Prediction is further hampered by the influence of atmospheric conditions. The effect of atmospheric conditions in modifying the operation of the inverse square law has already been discussed and it is particularly important that, where there is a well-established prevailing wind direction, noise sources are not located directly upwind of sensitive external areas.

In many countries, noise nuisance is limited by the use of a zoning concept. Industrial and other noisy developments are directed to zones remote from residential and office areas. Mining is less amenable to such planning techniques than most industries since the mineral can only be worked where it is found, but it may on occasion be practicable to locate processing plant, workshops and other service activities away from the mine site in areas where noise nuisance is unlikely to occur.

The nature of the ground surface between the source and the listener can have a small effect upon noise levels. A hard, reflective surface such as concrete causes a very slight increase in sound level whereas an absorbent surface, such as grass, can reduce the sound level by a maximum of 1–3 dB per 30 m (100 ft).

A final technique for reducing noise levels is to erect some form of barrier or screen between the source and the receiver (Fig. 49). Sound is diffracted over the top edge of a screen and hence a sound 'shadow' does

Fig. 49. Sound baffle embankments used to screen a church and a school beside an open cast mine.

not coincide with a geometrical shadow. More diffraction occurs at low frequencies than at high and hence the higher frequencies are more effectively screened. This can result in a subjective decrease in loudness level greater than the reduction in overall sound pressure level. Because of diffraction effects, attenuation increases as the height of the screen increases and, in order to achieve significant reduction in noise levels, the screen should be located close to the source or the receiver, as illustrated in Fig. 50. Moore[6] quotes the following formula for calculating the reduction in sound pressure level due to a screen wall:

$$R = 8 \cdot 2 \log_{10} \left(\frac{44H}{WL} \tan \frac{\theta}{2} \right)$$

where R = reduction in sound pressure level (dB); H = effective height of screen wall (m); WL = wavelength of the sound (m); θ = angle of diffraction.

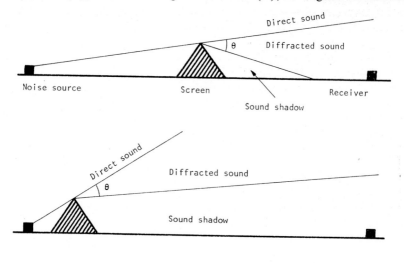

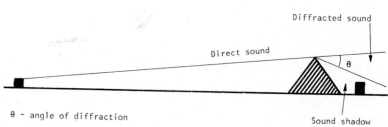

Fig. 50. Effects of screen location on the noise experienced by a receiver.

In practice, the reduction of noise is often less than that calculated because sound is scattered over the edge of the screen by air turbulence.

Screens used in practice include boundary walls, waste banks, trees and specially designed acoustic barriers. Walls, banks and trees are seldom placed for acoustic purposes only. Walls normally fulfil a security function, trees and banks provide visual screening and banks additionally are a convenient and environmentally acceptable means of solid waste disposal. It is often difficult to locate banks close to the source or the listener and hence their effectiveness is limited. However, major internal haul routes can sometimes be located close to the internal face of a bank and, under these circumstances, the bank may be effective in reducing mobile plant noise, which is usually more difficult to control than noise from permanent installations.

Tree screens are relatively ineffective in reducing noise levels. Attenuation increases with frequency and different authorities quote varying figures up to a maximum of about 7 dB per 30 m (100 ft) at 2000 Hz for very dense foliage. However, even if the maximum figure is applicable, sound travelling near ground level is little affected unless hedges and shrubs are planted and attenuation is only apparent when trees are in leaf. Hence it is probable that a wide belt of mature trees and dense undergrowth is necessary to cause significant attenuation. However, since noise is a subjective phenomenon, it is possible that tree screens and banks which prevent the listener from seeing the source of the noise may achieve a reduction in complaints out of proportion to the attenuation effected.

Purpose-built acoustic screens may achieve substantial reductions in noise levels but are also liable to be very expensive to install and for that reason their widespread use seems impractical.

6.5. CONCLUSIONS

When an existing mining operation is giving rise to complaints of noise nuisance, the problem should be approached in a logical manner as follows:

(a) Measure noise levels at the locations where nuisance is claimed and compare with relevant standards to assess the validity of the claims and the amount of attenuation required.

(b) Identify the most prominent noise sources. This may require further measurement and the use of special techniques such as frequency analysis.

(c) Consider the remedial measures available and the attenuation which each will yield.

(d) Select the techniques to be used having regard to cost, operational convenience and effectiveness.

(e) Implement the selected control measures.
(f) Measure the reduced noise levels to ensure that the nuisance has been removed.

In planning a new mining operation, the approach to noise control should be as follows:

(a) Measure existing ambient noise levels at locations where noise nuisance could arise.
(b) Consider the relevant standards and assess the maximum noise levels permissible from the mine without nuisance being caused.
(c) Estimate the noise levels arising from each major source within the planned mining operation.
(d) Predict the noise levels which will be caused at sensitive locations and compare with the maximum permissible levels.
(e) If necessary to prevent nuisance, consider the remedial measures available and their effectiveness.
(f) Select the techniques to be used having regard to cost, operational convenience and effectiveness.
(g) Implement the control measures at the appropriate construction stage.
(h) Measure the actual noise levels at sensitive locations and compare with the predicted and the maximum permissible levels.
(i) Implement further control measures if necessary to avoid nuisance.

Many facets of noise control require specialist knowledge and where actual or potential problems exist, the employment of an acoustic specialist is likely to yield optimum results at minimum cost and operational disruption.

REFERENCES

1. Bell, A. (1966). *Noise—an Occupational Hazard and Public Nuisance*, World Health Organisation, Geneva.
2. *Method of Rating Industrial Noise Affecting Mixed Residential and Industrial Areas* (1967). BS 4142: 1967, British Standards Institution, London.
3. Sisson, C. H. (1972). *Code of Practice for Reducing the Exposure of Employed Persons to Noise*, Her Majesty's Stationery Office, London.
4. Wilson, A. (1963). *Noise—Final Report of the Committee on the Problem of Noise*, Her Majesty's Stationery Office, London.
5. Beranek, L. L. (1971). *Noise and Vibration Control*, McGraw-Hill, New York.
6. Moore, J. E. (1966). *Design for Noise Reduction*, Architectural Press, London.
7. *Assessment of Noise with Respect to Community Response* (1971). I.S.O. R 1996, International Organisation for Standardisation.

8. Scholes, W. E., Salvidge, A. C. and Sargent, J. W. (1971). Field performance of a noise barrier, *Journal of Sound and Vibration*, **16**(4), 627–42.
9. *Code of practice for noise control on construction and demolition sites* (1975). BS 5228: 1975, British Standards Institution, London.
10. Botsford, J. H. (1971). Noise hazard meter, *Journal American Industrial Hygiene Association*, **32**(2), 92–5.
11. *First Report of the Royal Commission on Environmental Pollution* (1971). Her Majesty's Stationery Office, London.
12. Wilmot, T. J. (1972). The meaning of modern audiological tests in relation to noise induced deafness, *British Journal of Industrial Medicine*, **29**(2), 125–33.
13. Michael, P. L. (1972). Hearing conservation, *Mining Congress Journal*, **58**(6), 74–82.
14. Swift, R. L. (1971). Noise regulations and the individual, *Mining Congress Journal*, **57**(12), 50–6.
15. Vitunac, E. A. and Caughey, R. H. (1972). Relative noise levels in a strip mine operation and coal preparation plant, *Mining Engineering*, **24**(8), 29.
16. Muldoon, T. L. (1972). Noise levels and controls in coal cleaning plants, *Mining Engineering*, **24**(8), 29.
17. Special noise report released by EPA (1972). *Mining Congress Journal*, **58**(4), 126.
18. Rowe, R. H. (1973). Noise, *Quarry Manager's Journal*, **57**(3), 92–9.
19. Wood, C. H. (1970). Industrial noise control, *Canadian Mining Journal*, **91**(10), 67–8.
20. Chakravatti, J. L. (1973). Noise—its limits in fan selection, *Canadian Mining Journal*, **94**(9), 65–8.
21. Williams, H. (1973). Noise measurement in planning, *Journal of the Royal Town Planning Institute*, **59**(1), 7–9.
22. Kolp, R. J. (1974). Dust and noise, *Quarry Management and Products*, **1**(1), 10–14.
23. Peterson, G. (1974). Noise control in coal preparation plants, *Mining Congress Journal*, **60**(1), 30–6.
24. Murphy, J. N. (1972). Progress in noise abatement, *Mining Congress Journal*, **58**(9), 59–63.
25. Davies, G. M. (1973). Noise in industry, *Metallurgist*, **5**(10), 513–18.
26. Adamson, M. G. (1972). Noise, *Mining Magazine*, **127**(2), 94.
27. Richards, E. J. (1973). The noise environment in the United Kingdom, *Environmental Engineering*, **12**(56), 2–9.
28. Walker, A. (1963). Noise—its effects and control in mining operations, *CIM Bulletin*, **56**(619), 820–34.
29. Webb, R. S. (1974). Noise and the extraction of minerals, in *Minerals and the Environment Symposium*, Institution of Mining and Metallurgy, London, pp. 589–603.
30. Greenland, B. J. and Knowles, J. D. (1970). Environmental considerations of quarry blasting, *Quarry Manager's Journal*, **54**, 371–81.

7

Ground Vibrations from Blasting

7.1. THE NATURE OF GROUND VIBRATIONS FROM BLASTING

In blasting operations, the potential energy contained in an explosive is suddenly released, normally with the primary intention of fragmenting rock. A secondary, and undesirable, result of explosive detonation is that the surface of the ground in the vicinity of the blast undergoes displacement, the amplitude of which depends upon the distance from the blast, the energy released in the explosion and the local geological conditions.

When an explosive is detonated, rock in the immediate vicinity is crushed and shattered and an oscillatory wave is propagated through the rock mass causing particles along its path to move backwards and forwards longitudinally along the line of advance of this primary wave, which is normally designated the P-wave. Where the P-wave strikes a free surface or change of material at any angle other than 90°, complex displacements occur which give rise to secondary or shear waves usually termed S-waves.

The P- and S-waves are called body waves because they travel through the body of the materials which transmit them. At the free surface between ground and air, the body waves generate a number of surface waves, each of which is characterised by the motion through which a particle in its path goes as the wave passes.

The particle motions associated with each of the major surface waves are illustrated in Fig. 51. The Rayleigh or R-wave is longitudinal and causes mainly vertical retrograde motion. It is the most commonly observed surface wave, carries the major part of the surface ground energy and consequently is most likely to cause damage. The Love or Q-wave (from the German querwellen) causes transverse vibration in the horizontal plane with no vertical displacement. The displacement of particles by coupled or C-waves is elliptical and inclined, having components in both vertical and horizontal directions. The use of the term coupled implies combined P- and S-type motions. The H-wave moves particles in an elliptical orbit similar to the R-wave but in the reverse direction. It has only been detected in nuclear blasting. P-waves have the highest velocity, usually in the order of 3000–6000 m/s (10 000–20 000 ft/s) in hard rock formations. For surface waves the following order generally obtains:

$$\xrightarrow{\hspace{3cm}}$$
Decreasing velocity
$$C \longrightarrow H \longrightarrow Q \longrightarrow R$$

As waves travel outwards from the source of the explosion, higher frequencies are damped out and the frequency range thereafter is of the order of 3–70 Hz.

Explosions in the ground are generated under a wide variety of conditions and this may lead to the formation of one type only, all types or combinations of surface waves. In dealing with environmental problems it is the net resultant motion to which a structure or person is subjected which is of interest and hence it is usual to measure the net effect of all surface waves and not to attempt differentiation between the motion attributable to each type.

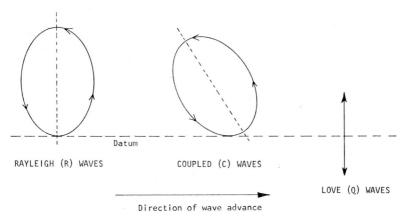

Fig. 51. Particle motions associated with R, C and Q surface waves.

The material involved in the transmission of surface waves is a zone about one wavelength in thickness and for dynamite blasts this does not normally exceed about 100 m (330 ft).

All surface waves are generated at approximately the same time and, in the immediate vicinity of the blast, the total surface displacement is controlled by the total energy contained within the waves. However, as the waves travel outwards at differing velocities, they quickly separate and maximum ground motion is then controlled by the energy contained within each individual wave. Hence maximum displacement decreases very rapidly at first but then diminishes more slowly as individual waves die out from loss of energy and dispersion. The rate at which the waves die out is dependent upon the nature of the materials through which they pass. The wave-forms are elastic and are more readily transmitted through competent rock which has a relatively high elasticity, than through clays, sand and similar unconsolidated material which rapidly convert wave energy into heat by friction.

7.2. MEASUREMENT AND RECORDING

The factors normally considered to be of importance in assessing the effects of ground vibrations are the maximum amplitude (particle displacement), the peak particle velocity, the peak particle acceleration and the frequency of vibration. It is generally assumed that surface waveforms are sinusoidal and for practical purposes this assumption is sufficiently accurate. Consequently the four main parameters are related as follows:

$$A = wV = w^2 U \tag{1}$$

where $w = 2\pi f$ (angular frequency); f = frequency; A = peak acceleration; V = peak velocity; U = maximum displacement.

Until recently mechanical vibrograph systems were commonly used to record ground vibrations. Various designs were available but essentially ground movements were transmitted to a mass whose movement actuated a mechanical or photographic recording system giving a record of maximum displacement.

At present most measurements are made with a seismic transducer in which ground movement generates a voltage. Commercially available systems mostly record peak velocity with a self-generating velocity pick-up or an accelerometer followed by an integration stage from which velocity is derived. Frequency may normally be determined by measuring the peak to peak distance on the recording and relating this to the speed at which the recording chart is driven. From a knowledge of peak particle velocity and frequency, maximum displacement and peak acceleration may be calculated from eqn. (1) if required. The equation in the form stated is only valid for peak values and it is thus essential to ensure that the instrumentation displays peak and not average or root mean square (r.m.s.) values. If this is not the case, peak values may be established by multiplying the r.m.s. value by 1·414 or the average value by 1·57.

Electronic transducers give a voltage reading and hence must be calibrated before use in the field. In addition to calibration, the transducer may be tested for linearity of response in order to determine the usable frequency range, and for accuracy. When measuring vibrations the transducer must be adequately coupled to the ground and this is often effected by firmly attaching it to a machined metal block of about 5 kg (10 lb) weight which is placed on the ground and acts as a reaction mass.

Vibration levels may be required in the vertical (compressional), horizontal (shear), or transverse directions and the pick-ups are oriented as required. There is at present no instrument available which automatically derives the resultant of all vibration vector components and, if required, this must be calculated from the readings of three mutually perpendicular transducers.

Frequently a number of readings are made simultaneously and the commercial instrumentation available ranges from single units with one channel recording to sophisticated multi-channel packages.

7.3. PREDICTION OF GROUND VIBRATION LEVELS

The propagation of ground vibration waves through the earth's crust is a complex phenomenon. Even over small distances rocks and unconsolidated material are anisotropic and non-homogeneous, and close to the rock/air interface at the ground surface, complex boundary effects may occur. These difficulties restrict theoretical analysis and derivation of a propagation law, and consequently research workers have concentrated upon empirical relationships based on field measurements.

The primary concern of most investigation programmes has been to derive a formula relating the weight of explosive charge with the distance from the blast and the magnitude of the ground vibration. Other factors such as local geology, type of explosive and method of initiation have also been studied.

Although many propagation equations have been proposed,[9] the most widely accepted general formula is of the type:

$$V = KW^a D^b \qquad (2)$$

where W is the weight of explosive charge; D is the distance from the blast; V is the magnitude of the vibration; and K, a, b are constants whose values depend upon individual site conditions. The quantity V may be expressed as the peak value of displacement, particle velocity or acceleration with consequent variation in the values of the constants.

Although for many years peak amplitude was in use to assess damage potential, recently peak particle velocity has gained more general acceptance as the parameter most closely related to structural damage and hence eqn. (2) is most frequently expressed with V as peak particle velocity and the appropriate values of the constants.

The actual formulae in use for prediction vary considerably. In 1950, Morris[5] suggested the simple relationship:

$$A = K \frac{(W)^{\frac{1}{2}}}{D} \qquad (3)$$

where A = maximum amplitude (in); K = site factor; W = explosive charge (lb); D = distance (ft) and the value of K was found to vary from 0·05 for hard competent rock to 0·30 for clay and up to 0·40 or 0·50 for completely unconsolidated material.

More recently in 1973 Gustafsson[6] citing the work of Langefors and Kihlstrom[7] proposed the equation:

$$V = K\left(\frac{W}{D^{3/2}}\right)^{\frac{1}{2}} \qquad (4)$$

where V = peak particle velocity (mm/s); K = constant (approx. 400 for hard Swedish rock); W = instantaneously detonated charge weight (kg); D = distance (m).

A comprehensive study by the U.S. Bureau of Mines[9] recorded blasts at 26 sites and analysed the results statistically to derive a propagation formula of general applicability in the form $V = KW^aD^b$ (see eqn. (2) above). Vertical, radial and transverse measurements were recorded and the general formula obtained was:

$$V = K\left(\frac{D}{W^a}\right)^b \qquad (5)$$

Values of K and b differed significantly for the three components but the variations in the constant a were small and it was concluded that a value of 0·5 may be used for all three components with sufficient accuracy.

Equation (5) may thus be rewritten:

$$V = K\left(\frac{D}{W^{\frac{1}{2}}}\right)^b \qquad (6)$$

The U.S. Bureau of Mines study indicated that a plot of peak particle velocity as a function of scaled distance $(D/W^{\frac{1}{2}})$ on log–log co-ordinates gives a straight line for each site and vibration component. Thus in order to predict the vibration levels at any site using this formula, it is necessary to monitor test shots and from these recordings plot peak particle velocity against scaled distance to a log–log scale as in Fig. 52.

The value of K in eqn. (6) is then the particle velocity intercept at $(D/W^{\frac{1}{2}}) = 1·0$ and b is the slope of the best fit line.

The U.S. Bureau of Mines also studied the effects of initiation method and geology. It was concluded that there is no significant error if comparisons of vibration levels in various blasts are based on maximum charge weight per delay, or total charge weight for instantaneous blasts. Detonating fuse incorporating delay connectors was found to give slightly higher vibration levels than equivalent instantaneous blasts, and electric delay detonators gave the lowest levels for a given weight of explosive.

Geology may cause variations in propagation in different directions and these may be large or small, but are difficult to predict. Hence in

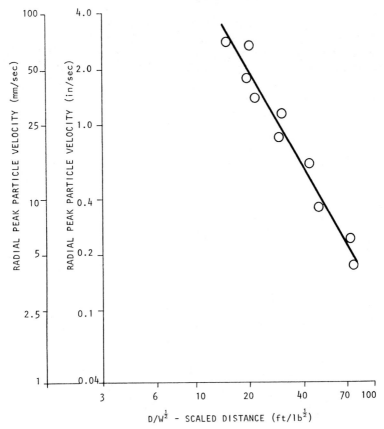

FIG. 52.   Log–log plot of peak particle velocity versus scaled distance.

deriving values of K and b for use in eqn. (6), it is desirable that test shots are measured in at least two directions, particularly where folding, faulting, jointing or other geological complexities are known to exist. On sites with thick overburden, displacements were found to be higher, and frequencies lower, than when little or no overburden was present.

Under normal circumstances the mine operator has little control over the distance between blasting sites and the nearest buildings and hence the usual purpose of ground vibration predictions is to determine the maximum instantaneous charge which may be detonated without causing structural damage and/or nuisance. The precise details of calculations at a specific site depend in part on the propagation formula adopted and the type of local standard in force.

In order to illustrate the principles involved, the hypothetical case shown in Fig. 53 is considered. It is necessary to protect the buildings A, B and C from vibrational damage resulting from blasting and it is assumed that a peak particle velocity standard of 50 mm/s (2·0 in/s) is in force and that propagation follows the U.S. Bureau of Mines relationship[9] in eqn. (6).

This has been verified by monitoring two test shots at the locations shown, in which it was established that there are no marked directional

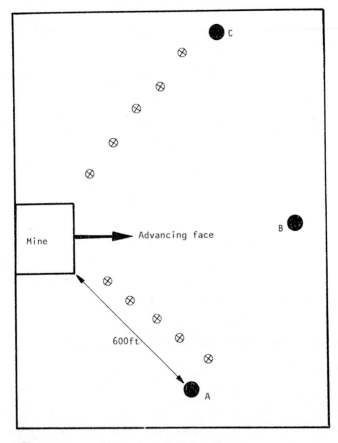

● A,B,C - Buildings to be protected

⊗ - Monitoring points for test shots

FIG. 53.   Protection of buildings.

propagation effects and the radial component of velocity was significantly higher than the vertical or transverse components. The results of the test shots have been plotted in Fig. 52 and the close approximation of the data to a straight line verifies the validity of using eqn. (6).

From Fig. 52 it can be seen that a peak particle velocity of 50 mm/s (2·0 in/s) is equivalent to a scaled distance of 20 ft/lb$^{\frac{1}{2}}$. The maximum permissible instantaneous charge may now be expressed in one of two ways. First the distance between the advancing mine face and the nearest building may be used to derive a precise value for the maximum instantaneous charge weight (W) as shown below:

$$\text{Scaled distance (S.D.)} = 20 \text{ ft/lb}^{\frac{1}{2}}$$

$$\text{Distance to nearest building } (D) = 600 \text{ ft}$$

$$W \text{ (lb)} = \frac{D^2}{(\text{S.D.})^2} = \frac{(600)^2}{(20)^2} = 900 \text{ lb}$$

Alternatively, a series of safe limit contours may be calculated and plotted on the plan as shown in Fig. 54. In this case the safe distance (D) is given by the expression:

$$D^2 = W \cdot (\text{S.D.})^2$$

Hence, for $W = 200$ lb, 400 lb, 600 lb, 800 lb (90 kg, 180 kg, 270 kg, 360 kg), $D = 280$ ft, 400 ft, 490 ft, 565 ft (85 m, 120 m, 150 m, 170 m).

7.4. EFFECTS OF GROUND VIBRATIONS

7.4.1. Structural Damage
The most common complaint concerning blasting is probably of damage to buildings. As the severity of blasting vibrations increases, the order in which damage occurs is usually:

(1) Dust falling from old plaster cracks.
(2) The extension of old plaster cracks.
(3) The formation of new plaster cracks.
(4) Flaking of plaster.
(5) Large areas of plaster drop.
(6) Masonry cracks form and partitions separate from exterior walls.
(7) Further severe damage and ultimate collapse of the building.

The overwhelming preponderance of complaints relate to categories (1)–(4). However, plaster cracking and other minor damage is commonly experienced in many buildings not subjected to blasting vibrations and the Architects Small House Service Bureau of the U.S.A. has listed over 40

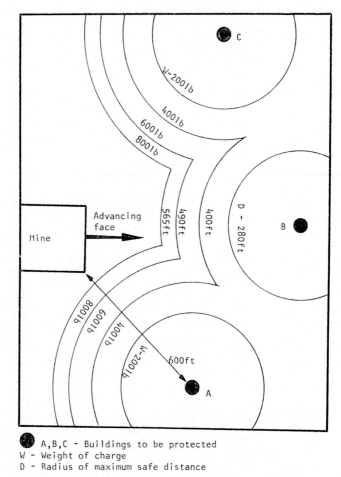

A,B,C - Buildings to be protected
W - Weight of charge
D - Radius of maximum safe distance

FIG. 54.   Maximum safe charge contours.

reasons other than vibrations why ceilings crack. These generally relate to poor building practice such as inadequate foundations and incorrect preparation or application of materials. There is thus a shortage of data concerning damage unequivocally attributable to blasting.

Past work has attempted to relate damage to one or more of the important blasting vibration parameters, *i.e.* frequency, amplitude, velocity and acceleration.

Field studies from 1935 to 1940 by the U.S. Bureau of Mines[10] in the frequency range 4–40 Hz and amplitude range 0·0025–12·0 mm (0·0001–0·5 in) related damage to acceleration as shown in Table 39.

TABLE 39

EFFECT OF GROUND VIBRATIONS UPON STRUCTURES
(after U.S. Bureau of Mines[10])

Peak particle acceleration	Damage
<0·1 g	Nil
0·1–1·0 g	Minor (fine plaster cracks)
>1·0 g	Major (fall of plaster, serious cracking)

However, later work by the Bureau[12] indicated that, using this data, major damage correlated best with velocity and minor damage with acceleration.

A report by Langefors et al.[11] in 1958 described the relationship between ground vibrations from blasting and structural damage during a reconstruction project in Stockholm. Frequencies measured ranged from 50 to 500 Hz and amplitudes from 0·02 to 0·5 mm (0·0008 to 0·02 in). They concluded that particle velocity gave the best guide to damage potential and derived the results shown in Table 40.

TABLE 40

EFFECT OF GROUND VIBRATIONS UPON STRUCTURES
(after Langefors et al.[11])

Peak particle velocity (in/s)	(mm/s)	Damage
2·8	70	Nil
4·3	110	Fine cracking and fall of plaster
6·3	160	Cracking
9·1	230	Serious cracking

Investigations by Edwards and Northwood[13] published in 1960 for the frequency range 3–30 Hz and amplitude range 0·25–9 mm (0·01–0·35 in) concluded that damage was more closely related to velocity than displacement or acceleration and that minor damage was likely to occur with a peak velocity of 100–125 mm/s (4–5 in/s).

Many other studies have been undertaken and Table 41 summarises some of the conclusions reached.

The apparent discrepancies disclosed in Table 41 are hardly surprising. The response of a structure to ground vibrations depends upon the nature of the building as well as the characteristics of the vibration. Steffens[16]

TABLE 41

SUMMARY OF THRESHOLD LEVELS FOR DAMAGE DERIVED BY VARIOUS AUTHORITIES

Authority	Criterion	Threshold level at which minor damage occurs
U.S. Bureau of Mines[10]	Acceleration	0·1 g
Crandell[14]	Energy ratio (a^2/f^2)	3·0 $(a, ft/s^2)$ (f, Hz)
Langefors et al.[11]	Velocity	70 mm (2·8 in)/s
Dvorak[15]	Velocity	10–30 mm (0·4–1·2 in)/s
Edwards and Northwood[13]	Velocity	100 mm (4 in)/s
American Society of Civil Engineers	Velocity	135 mm (5·36 in)/s radial 175 mm (6·86 in)/s vertical 45 mm (1·71 in)/s transverse

discusses the relationship between structural vibration and damage for many sources, including blasting. The age, condition, method of construction and geometry of the building are important as are the natural resonant frequency of the structure and its component elements. It is generally assumed that buildings are most susceptible to damage from horizontal ground movements and frequently only one component of vibration is recorded. However, the actual movement of a building depends upon the vector summation of all three components.

It therefore seems likely, as suggested by Oriard,[17] that a single value of one vibration parameter is inadequate to describe accurately the damage potential of a surface blast wave. However, peak particle velocity appears to be more closely related to structural damage over a wider frequency range than peak amplitude or acceleration.

7.4.2. Human Response to Vibrations

Vibration levels well below those necessary to cause damage to structures are readily detectable by humans and can cause annoyance. Pioneering work by Reiher and Meister[1] published in 1931 established that human response to vibration is complex, depending amongst other factors upon frequency and amplitude and that sensitivity is greatest towards vibrations parallel to the long axis of the body, *i.e.* vertical vibrations when standing and horizontal vibrations when lying down. Figure 55 summarises their findings.

Dieckmann[2] suggested the use of a K factor to determine response. The K factor is calculated as shown in Table 42 and the interpretation suggested by Dieckmann in Table 43.

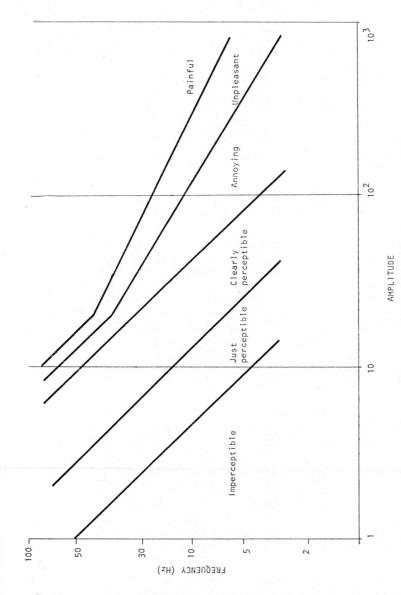

Fig. 55. Human sensitivity to vertical vibrations (after Reiher–Meister[1]).

TABLE 42

CALCULATION OF DIECKMANN'S K FACTOR
(A, amplitude in μm, f, frequency in Hz)

Vertical vibrations		Horizontal vibrations	
Frequency range	Value of K	Frequency range	Value of K
Below 5 Hz	$0{\cdot}001\ Af^2$	Below 2 Hz	$0{\cdot}002\ Af^2$
5–40 Hz	$0{\cdot}005\ Af$	2–25 Hz	$0{\cdot}004\ Af$
Above 40 Hz	$0{\cdot}2\ A$	Above 25 Hz	$0{\cdot}1\ A$

Other research workers have suggested a range of criteria for assessing human response, using particle acceleration, velocity or amplitude in conjunction with frequency.

Frequency ranges generated in blasting are predominantly in the range 5–60 Hz and a review by Goldman[3] of the range 1–70 Hz produced responses shown in Fig. 56.

Much of the research in this field has been concerned with soundless vibration of relatively long duration. Blasting vibrations are normally accompanied by an air blast wave causing a primary noise itself and also giving rise to secondary noises such as the rattling of windows. Furthermore the vibrations are of very short duration and in these circumstances it is probable that subjective response is governed at least as much by reaction to the air blast wave as by the body's sensitivity to vibration.

TABLE 43

INTERPRETATION OF DIECKMANN'S K FACTOR

Value of K	Human response
0·1	Lower limit of perception
1	Allowable in industry
10	Allowable for brief periods
100	Limit of tolerance

Consequently it is probable that criteria derived for soundless, steady-state vibration are of limited value in assessing the response to blasting and the available evidence suggests that these criteria are conservative when applied to the human reaction to blasting.

A survey of complaints published by Power[4] following the Salmon Underground nuclear detonation in the U.S.A. has obvious parallels with mine blasting operations and the graph in Fig. 57 is probably a good guide to the human response to ground vibrations from mining.

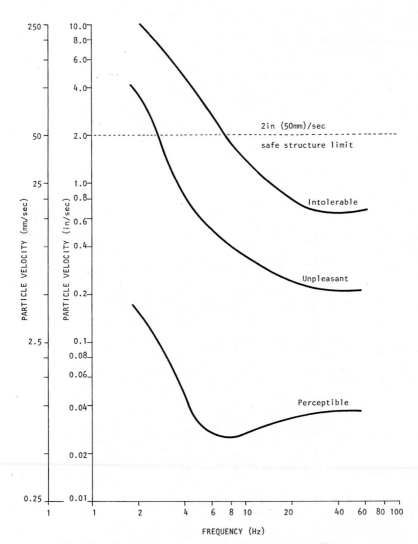

Fig. 56.   Human response to vibrations (after Goldman[3]).

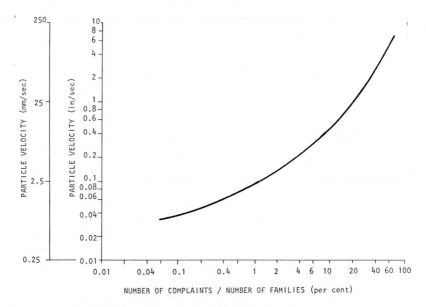

FIG. 57.   Survey of complaints after the Salmon nuclear event.

7.5. DAMAGE CRITERIA AND STANDARDS

In order to conduct blasting operations without causing damage, it is desirable to have some standard, compliance with which ensures no detrimental effect upon structures. Such a standard would ideally be practical, readily monitored and thus easily enforced and this implies a single value of one of the major vibration wave characteristics.

Since authorities in different countries and at various times have related structural damage to various parameters, it naturally follows that a range of standards and criteria have been used.

The difficulties of relating damage potential to any single criterion have been discussed above. Nonetheless, because of simplicity of measurement and enforcement, it has been relatively common practice to adopt a maximum of the peak value of amplitude, velocity or acceleration as the standard and to allow for any anomalies in this approach by taking a conservative value well below the minimum at which damage has actually been observed. However, some authorities have used more complex standards combining frequency with amplitude or velocity.

In the U.K. there is no universal standard. For many years a maximum amplitude of 0·1 mm (0·004 in) for blasting vibrations received general

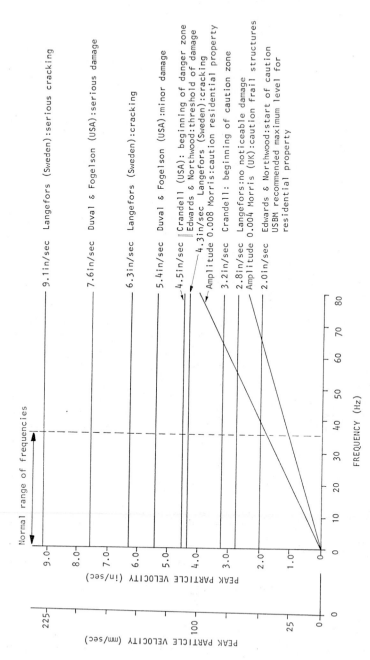

FIG. 58. Comparison of vibration criteria (after Grimshaw[18]).

acceptance but recently, following the work of the U.S. Bureau of Mines, peak particle velocity has come into increasing use and the values stipulated by local planning authorities have ranged from 5 to 50 mm (0·2–2·0 in)/s.

In the U.S.A. a variety of standards have been adopted in different States and these are reviewed by Berger.[19] Some have used maximum amplitude and others more complex criteria combining amplitude and frequency. Thoenen and Windes[10] recommended a standard based on the quantity Af^2 (A = amplitude, f = frequency) with a suggested maximum of 25 mm (1 in)/s² whilst Crandell[14] used the quantity Af and specified a 'caution' value of 13·4 mm (0·5 in)/s (equivalent to a peak velocity of 84 mm (3 in)/s).

Recently both Edwards and Northwood[13] and the comprehensive study by the U.S. Bureau of Mines[9] have suggested a peak particle velocity standard of 50 mm (2·0 in)/s. The graph in Fig. 58 based on Grimshaw,[18] compares various damage levels and criteria.

The criteria described above generally refer to a single component of vibration and the values specified are the maximum permissible for any individual component. Standards seldom specify the precise location or orientation of measurements although this can affect the magnitudes recorded. Nicholls *et al.*[9] cite an investigation by the American Society of Civil Engineers in which measurements at the roof level of a house showed an amplification of up to 2·0 compared to ground level response.

Although adopting a single criterion, maximum velocity, the German draft DIN 4150 (1970) standard is otherwise very specific. The maximum allowable velocity is the combined vectorial component of the vertical and two horizontal directions, and measurements should be made at ground level on the external foundation. One of the horizontal measurements should be directed towards the source of the vibration or parallel to a sidewall of the building. The proposed levels are shown in Table 44.

TABLE 44

MAXIMUM ALLOWABLE VELOCITIES SPECIFIED IN DRAFT DIN 4150 (1970) FOR TRANSIENT VIBRATIONS OF 8–80 Hz
(after Steffens[16])

Class	Description	Maximum velocity mm/s	in/s
1	Ruins and buildings of historic value	2	0·08
2	Buildings with existing visible cracks in brickwork	5	0·2
3	Undamaged buildings in good condition apart from minor defects	10	0·4
4	Strong buildings in steel, reinforced concrete, etc.	10–40	0·4–1·6

7.6. NUISANCE CRITERIA AND STANDARDS

Almost all the research into blasting vibrations has concentrated on deriving damage rather than nuisance criteria. Evidence from the Salmon nuclear event indicates that at a peak velocity of 50 mm (2 in)/s more than 35% of families complained, and that a standard of 10 mm (0·4 in)/s would be required to keep complaints to below 8%.

Steffens[16] cites German proposals for a revision to DIN 4150 which would set new standards for vibration nuisance. This requires the calculation of a K value:

$$K = \frac{0\cdot005Af^2}{(100 + f^2)^{\frac{1}{2}}} = \frac{0\cdot8Vf}{(100 + f^2)^{\frac{1}{2}}} = \frac{0\cdot125a}{(100 + f^2)^{\frac{1}{2}}}$$

where f = frequency (Hz); A = maximum amplitude (μ); V = maximum velocity (mm/s^2). K is evaluated as shown in Table 45.

TABLE 45

SUBJECTIVE EVALUATION OF K VALUES

K value	Subjective effect
<0·1	Not felt
0·1	Threshold of perception
0·25	Barely noticeable
0·63	Noticeable
1·6	Easily noticeable
4	Strongly detectable
10	Very strongly detectable

The section of the proposed standard applicable to blasting is shown in Table 46.

The present plethora of criteria and standards in different countries, and even in different parts of the same country, is understandable in view of the complexities of assessing damage and nuisance potential. However, such confusion is conducive neither to economic mineral working nor to effective environmental control and hence the publication of a realistic, workable and generally accepted International Standard would probably be advantageous to both the mineral industry and the general public.

7.7. COSTS

Whilst it is desirable that buildings and people should be protected from the adverse effects of blasting vibrations, this normally entails an increase

TABLE 46
DRAFT DIN 4150 (1970)—PROPOSED LEVELS OF
VIBRATION ACCEPTABLE FOR VARIOUS SITUATIONS

Building areas	Time	Permissible K-value for seldom occurring shocks
Health resorts	Day	2·5
Hospitals	Night	Threshold of perception
Residential areas	Day	4
University areas	Night	Threshold of perception
Village areas	Day	8
Mixed areas	Night	Threshold of perception
Business areas	Day	12
Industrial areas		
Port areas	Night	0·4

in operating costs which is reflected in the price paid by the consumer for the mineral. Thus it is in the interests of both the mine operator and the general public that standards should be realistic and not excessively restrictive.

The additional cost factors may be divided into those wholly attributable to vibration control, and those in which the imposition of controls causes an increase in existing costs. The former class includes costs associated with vibration measurements, building inspections, insurance and consulting fees, whilst the second category includes increased drilling, blasting (and possibly loading and haulage) costs because of the smaller scale of working and higher planning and supervision costs.

It is almost impossible to indicate the relative increases in costs which may arise since these are very site specific. Gustafson[6] quotes actual examples of blasting cost increased by factors of up to several hundred per cent, but these refer to city centre work within 30 m (100 ft) or less of buildings. In normal mining situations it is unlikely that drilling and blasting costs would increase by more than 25% because of vibration control and frequently the figure would be much lower.

7.8. CONCLUSIONS

Where blasting vibrations are a potential source of damage or nuisance, it is recommended that the following actions are taken, in addition to those described in Chapter 8 for Air Blast.

(1) Determine which standards or criteria and propagation formula are applicable.

(2) Monitor the vibrations from test shots to determine site factors and the directional influence (if any) of local geology.

(3) Calculate the maximum instantaneous charge permissible.

(4) Design blasting patterns to conform with (3).

(5) Inspect buildings likely to be affected by blasting, noting age, condition, etc.

(6) Carefully supervise drilling and blasting—there is evidence that many complaints arise from 'maverick' blasts where drilling deviation or overcharging of holes invalidate design controls.

(7) Check monitor the vibration levels regularly, especially at sensitive locations.

Items (3)–(7) above fall within the ambit of operating engineers and geologists. However, in deciding upon the use of standards and propagation laws, the layout and interpretation of test shot monitoring, the use of specialist assistance is advised.

REFERENCES

1. Reiher, H. and Meister, F. J. (1931). Human sensitivity to vibration, *Forsch Geb. Ing. Wes.*, **2**(11), 381–6.

2. Dieckmann, D. (1958). A study of the influence of vibration on men, *Ergonomics*, **1**(4), 347–55.

3. Goldman, D. E. (1948). *A Review of Subjective Responses to Vibrating Motion of the Human Body in the Frequency Range 1 to 70 Cycles per Second*, Report No. 1, Project NM 004001, Naval Medical Research Institute.

4. Power, D. V. (1966). A survey of complaints of seismic related damage to surface structures following the Salmon Underground Nuclear Detonation, *Bulletin of the Seismic Society of America*, **56**(6), 1413–28.

5. Morris, G. (1950). Vibrations due to blasting and their effects on building structures, *The Engineer*, **190**, 394–414.

6. Gustafsson, R. (1973). *Swedish Blasting Technique*, SPI, Gothenburg, Sweden.

7. Langefors, U. and Kihlstrom, B. (1963). *Rock Blasting*, John Wiley, New York.

8. Cummins, A. B. and Given, I. A. (1973). *SME Mining Engineering Handbook*, Section 11, Port City Press, Baltimore.

9. Nicholls, H. R., Johnson, C. F. and Duvall, W. I. (1971). *Blasting Vibrations and their Effects on Structures*, U.S. Bureau of Mines Bulletin 656, U.S. Department of the Interior.

10. Thoenen, J. R. and Windes, S. L. (1942). *Seismic Effects of Quarry Blasting*, U.S. Bureau of Mines Bulletin 442, U.S. Department of the Interior.

11. Langefors, U., Kihlstrom, B. and Westerberg, H. (1958). Ground vibrations in blasting, *Water Power*, **10**, 335–8, 390–5, 421–4.

12. Duvall, W. I. and Fogelson, D. E. (1962). *Review of Criteria for Estimating Damage to Residences from Blasting Vibrations*, U.S. Bureau of Mines Report of Investigations RI 5968, U.S. Department of the Interior.

13. Edwards, A. T. and Northwood, T. D. (1960). Experimental studies of the effects of blasting on structures, *The Engineer*, **210**, 538–46.
14. Crandell, F. J. (1949). Ground vibration due to blasting and its effect upon structures, *Journal Boston Society of Civil Engineers*, **36**, 222–45.
15. Dvorak, A. (1962). Seismic effect of blasting on brick houses, *Prace geofyrikeniha Ustance Ceskoslovenski Akademie Ved, Geofysikal Sbornik*, Number 169, 189–202.
16. Steffens, R. J. (1974). *Structural Vibration and Damage*, Building Research Establishment Report, Her Majesty's Stationery Office.
17. Oriard, L. L. (1972). Blasting operations in the urban environment, *Bulletin Association of Engineering Geologists*, **IX**(1), 27–46.
18. Grimshaw, G. B. (1971). The current scene in quarry blasting, *Quarry Managers' Journal*, **55**(4), 1–13.
19. Berger, P. R. (1973). Blasting controls and regulations, *Mining Congress Journal*, **59**(11), 48–51.
20. Berger, P. R. (1971). Blasting seismology, *Quarry Managers' Journal*, **55**(6), 187–90.
21. Duvall, W. I., Devine, J. F. and Johnson, C. F. (1963). *Vibrations from Blasting at Iowa Limestone Quarries*, U.S. Bureau of Mines Report of Investigations RI 6270, U.S. Department of the Interior.
22. Duvall, W. I., Johnson, C. F. and Mayer, A. V. C. (1963). *Vibrations from Instantaneous and Millisecond Delayed Quarry Blasts*, U.S. Bureau of Mines Report of Investigations RI 6151, U.S. Department of the Interior.
23. Defining vibration discomfort—new draft for development to improve the working environment (1974). *British Standards Institute News*, April, 12–13.
24. Broadbent, C. D. (1974). Predictable blasting with in-situ seismic surveys, *Mining Engineering*, **26**(4), 37–41.
25. Isham, A. R. (1974). An investigation into primary blasting, *Quarry Management and Products*, **1**(3), 103–9.

8

Air Blast

8.1. INTRODUCTION

Air blast is the term used to describe the air vibrations generated by blasting operations. An explosion causes a diverging shock-wave front which quickly reduces to the speed of sound and the air blast is then propagated through the atmosphere as sound waves.

The principal causes of air blast are:

(a) The release to atmosphere of the gases from unconfined or partially confined explosions. Inadequate confinement liberates large quantities of expanding gases which do little useful work and the dissipation of this energy in the atmosphere generates shock waves of large amplitude. These can arise from secondary blasting (particularly plaster shooting), overcharging and poor stemming.
(b) The release to the atmosphere of the gases from an exposed detonating fuse.
(c) Ground vibrations resulting from blasting. Ground movement generates air vibrations but these are small compared to those created by the venting gases described in (a) and (b).

Air blast consists of an initial concussion wave which lasts a few milliseconds rising quickly to its peak and falling off more slowly, followed by a rarefaction of longer duration but lower pressure change. Research indicates that damage is closely related to the concussion wave and hence it is usual to characterise air blast by the peak overpressure created. Overpressure measurements from 0·7 to 70 g/cm^2 (0·01–1 lb/in^2) may be undertaken with microphone sound recorders but above 70 g/cm^2 (1 lb/in^2) a piezoelectric, direct reading pressure gauge is more satisfactory. When recording shock waves from millisecond delay blasting there are a series of peaks which are separated by a few milliseconds only, and it is important to ensure that the recording equipment has response characteristics which permit resolution of each peak.

The principal factors governing air blast effects are: (a) the type and quantity of explosive; (b) the degree of confinement; (c) the method of initiation; (d) local geology and topography; (e) the distance and condition

185

of structures; (f) atmospheric conditions. Factors (a), (b) and (c) are variables within the control of the mine operator whereas (d), (e) and (f) are essentially uncontrollable at any particular site, although by varying the timing of a blast the operator may in effect achieve partial control over atmospheric conditions.

It is normally the objective in mine blasts to use the explosive energy to do useful work and thus energy escaping as an air blast wave represents blasting inefficiency. It is difficult to derive a working formula for predicting the air blast at a given distance from the explosion of a known

FIG. 59.   This view of a quarry blast demonstrates the possible problems of air blast, fly rock and dust dispersion which can arise from this source.

weight of explosive since the energy available for dissipation as air blast waves depends in part upon the design of the blast round. A badly designed round, for instance with inadequate burdens or stemming, may liberate a substantially greater percentage of the explosive energy as air blast waves than an apparently similar but well designed round at the same site.

8.2. PREDICTION

A programme of research by the U.S. Bureau of Mines in the early 1940s[1] studied the decay of amplitude of air blast with distance by detonating charges in air and measuring the increase in air pressure due to the passage of the blast wave. The damaging effects upon glass window frames were also studied. A plot of overpressure against scaled distance was derived, where scaled distance is defined as the distance in feet divided by the cube root of the charge weight in pounds. Cube root scaling of this type has become fairly common practice and implies spherical propagation of blast waves from a point source. Normal mining blasts do not conform to this

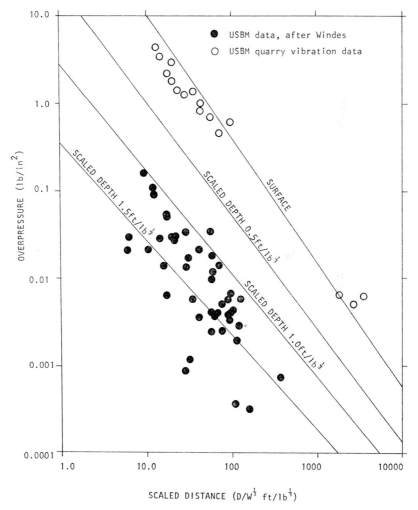

FIG. 60.   Relationship between overpressure and scaled distance.

model, since usually a number of charges are buried at some depth below the surface.

Research by the U.S. Bureau of Mines and Ballistics Research Laboratories[2] has produced the graph shown in Fig. 60 which incorporates the concept of cube root scaling for distance and for depth of burial. This type of graph can be used to predict the overpressure at a given distance from a blast or to determine the maximum charge weight permissible without exceeding a standard overpressure. In practice it is found that the maximum charge size for safe air blast limits obtained from such calculations is very much higher than the maximum charge permissible for safe ground vibrations. Hence, in normal blasting the measures taken to ensure safe vibration levels automatically limit overpressures to levels at least an order of magnitude below those necessary to cause damage. Air blast problems are only likely to arise where it is necessary to blast relatively unconfined charges, where unusual atmospheric conditions prevail or in a badly designed blast. Lack of stemming or exposed detonating fuse can increase the air blast pressure predicted for a given blast by a factor of 10 or more.

Atmospheric conditions affect the propagation of air blast waves in the same manner as sound waves, as described in Chapter 6. Grant et al.[3] found that the significant variables in order of importance were wind velocity and direction, barometric pressure, and temperature. Other conditions including low cloud ceiling, rain, mist and fog may cause reflection and enhance air blast levels at specific locations. In extreme cases, atmospheric conditions may cause a local increase in blast overpressure by up to a hundredfold.

8.3. EFFECTS OF AIR BLAST

Air blast waves may give rise to damage or nuisance. The damage potential of overpressure has been studied by many authorities and there is universal agreement that window glass failure occurs before any other type of structural damage. Table 47 summarises the effects of air blast overpressure upon structures.

The precise overpressure required to crack a window pane depends on many factors including the size and thickness of the glass, the standard of manufacture and installation, the rigidity of the frame and the time taken to reach peak overpressure (i.e. the rate of loading). Plaster cracks are generally caused by structural flexing induced by building vibrations arising from the air blast wave.

Normal mine blasting results in general overpressures well below those necessary to cause failure of correctly installed window panes. Hence the damage potential of air blast is largely confined to structurally unsound

TABLE 47
EFFECTS OF OVERPRESSURE UPON STRUCTURES

Overpressure lb/in²	g/cm²	Structural effect
0·03–0·05	(2–4)	Loose window sash rattles
over 0·1	(7)	Failure of stressed or badly installed window panes
over 0·75	(52)	Failure of correctly installed window panes begins
2·0	(140)	All window panes fail
over 2·0	(140)	Plaster cracks begin and, at higher pressures, masonry cracks may be evident

windows, bad blasting practice (such as the detonation of large, unconfined charges), and unusual atmospheric effects.

Nuisance can arise at overpressures well below those necessary to cause damage. The sound level at an overpressure of 0·7 g/cm² (0·01 lb/in²) is equivalent to a pneumatic riveter at 1·2 m (4 ft), and at 0·07 g/cm² (0·001 lb/in²) to a car horn at a distance of 1 m (3 ft). The problems of assessing nuisance are discussed in Chapter 6 and, in the case of air blast, these are compounded by the intermittent nature of blasting operations and by the adverse subjective response of most individuals to a phenomenon deriving from explosions.

There have until recently been no attempts to lay down quantified standards for air blast overpressure. However, in 1971 the U.S. Bureau of Mines[4] recommended that a standard of 35 g/cm² (0·5 lb/in²) be adopted. This would not prevent the rattling of loose sashes or the failure of badly installed panes but would be adequate with a reasonable margin of safety to preclude major damage. Complaints would still arise from minor damage and nuisance.

8.4. REMEDIAL MEASURES

Air blast overpressures can be minimised by correct design of blasting rounds, avoiding the detonation of large unconfined charges and considering atmospheric conditions before blasting. The important controls available to the mine operator are summarised below:

(a) Avoid overcharging by considering depth, burden and spacing when calculating charge weight per hole.
(b) Limit the maximum instantaneous charge detonated by the use of delays.

(c) Ensure adequate stemming.
(d) Consider replacing detonating fuse with electric initiation or
(e) use low energy rather than high energy detonating fuse or
(f) cover detonating fuse with at least 6 in of unconsolidated material (*e.g.* drilling dust and chippings).
(g) Consider alternatives to blasting for secondary breakage on surface, *e.g.* drop ball or impact hammer.
(h) Avoid blasting when unfavourable atmospheric conditions prevail, *e.g.* temperature inversions, adverse wind direction, low cloud ceiling.

Implementation of the appropriate measures from (a) to (h) should ensure that air blast causes no damage. However, the experience of many operators indicates that complaints are still likely to arise from the general public. These may be reduced, if not eliminated, by a programme of public relations. The measures discovered to be most effective in reducing public disquiet are:

(a) Blast at a regular time whenever possible and publicise the timing of blasts to reduce the element of surprise.
(b) Publicise widely the measures adopted to reduce the impact of blasting.
(c) Make a preblast survey of structures close to the blast site.
(d) Monitor at strategic points.
(e) Settle promptly any legitimate damage claim.
(f) Avoid blasting at night, weekends and public holidays.

REFERENCES

1. Windes, S. L. (1942). *Damage from Air Blast*, U.S. Bureau of Mines Report of Investigations RI 3622 and RI 3708, U.S. Department of the Interior.
2. Kingery, C. N. and Pannill, B. F. (1964). *Peak Overpressure versus Scaled Distance for TNT Surface Bursts*, Report No. 1518, U.S. Ballistics Research Laboratories.
3. Grant, R. L., Murphy, J. N. and Bowser, M. L. (1967). *Effect of Weather on Sound Transmission from Explosive Shots*, U.S. Bureau of Mines Report of Investigations RI 6921, U.S. Department of the Interior.
4. Nicholls, H. R., Johnson, C. F. and Duvall, W. I. (1971). *Blasting Vibrations and their Effect on Structures*, U.S. Bureau of Mines Bulletin 656, U.S. Department of the Interior.
5. Siskind, D. E. and Summers, C. R. (1974). *Blast Noise Standards and Instrumentation*, U.S. Bureau of Mines Technical Progress Report 78, U.S. Department of the Interior.
6. McKinley, R. W. (1964). Response of glass in windows to sonic booms, *Materials Research and Standards*, **4**(11), 594–600.

7. Perkins, B. and Jackson, W. F. (1964). *Handbook for Prediction of Air Blast Focusing*, Report 1240, U.S. Ballistics Research Laboratories.

8. Poulter, T. C. (1955). *The Transmission of Shock Pulses in Homogeneous and Non-homogeneous Air and Possible Damage to Building Structures from Moderately Small Explosive Charges*, Report 052710, Stanford Research Institute.

9. Reed, J. W. (1969). *Acoustic Wave Effects Project: Airblast Prediction Techniques*, Report SC-M-69-332, Sandia Laboratories.

10. Review: Airblast effects (1970). In *Symposium on Engineering with Nuclear Explosives*, American Nuclear Society, Las Vegas, 1484–1507.

11. Viksne, A. (1972). *Measurement and Reduction of Noise from Detonating Cord used in Quarry Blasting*, U.S. Bureau of Mines Report of Investigations RI 7678, U.S. Department of the Interior.

12. Van Dolan, R. W., Gibson, F. C. and Hanna, N. E. (1964). *Abatement of Noise from Explosives Testing*, U.S. Bureau of Mines Report of Investigations RI 6351, U.S. Department of the Interior.

13. Kenney, G. F. (1962). *Explosive Shocks in Air*, MacMillan, New York.

9
Transport

9.1. INTRODUCTION

There are various interfaces between a mining operation and the general public, the most obvious of which is the boundary between the mine site and the adjoining property. It is self-evident that the close juxtaposition of mining activity and other community functions maximises opportunities for conflict, and from this point of view it is fortunate that many aspects of mining are frequently concentrated on a single site thus reducing the chances of conflict.

Mining, like other forms of industrial activity, requires a two-way flow of goods, material, personnel and products with the external community. The transport associated with this exchange inevitably represents an extension of mining activity into the external environment, and therefore often directly confronts the general public with environmental problems. It is thus hardly surprising that transport is commonly a major source of public criticism. There are, however, particular difficulties in assessing the environmental impact of mineral traffic. These arise principally for the following reasons:

(i) The total volume of traffic may be unknown—this is seldom the case for outgoing valuable mineral products, but the traffic generated by waste disposal and the flow of men and supplies can be harder to quantify. Even for mineral products, whilst the total tonnage despatched is normally known, the ultimate destination is often less certain and hence the total transport load, in tonne-kilometres, can be difficult to determine. For instance, roadstone aggregates are often sold ex-mine and the mineral producer may have limited data on the ultimate destination.

(ii) Once outside the mine boundary it is often difficult to identify mineral traffic from the general flow within an overall transport network. For example a freight train or road vehicle containing mineral products is one element in a system transporting goods from many other sources. Unless mineral traffic represents a major proportion of the total transport load, it can at best be identified as one factor contributing to a general community problem.

Because of these problems there are seldom adequate statistics available to quantify mineral transport even within a limited area. This fact is readily illustrated by the experience of Local Planning Authorities in the U.K. Despite the fact that Britain is a relatively small, developed country with sophisticated central control of industrial activity, Planning Authorities concerned with developing minerals extraction policies have experienced grave difficulties in quantifying the markets served by minerals produced within their region. Until much more comprehensive data is available a general discussion of the environmental impact of mineral transport is thus necessarily qualitative rather than quantitative.

Complaints relating to mineral traffic cover a wide range of topics. The principal areas of concern to the general public are:

(i) Public safety.
(ii) Noise and vibration.
(iii) Air pollution.
(iv) Water pollution.
(v) Visual intrusion.
(vi) Interference with other community activities (*e.g.* private transport).

The relative importance of these impacts cannot be assessed in abstract since it depends first on the type of mineral transport system in use and second on the nature of the local environment.

9.2. RANGE OF TRANSPORT METHODS

The selection of transport method is principally determined by technical and economic factors. A limited number of environmental constraints have been in force for many years. For instance most countries restrict the size of road vehicle which may be used on public highways. However, it is mainly in the last decade that environmental factors have begun to play a significant role in the choice of a mineral transport system, but inevitably such constraints give only limited flexibility within the framework of what is technically and economically feasible.

Thus before discussing environmental aspects of mineral transport it is important to consider the practical constraints which apply to the principal transport methods. In the following section the seven most important types of mineral transport are examined. In practice mineral producers often use a variety of transport methods in combination in order to take advantage of the strong points, or overcome the weak points, of particular methods.

9.2.1. Road

The overwhelming advantage of road transport is its very great flexibility. Every developed country has a road network extending to the smallest communities and even the most underdeveloped countries normally have roads connecting important centres. Road is the major system in use for general transport requirements in almost every country in the world. The tonnages despatched or received are infinitely variable between zero and the maximum capacity of the road network. Because of the widespread use of road haulage for other purposes, the mineral producer can normally rapidly vary his transport capacity by purchasing and hiring (or selling) vehicles. Road vehicles are subject to few gradient constraints and hence can cope with a wide range of terrain.

In most countries the provision of a road network is a community responsibility. Hence the capital investment of a mineral producer is frequently limited to the construction of private roads in the immediate mine area, maintenance and load-out facilities, and the purchase of vehicles (although even the latter may be reduced by hiring or leasing).

Because of the above advantages, road transport is unrivalled for movement of personnel, receiving small tonnages of materials from a variety of suppliers, and the despatch of products in small batches to a range of geographically distinct customers. No other existing system has the flexibility to cope with these requirements.

The principal technical and economic disadvantages of road transport are relatively high manpower requirement and maintenance costs (because of limitations on the size of load), and high fuel consumption per tonne-kilometre.[9,23] A further disadvantage in an undeveloped area may be the high capital cost of constructing a new road system which, depending upon location, may be the responsibility of the mine operator or the government.

Because of these disadvantages, other transport methods may be preferred to road for the bulk movement of products from the mine to a single, or relatively few, outlets.

The main advantages and disadvantages of road transport are summarised in Table 48.

9.2.2. Rail

As with road transport, most countries have a rail network operating at least between major urban and industrial centres. However, this network is never as extensive as a road system and there is usually a limit to the number of rail despatch points. Consequently a mining operation and its suppliers and customers are much less likely to be located adjacent to rail than to road facilities. Hence the use of rail transport frequently entails construction of branch lines in addition to loading facilities and the purchase of rolling stock.[17,18,21]

Freight cars for mineral transport are often custom built and not

TABLE 48

OPERATING ADVANTAGES AND DISADVANTAGES OF ROAD TRANSPORT

Advantages	Disadvantages
1. Great geographical flexibility	1. Restriction on maximum size of unit
2. Few gradient constraints	2. Low to medium manpower productivity
3. Low capital cost of transport unit	3. High fuel costs
4. Technology and costs well-established and predictable	4. Medium to high maintenance costs
5. Rapid response to market requirements	5. Capital costs high in undeveloped areas

available off the shelf (unlike road vehicles) and there is in practice only limited flexibility in output if an adequate return on capital investment is to be assured. Payload, efficiency and safety reduce rapidly with increasing gradient so that it is advisable to restrict inclines to a maximum one per cent gradient. This limits the use of rail in hilly terrain.

A major advantage of rail is the low fuel cost per tonne-kilometre. Maintenance costs can also be low but this depends partly on whether track must be maintained to passenger or freight safety standards.

Because of these advantages and disadvantages, rail is of greatest interest for shipping relatively constant, preferably medium to large, tonnages between a few fixed locations. Examples of its use include transport of concentrates to smelter or port, and bulk transport of aggregates from quarry to urban depot for onward distribution to customers by road.

For new mines remote from developed areas where neither road nor rail networks exist, the high capital cost of railway construction may be matched by the cost of building roads and in these circumstances economics might favour the use of rail.

Table 49 summarises the main advantages and disadvantages of rail.

Every other transport method can be (and frequently is) under the

TABLE 49

OPERATING ADVANTAGES AND DISADVANTAGES OF RAIL

Advantages	Disadvantages
1. Low fuel cost	1. Gradient limitations
2. Low operating cost for constant large tonnages between fixed points	2. High capital cost of infrastructure, transport units and bunkerage
3. Medium to good productivity	3. Limited flexibility in size of unit and distribution points
4. Technology and costs well-established and predictable	

complete control of the mining company. With the exception of relatively few mines where the traffic volume has enabled private railways to be economically justified, use of rail entails a common-carrier railway and thus the mining company is completely in the hands of another party for its transport.

9.2.3. Sea

The most obvious constraint and principal disadvantage of sea transport is that both loading and distribution point must be on the coast. Since only infrequently are both mineral producer and customer at coastal locations, sea transport is normally used in conjunction with other methods.[16,22,24,26]

The main advantages of sea transport are that manpower productivity is high and fuel costs per tonne-kilometre are very low. Thus operating costs are very cheap. In contrast capital costs are high and there is very little output flexibility unless shipping can be diverted to and from other cargoes as required.

Sea transport can conveniently be divided into two major classes. First there is the international trade in a wide range of minerals for which land based transport methods are inappropriate. Here there is no viable alternative to sea transport and the only points at issue are the size and type of vessel to be used. In general, low value minerals such as roadstone cannot bear the costs of long distance international transport unless there is an acute shortage in the importing country. For medium value minerals such as coal, potash and iron ore, the transport element of cost is an important factor. They are commonly shipped in bulk carriers of 25 000 tonnes dwt and larger and for major mineral routes vessels of 100 000 or even 200 000 tonnes dwt may be used. The more valuable minerals and concentrates are not transported in such quantities and either use smaller ships or form one part only of the cargo of a large vessel.

Secondly there is coastal shipping of minerals for which sea transport must often compete with land based methods. In this case, unless producer or customer or both are located close to port facilities, the multi-handling costs when transferring from one transport method to another may easily offset the low operating costs of ships.

In either case, the economics of sea transport are much more attractive if loads are available to obviate ships returning empty. Return loads can likewise reduce the costs of rail and road transport.

The main advantages and disadvantages of sea transport are summarised in Table 50.

9.2.4. Inland Waterways

Most of the advantages and disadvantages of sea transport are also apparent in the use of rivers, canals and lakes. Although some inland

TABLE 50

OPERATING ADVANTAGES AND DISADVANTAGES OF
SEA TRANSPORT

Advantages	Disadvantages
1. Low fuel cost	1. High capital cost
2. Low maintenance cost	2. Limited output flexibility
3. High manpower productivity	3. Severe loading and distribution limitations

waterways, such as the Great Lakes in North America, carry important tonnages of mineral, in general they do not provide an adequate network between producer and consumer. Hence their principal use is as one element in a transport system.[20,31]

Normally inland waterways must compete with land-based transport methods. The principal advantages of the method are low fuel cost and relatively high manpower productivity, although the latter may be restricted by the maximum size of vessel, particularly for canals. The main disadvantages of inland waterway transport are the multihandling costs if producer and customer are not located close to the waterway; the slowness and difficulty of negotiating rapid changes in gradient; the inflexibility of output; and the high capital cost which may be required for sophisticated storage and loading facilities and vessels.

Recently the British Waterways Board has been seeking ways to increase the use of inland waterway transport by overcoming some of the disadvantages. Three systems have been investigated to increase flexibility. These are trains of up to nineteen compartment boats, barge–aboard–catamaran (BACAT), and lighter–aboard–ship (LASH). Both the BACAT and LASH concepts enable canal barges to be loaded and carried by mother ships along appropriate sea lanes to a canal head for dispersal without any break of bulk. The cost implications of these systems are not

TABLE 51

OPERATING ADVANTAGES AND DISADVANTAGES OF
INLAND WATERWAY TRANSPORT

Advantages	Disadvantages
1. Low fuel cost	1. Medium to high capital cost
2. Low maintenance cost	2. Severe loading and distribution point limitations
3. Medium to high manpower productivity	3. Limited output flexibility
	4. Gradient restrictions

yet known although it is self-evident that greater flexibility is achieved at considerably increased capital cost.

The principal advantages and disadvantages of inland waterway transport are summarised in Table 51.

9.2.5. Aerial Ropeway

Aerial ropeways are quite commonly used in mining for internal transfer of material, particularly for transport of ore from mine to treatment plant or waste to dumps. Their principal advantages are the ability to negotiate hilly or difficult terrain, relatively high manpower productivity and medium to low operating cost.

The main disadvantages of the method are the inflexibility of output and distribution points and the high capital cost. For the latter reason, if no other, long-distance use of aerial ropeways is unlikely and at best they may be expected to form one element of an external transport system.

TABLE 52

OPERATING ADVANTAGES AND DISADVANTAGES OF
AERIAL ROPEWAYS

Advantages	*Disadvantages*
1. Ability to negotiate difficult terrain	1. High capital cost
2. High manpower productivity	2. Output inflexibility
3. Medium to low operating cost	3. Loading and distribution point limitations

9.2.6. Conveyers

Conveyers are widely used for internal transport of material within a mining operation. They are well suited to handle steady tonnages moving between fixed points and hence are ideal for transfer of material from one section of a mine or treatment plant to another.[10]

Their principal advantages are the ability to negotiate considerable changes in gradient, high manpower productivity and medium to low operating cost if used in correct applications. The main disadvantages are high capital cost, requiring good utilisation if costs per tonne are to be acceptable; severe loading and distribution point inflexibility; limitations on the size of material transported; inability to change direction rapidly without transfer points; and restrictions on the maximum length of a single flight. The latter limitation has been the subject of much research in the last decade. Several methods for increasing single flight length have been evolved. The best known of these is the cable belt in which the

TABLE 53

OPERATING ADVANTAGES AND DISADVANTAGES OF CONVEYERS

Advantages	Disadvantages
1. Moderate changes in gradient negotiable	1. High capital cost
2. High manpower productivity	2. Lump size limitation
3. Medium to low operating cost if utilisation is high	3. Inability to change direction rapidly
	4. Loading and distribution point limitations
	5. Output inflexibility
	6. Maximum length of single flight limited

driving tension is transmitted through steel cable, rather than the belt fabric, and single flights of several kilometres length are now commonplace.[30]

9.2.7. Pipelines

Pipelines are in widespread use in the mineral industry for the transport of liquids and slurries. No other method can rival the low operating cost of pipelines for routine transfer of medium to large tonnages of these materials between fixed points.[15] Pipelines up to 100 km (60 miles) in length, and occasionally more, are used to convey water, slurried mineral products and wastes, particularly tailings. The main disadvantages of pipelines are medium to high capital cost, and inflexibility in output, loading and distribution points.[12]

A further disadvantage of pipelines has been the inability to carry solid material (of density greater than one) except as finely ground particles in suspension. Although some material is pumped in saltation (rather than suspension) over short distances, high energy cost and excessive pipe wear have precluded long distance transport in this manner. In the last few

TABLE 54

OPERATING ADVANTAGES AND DISADVANTAGES OF PIPELINES

Advantages	Disadvantages
1. High manpower productivity	1. Medium to high capital cost
2. Low fuel cost	2. Loading and distribution point restrictions
3. Low maintenance cost	3. Limited to liquids and slurries at present
4. Ability to negotiate changes in gradient	4. Output inflexibility

years there has been research in a number of countries (notably the U.S.S.R., Canada, Germany and Britain) into the possibility of encapsulating bulk minerals and conveying them through pipelines using either a hydraulic or pneumatic medium. Neither the technology nor economics of this transport method are yet adequately defined and hence at present it remains no more than a future possibility.[25,27-29]

9.2.8. Combined Methods

In practice, because of the limitations of each method, a mineral distribution network seldom comprises only one type of transport. A typical system consists of one or more trunk line methods with feeder and/or distribution elements. The example given in Table 55 illustrates the complexity of many mineral transport systems.

TABLE 55

TRANSPORT NETWORK OF LIMESTONE QUARRY PRODUCING 5 MILLION TONNES/A

Product	From	To	Transport method	Annual tonnage
Limestone	Quarry	Customer	Road	One million
Limestone	Quarry	Urban depots	Rail	Three million
Limestone	Urban depots	Customer	Road	Three million
Limestone	Quarry	Adjacent dock	Conveyer	One million
Limestone	Dock	Overseas port	Ship	One million
Limestone	Overseas port	Customer	Road	One million
Solid waste	Quarry	Tip	Road	250 000
Slurry waste	Quarry	Settling ponds	Pipeline	50 000

Table 56 indicates the suitability of the various transport methods for use as branch and trunk line elements in a system. The table gives general guidance only. For reasons specific to individual operations, there are exceptions to the general pattern shown.

In any practical situation the operator is often faced with a choice of both branch and trunk line systems. Considerable interchangeability exists and decisions are usually taken on grounds of cost and operational convenience.

9.3. ENVIRONMENTAL PROBLEMS OF MINERAL TRANSPORT

9.3.1. Road

Road transport almost certainly carries more mineral, in terms of total tonne-kilometres, than any other method and is particularly dominant in

TABLE 56

MAIN USES OF INDIVIDUAL TRANSPORT
METHODS AS ELEMENTS IN TRANSPORT
NETWORK

Branch lines	Trunk lines
Road	Road
Rail	Rail
Conveyer	Sea
Aerial ropeway	Pipeline
Inland waterway	Inland waterway

the field of low-value industrial minerals. It is unfortunate, in view of this widespread operational convenience, that it is the transport method which normally occasions the greatest number and range of environmental complaints. The principal areas of concern are public safety and inconvenience; noise and vibration; air pollution; and visual intrusion.

Wynn[1] cites evidence that goods vehicles in the U.K. have a lower rate of involvement in accidents than any other type of vehicle, but that accidents in which they are involved lead to a relatively high proportion of deaths. From the point of view of public safety and inconvenience, there seems no reason to single out minerals transport from other heavy road vehicles. However, most heavy traffic occurs in urban or industrial areas and on major trunk highways. In contrast many mineral operations are located in remote rural areas, since the mineral must obviously be mined where it is found. Consequently mineral traffic is often readily identifiable as such and may be required, especially in the vicinity of the mine, to traverse small roads unsuited to heavy traffic flows. The main methods of reducing public hazard and inconvenience from heavy minerals road traffic are:

(a) Transfer some of the traffic to a more acceptable transport method, such as rail or inland waterway. This course of action may be possible in some cases but, apart from any economic implications, other transport methods often lack the flexibility to be an adequate substitute for road.

(b) Limit the maximum vehicle size on public roads—this is commonly done in many countries (*e.g.* maximum 32 tonnes gross vehicle weight in the U.K.). However, restrictions on vehicle size not only adversely affect transport economics but also increase the number of units required to move a given tonnage. Hence a balance must be struck between an unacceptably large number of vehicle movements and the maximum vehicle size suitable for local roads.

(c) Vehicle routing—this may be achieved directly, or indirectly, by placing weight or width restrictions on some roads. Where adequate alternate routes exist, it is a satisfactory method of avoiding narrow, winding, unsuitable roads and sensitive population centres such as rural villages.

The noise levels of heavy road vehicles with engines exceeding 200 h.p. are often 90 dB(A) or higher at the kerbside. At this level, unacceptable nuisance can certainly be caused in property adjacent to mineral traffic routes.[14] The extent of the nuisance depends upon the volume of traffic, the nature of the locality and the time of day. Obviously the problem is most severe for heavy traffic flows in rural areas at night. The main control measures are:

(a) Better vehicle design to reduce emitted noise levels—increasing research efforts are concentrating on the design of quieter vehicles and there are hopeful signs that kerbside noise levels can be reduced to 80–85 dB(A) at relatively low cost. It is uncertain whether design factors could further reduce noise levels without escalating cost unacceptably.

(b) Vehicle routing and scheduling—routes may be selected to avoid sensitive areas and night traffic may be banned or limited. As often occurs in environmental matters, there is direct conflict between one consideration and another. Night transport is desirable to reduce public hazard and inconvenience, but may cause extra noise problems.

(c) Screening of roads—the use of tunnels, cuttings, banks or even vegetative screens can reduce received noise levels. However, unless these are required for road construction, any noise benefit is achieved at considerable cost and probably by the unnecessary sterilisation of land adjoining highways.

Vehicles can give rise to ground vibrations of a similar nature to those caused by blasting. There are two main categories of force which occasion these vibrations. First, out of balance and other forces from the engine, wheels and transmission may be reacted by the suspension–roadwheel system. Secondly, roughness or discontinuities in the road surface cause abrupt dynamic loads between the wheels and the road.

Vibrational energy is propagated by surface waves in the manner described for blasting in Chapter 7, and similar criteria are applicable both for structural damage and nuisance. An important difference between vibrations from blasting and from traffic is frequency rate. Blasting seldom occurs more often than once or twice per day whereas heavy traffic may

TABLE 57

LIFE REDUCTION IN HOUSES DUE TO TRAFFIC

Life decrement (%)	Decrease code[a]	Vehicle flow units per 24 hours[b]
Zero	1	0–260
4·0	2	260–600
7·5	3	600–960
10·0	4	960–1540
15·0	5	1540–2660
20·0	6	2660–3440
25·0	7	3440–4660
35·0	8	4660–7440
50·0	9	7440+

[a] See Table 58.

[b] Only the following are counted: vehicles over 5 tonnes at 1 unit each; lorries under 5 tonnes at 0·4 units each; trams on track with poor foundations at 1·5 units each.

have an incidence rate several hundred times as great. There is very limited data available on fatigue effects in buildings. According to Steffens[2] fatigue strengths of masonry and mortars are of the order of 0·6 of the static values. This is based on laboratory work over several millions of cycles.

The effect of traffic upon houses has been studied in Czechoslovakia[3] and the results are cited by House[4] as shown in Tables 57 and 58.

TABLE 58

ADJUSTMENTS TO DECREASE CODE IN TABLE 57

Condition	Change to code
Older buildings, frontage up to 10 m (30 ft) apart, paved surface	No change
Bituminous road surface	Add 1
Concrete surface	Add 3
New brick buildings	Add 2
Steel or concrete framed structures	Add 4
Mean traffic speed below 20 km/h (12 mi/h)	Subtract 2
For each increase of 8 m (25 ft) frontage separation	Subtract 1
Footpath 1·5 m (5 ft) wide or less	Subtract 1
No footpath	Subtract 2

The principal methods of reducing problems caused by traffic vibrations are:

(a) Route heavy vehicles to avoid passing close to structures, particularly those which are old and delicate.
(b) Improve vehicle suspension.
(c) Improve road construction to avoid abrupt impulsive loading.
(d) Impose weight restrictions on vehicles.

Air pollution arises principally from exhaust emissions and dust. There is no reason to single out mineral traffic from other heavy diesel vehicles on the subject of exhaust gases. Increasing general environmental concern is leading to the development of more efficient exhaust scrubbers for all vehicles, and mineral lorries will benefit from these improvements. There is little sign that any alternative to the internal combustion engine is likely to be available at acceptable performance and cost in the foreseeable future. Hence the emphasis will be on improved fuel utilisation and more effective scrubber design, probably by the use of catalysts.

Dust is created by wind blow from the lorry load and the movement of the vehicle. The problem is mainly one of nuisance from non-toxic

FIG. 61.   An automated vehicle washer at an open cast coal site ensures that the entire underside of the truck is cleaned before it joins the public highway.

Fig. 62. At the weighbridge at an open cast coal site a grab is employed to prevent the lorry carrying more than its legally permitted weight, and also trims the load. The small stockpile accumulated by the grab is used to top up lorries carrying less than the legal maximum load.

industrial minerals, but serious pollution can be caused by the toxic dust of metal concentrates. Since the latter also causes financial loss, measures would normally be taken to avoid its occurrence. Where dust is emitted, it can cause soil and water pollution when it settles.

The principal means used to avoid air pollution are:

(a) Fitting efficient exhaust scrubbers.
(b) Coating internal and external roads to avoid creating dust plumes.
(c) Regular sweeping of internal roads and, sometimes, external roads close to the minerals operation.
(d) Sheeting of loads.
(e) The use of tankers or other specialist vehicles for the transport of products in powder form.
(f) Regular washing of vehicles.

9.3.2. Rail
In terms of overall freight movements, rail is comparable to road in importance. In 1970 the U.S. railroads carried 1230 billion tonne-km (768 billion ton-miles) in contrast to 960 billion (596 billion) on the waterways and 800 billion (496 billion) on the roads. However it is probable

that rail is less important to mineral producers than road because of its lower flexibility. In Britain in 1972 only 6% of mineral products (by tonnage) travelled by rail although the percentage was probably larger when expressed in tonne-kilometres, because of the greater average haul with rail. There is evidence that increasing fuel and labour costs, environmental pressures, and the continuing trend to centralisation of mineral production are combining to induce some shift of traffic away from road onto rail.

The main environmental disadvantages of rail are noise, vibration and dust. Additionally, the necessity for rail depots in urban areas can cause further problems.

Noise and vibration are similar problems to those caused by road transport. Generally they are less severe because the levels generated are normally lower, the frequency of occurrence is less and railroads less often run adjacent to residential areas. Where problems occur, the principal control measures are:

(a) A high standard of track and rolling stock maintenance; long-welded rails to eliminate rail joint noise.
(b) Speed and weight restrictions.
(c) Scheduling to avoid excessive night movements in residential areas.

Loss of dust in transit has recently been a problem highlighted because of the pollution which can arise, and the financial loss where valuable metal concentrates are concerned. Schwartz[5] reports interesting work on transit losses undertaken by the Canadian National Railways. Actual dust losses depend not only on the type of rolling stock but also on a wide range of other factors such as transit speed, time and distance and the preventive measures adopted. Tests on zinc concentrate movements from Quebec to Pennsylvania [1700 km (1060 miles)] in open top cars protected with only a light water spray disclosed a loss of about 2·1% of concentrate. Similar movement of copper concentrate from British Columbia to Quebec [3860 km (2399 miles)] disclosed a 3·7% loss. Various control measures were tried with the following results:

(a) Gondola cars with tarpaulin covers—perfect protection from wind loss and moisture pick-up but high cost of tarpaulin replacement.
(b) Spraying with chemical binders—effective for products that do not slump but with some concentrates the crust breaks and strips off giving high losses. No protection is afforded against moisture pick-up.
(c) Polyethylene sheets held in place with wrought lumber—provides only limited protection due to cover failure.

(d) 'Fabrene' cover (woven polyolefin fabric)—good protection against dust losses and moisture pick-up.
(e) Fibreglass cover—perfect protection against dust loss and moisture pick-up.

Table 59 summarises the results for copper concentrate movements, at a concentrate value of $300 per tonne.

TABLE 59

COPPER TEST SUMMARY

Cover type	Product loss (%) Handling	Transport	Loss cost ($)	Cover cost (c)	Total cost ($/tonne)
Open top car	0·75	3·7	13·35	—	13·35
Fabrene	0·75	0·2	2·85	21·6	3·06
Fibreglass	0·75	—	2·25	56·0	2·81
Polyethylene	0·75	0·4	3·45	33·0	3·78

A theoretical study of four rail transport systems was undertaken by the Canadian National Railways. These were:

(i) 6% moisture content concentrate in standard gondola cars.
(ii) 6% moisture content concentrate in steel containers.
(iii) Bone-dry concentrate in pressure-differential covered hopper cars.
(iv) Pellets or briquettes in standard covered cars.

The results are summarised in Table 60.
It is of interest that those systems which minimise dust pollution also optimise transport costs.

TABLE 60

SYSTEM COST COMPARISON

System	Cost per tonne ($)
6% moisture content concentrate (gondola)	13·70
6% moisture content concentrate (container)	14·80
Bone-dry concentrate	12·70
Pellets or briquettes	12·00

The establishment of urban distribution depots for low value industrial minerals moved by rail causes several problems.[8,11,13,14] Principally these are dust, noise, visual intrusion and the generation of heavy road traffic movements locally. The solution to these problems lies first in careful site selection to minimise the potential for disruption. A totally enclosed depot design affords the greatest opportunity for minimising noise and dust emissions and for creating a generally acceptable visual profile. However, evidence from Britain suggests that such enclosed designs cost up to 50% more to construct than conventional open depots.

9.3.3. Sea

Almost all the environmental disadvantages of sea transport are apparent during loading and unloading, and in the limitations of the method itself.[7] Once at sea, there are virtually no environmental problems except the slight pollution potential caused by the chance of shipwreck or careless discharge of oil. However, there are few occasions when sea transport can directly connect mineral producer with consumer and hence the method normally forms one element only in a transport network.

The environmental problems at the loading and discharge points are similar to those of rail urban depots. The low cost of sea transport is dependent upon a quick turn-round time in port. This necessitates rapid loading and discharge facilities and stockpiling with consequent potential for visual intrusion, dust and to a lesser extent noise. The large ore ships of 100 000 tonnes dwt commonly are loaded at 3000–6000 tonnes/h and occasionally the rate is as high as 10 000 tonnes/h. The solution to the problems caused lies in sophisticated loading systems to minimise the generation of dust and, as far as practicable, enclosed dock areas to contain dust and to create a more acceptable visual profile. An example of the dust problems associated with rehandling at ports is given in Chapter 4.

The shipment of metal concentrates as briquettes may increase in the future. In this form, little dust is generated. Further advantages of briquettes are that the tendency to oxidation and spontaneous combustion during transit is minimised by the absence of voids, and the tendency of loads to shift during transit is also reduced.

In the immediate port area congestion is likely to arise from the onward transfer of minerals by other transport methods. This can only be eliminated by the careful siting of docking facilities adjacent to an adequate land-based transport network and, if necessary, limiting the sea tonnages to those which can be handled readily by this land network.

9.3.4. Inland Waterways

The environmental benefits and drawbacks of inland waterways are similar to those of sea transport. Once in transit minimal environmental disruption occurs and, according to the project director of the Port of Toronto, a waterway can transport far more freight than an equivalent width of road or rail. Where an adequate network exists, the cheapness of this transport method makes it popular. In the U.S.A. there is extensive use of the Great Lakes system and nationally 14% of freight travels by inland waterway. Equivalent figures elsewhere include Germany 25%, Holland 66%, Belgium 29% and France 31%. In Britain the percentage is less certain but is lower because of the present deficiencies of the canal and river systems.

The problems associated with loading and unloading are the same as for sea transport, although usually on a smaller scale, and the remedies are similar.

9.3.5. Conveyers

The use of conveyers is presently confined to internal movements and branch or feeder lines. Their principal environmental disadvantages are dust and visual intrusion.

Dust arises from wind blow and spillage. It may be reduced by regular maintenance, the use of water sprays or, preferably, by total enclosure. Visual intrusion is most pronounced when conveyers are used to elevate material for loading or processing. It is probable that the clean lines associated with enclosure are more generally acceptable than the network of gantries and struts common in open conveyers. However, these environmental advantages are only achieved at considerably increased capital cost.

9.3.6. Aerial Ropeways

The application and environmental advantages and disadvantages of aerial ropeways are very similar to those of conveyers. However, total enclosure is not an economically practicable solution and little can be done to reduce dust other than avoiding overloading, spraying with water, and regular maintenance to reduce spillage. Yet dust nuisance from aerial ropeways is seldom serious.

Visual intrusion can only be reduced by the careful selection of the ropeway route and minimising the height of supporting structures. The latter would normally be undertaken in any case to reduce capital costs.

9.3.7. Pipelines

Environmentally, the use of pipelines for mineral transport is probably the most satisfactory method. There is an absence of dust, noise, vibration, air pollution and public safety hazard, and visual intrusion may be very small or non-existent if the pipe is buried. The principal drawback is the very slight possibility of pollution from leaks.

The method is at present severely limited in that relatively constant throughput is necessary and only liquids or slurries may be transported over significant distances. Considerable research and development effort is now under way on the encapsulation of solids within pipelines and their transport using a pneumatic or hydraulic medium. It is too early to determine the technical and economic feasibility of such methods but their successful implementation would certainly hold the promise of very low environmental impact.

9.3.8. Combined Methods

In any transport system comprising a number of elements, the disadvantages and advantages of each element would be as described above. The principal advantage of combined systems is that they permit each element to be used in the circumstances for which it is economically and environmentally most appropriate. The main disadvantage is that trans-shipment is required between each element. In addition to cost considerations, it is at this transfer stage that environmental problems are frequently most apparent. The potential for spillage, dust, noise and visual intrusion is usually high during transfer and consequently careful site selection and costly design is necessary if transfer points are to be environmentally acceptable.

9.4. CONCLUSIONS

It is a curious coincidence that the transport methods which have the lowest environmental impact also have the lowest unit costs when the method is used in favourable conditions. Breach[6] cites energy efficiency research at the Oak Ridge National Laboratory, summarised in Table 61.

In a Wall Street Journal story in 1973 it was stated that 'a dollar will move a ton of freight 8 km (5 miles) by air, 24 km (15 miles) by truck, 106 km (66 miles) by rail and 530 km (330 miles) by water'. Nonetheless these figures mask the network deficiencies and operating disadvantages of the cheaper and environmentally preferable transport methods.

TABLE 61
ENERGY EFFICIENCY OF TRANSPORT METHODS

Transport method	Tonne-mile/gal	Tonne-km/litre	Btu/tonne-mile
Pipelines	300	110	450
Waterways	250	90	540
Railroads	200	70	680
Roads	58	20	2 340
Airways	3·7	1·3	37 000

The reduction of environmental impact from mineral transport in the future seems to require three strategies:

(a) Improved design of existing transport and loading facilities.
(b) Increased flexibility to permit transfer of freight from transport methods with a high environmental impact to those which cause less disruption.
(c) Development of new, economically practicable, methods such as hydro- and pneumo-capsule pipelines.

Much of the need to implement these strategies stems from general problems of transport and is not specific to the mineral industries.

REFERENCES

1. Wynn, N. R. (1974). Road delivery of materials, *Quarry Managers' Journal*, **58**(4), 111–19.
2. Steffens, R. J. (1965). Some aspects of structural vibration, in *Proc. Symp. Brit. Nat. Soc. International Soc. Earthquake Engng*, Imperial College, London.
3. Kapsa, L. (1971). *Protection of Housing Estates against Unfavourable Effects of Traffic*, Research Institute for Building & Architecture, Prague.
4. House, M. E. (1973). Traffic-induced vibrations in buildings, *The Highway Engineer*, **20**(2), 6–16.
5. Schwartz, P. L. (1974). Innovative rail transport systems cut concentrate losses, *World Mining*, **27**(7), 38–43.
6. Breach, I. (1973). Freightways of tomorrow, *New Scientist*, **59**(854), 68–70.
7. Yu, A. T. (1972). Environmental control in shiploading and unloading, *Mining Engineering*, **24**(8), 29.
8. A major quarry development in the Mendips (1970). *Quarry Managers' Journal*, **54**(9), 309–18.
9. Relleen, R. (1970). Economics of external transport, *Quarry Managers' Journal*, **54**(9), 321–34.
10. Taylor, H. J. (1970). Field belt conveyor systems, *Quarry Managers' Journal*, **54**(7), 240.
11. New local distribution centre for aggregates (1973). *Quarry Managers' Journal*, **57**(4), 119–24.
12. Baker, P. J. (1973). Pipeline transport of material, *Quarry Managers' Journal*, **57**(8), 281–9.
13. New aggregates depot to serve the south-east of England (1973). *Quarry Managers' Journal*, **57**(7), 231–4.
14. Foster Yeoman's Botley coating plant supplies the south-east (1973). *Mine & Quarry*, **2**(8), 15–18.
15. Bougainville's 17 mile concentrate slurry pipeline (1973). *World Mining*, **26**(5), 45.

16. Vander Laan, R. W. (1974). World's first 10,000 tph ore and coal unloading terminal, *Mining Engineering*, **26**(3), 33–6.
17. Why Tytherington transports aggregate by rail (1973). *Mine & Quarry*, **2**(10), 19–25.
18. Rail depot keeps rocks off the road (1974). *Mine & Quarry*, **3**(1), 15–16.
19. Williams, H. (1973). Noise measurement in planning, *J. Royal Town Planning Institute*, **59**(1), 7–9.
20. Stearn, E. W. (1973). Cement and aggregate ride the water-way, *Rock Products*, **76**(6), 62–6.
21. Stearn, E. W. (1973). Rock products and the railroad option, *Rock Products*, **76**(6), 67–71.
22. Saunt, T. (1974). Transport of aggregates by sea, *Quarry Managers' Journal*, **58**(4), 120–3.
23. The costliest operation in quarrying (1974). *Mine & Quarry*, **3**(3), 11–19.
24. Vickers, C. L. and Tonsley, A. H. (1963). Shiploading bulk materials at Long Beach, California, *Trans. A.I.M.E.*, **226**(3), 248–57.
25. Hydraulic transport of minerals (1972). *Mining Magazine*, **126**(4), 248–68.
26. Super ore carriers (1969). *Mining Magazine*, **120**(7), 3.
27. Nickeson, F. H. (1974). Hydraulic pipeline makes continuous haulage possible, *Coal Mining & Processing*, **11**(7), 43–53.
28. Berkowitz, N. and Jensen, E. J. (1963). On possibilities for pipelining coal in Canada, *Canadian Mining & Metallurgical (CIM) Bulletin*, **56**(615), 504–8.
29. Thornton, W. A. (1972). *The pneumatic transport of solids in pipes; a bibliography*, British Hydromechanics Research Association, Cranfield, 104 pp.
30. Walker, R. (1971). 5¼ mile single flight conveyor, *Canadian Mining & Metallurgical (CIM) Bulletin*, **64**(716), 46–64.
31. Jones, R. M. (1971). The impact of superfreighters and Nova Scotia deep-water ports on bulk material distribution and marketing, *Canadian Mining & Metallurgical (CIM) Bulletin*, **64**(715), 82–4.

10

Reclamation

10.1. INTRODUCTION

Mining land has hitherto passed through a cycle of land use which in general can be summarised as: the exploitation of virgin land for the particular minerals; abandonment of the land and subsequent dereliction; and, after an interval of decades, the treatment of that derelict land to suit it for some productive use. It is the intermediate derelict land phase which has frequently been responsible for mining's unfavourable public image, and only in comparatively recent years has any attempt been made to remedy the numerous evils which, it is widely acknowledged, arise from the existence of dereliction.

10.1.1. Definitions

In attempting to define the terminology, it must be recognised that much confusion exists. For example, in Britain the Commission on Mining and the Environment[130] accepted the following:

Restoration—recreating conditions suitable for the previous use of the area.
Rehabilitation—creating conditions for a new and substantially different use of the mine site.
Reclamation—returning a derelict site to some use.

The precision of these terms has not always found favour. Restoration is widely accepted to mean returning the mined-out area to its former condition and land use; as such, shallow strip mine workings for coal and ironstone, which have high proportions of overburden, are conventionally spoken of as being restored. Rehabilitation and reclamation have no well-accepted meanings although derelict land, in common parlance, is 'reclaimed'. In Britain, the sand and gravel industry has proposed the terms:

After-treatment—any process which leads to the 'after-use' or 'redevelopment' of the land.
Reconstituted—when after-treatment involves refilling the pit, the resulting land is 'reconstituted'.

In 1976 in Britain, the Committee on Planning Control over Mineral Working[88] proposed the following definitions:

Restoration—the whole complex of operations which follow the extraction of minerals up to the time at which the land is fully established once more in an acceptable environmental condition. It includes removal of buildings and plant, and after-treatment of the land. When restored, the area can either be suited to its old use, or be subject to special treatment for redevelopment.

After-treatment—treatment of land after mineral extraction by filling, contouring, topsoiling and seeding, to return it to an acceptable condition fit for further use.

After-care—management of land to ensure that its restored condition is maintained; for example, agricultural cultivation and drainage.

After-use—the use to which the restored land is put, whether it be the old use or a new use.

Special treatment for redevelopment—as a supplement to after-treatment, extra work may be undertaken for some uses, such as public parks, golf courses, etc.

It would be tedious to instance more examples (although there are many) but there is a need for simplification rather than increased complexity. Any too-specific definitions (as in some of the cases above) merely mean that a variety of different words have to be used to define the process at a particular site. The definitions used in this book are those most widely accepted:

Restoration—recreating the original topography and re-establishing the previous land use.

Reclamation—any treatment which is not restoration. This includes the most common forms of treatment at mineral extraction sites.

10.1.2. Derelict Land

Definitions of derelict land show a dichotomy between legal implications and the usage of the 'man in the street'. To the public in general, land is derelict if it looks unpleasant and uncared for. As such, this common-sense definition includes a whole range of mining and other land which is, in fact, in productive use but which, either because of the inherent nature of the activities carried out upon it or due to defects in environmental controls, has an objectionable appearance.

As embodied in legal terminology a more precise specification obtains. In England and Wales, derelict land is defined as 'land so damaged by industrial or other development that it is incapable of beneficial use without treatment'. This excludes land which has become derelict from natural causes such as marshland or neglected woodland, and also omits land in the following categories:

(i) Damaged land which is still subject to reclamation requirements under a planning consent or statute.
(ii) Land still in industrial use, such as an active tip, which is not subject to any requirement for reclamation.
(iii) Damaged land upon which planning permission for further development (such as coal stocking, backfilling, etc.) exists, or may be granted in the foreseeable future.
(iv) Damaged land which has blended into the landscape over a period of time, to the extent that it can reasonably be considered as part of the natural surroundings, or has been put to a use such as sailing, fishing, potential nature reserve, etc.
(v) War damaged land, sites awaiting clearance and urban clearance in the course of urban renewal.

These exclusions render the definition of derelict land exclusive to Britain, thus creating problems of comparisons with other countries, and also exclude extensive land which in layman's terms would be described as derelict. In quantitative terms, a 1954 survey suggested that England and Wales contained 51 000 ha (127 000 acres) of derelict land; in 1960 a rougher estimate suggested 50 000 ha (125 000 acres) derelict and the same area disturbed but not legally derelict; in 1963, a survey showed 40 000 ha (99 000 acres) of derelict land; and in 1970, the total was 19 800 ha (48 950 acres). Even at the highest estimate, less than 1% of the land area of England and Wales is derelict, but the overwhelming proportion of these totals is produced by mineral working or its adjuncts. The 1954 survey[87] showed that 40% of the dereliction was spoil heaps, 41% excavations and 19% other forms. Most of the heaps and excavations were caused by mining. The average site area was 3·6 ha (9 acres).

In the U.S.A., statistics are collected on the basis of areas 'disturbed' by mining.[69] Up to the end of 1964, the total area disturbed by mine excavation or tipping was 1 300 000 ha (3 212 000 acres), with a further 130 000 ha (321 200 acres) affected by mine access roads and exploration. Some 95% of the disturbance was caused by seven commodities: coal (41%); sand and gravel (26%); stone, gold, clays, phosphate and iron (28%); and the remainder 5%. Of the disturbed area, two thirds was assessed as requiring reclamation.

A U.S.A. survey covering the period 1930–1971 showed that nationally only 0·16% of the land area of the country had been used by mining (including surface disturbance of underground mines), in total 1 477 000 ha (3 650 000 acres). Of this, 40% [590 000 ha (1 460 000 acres)] had been reclaimed. In three states, the percentage of land area utilised by mining

FIG. 63.   Derelict land in a slate quarrying area in North Wales. Note the social decay indicated by the derelict houses. The juxtaposition of waste tip and flooded quarry indicates the method of reclamation which could be employed.

FIG. 64.   Very extensive areas of dereliction at Blaenau Ffestiniog, North Wales have been caused by the waste from underground slate mining. Tipping has buried the mountainside from its crest at an altitude of 450 m down to 200 m.

exceeded 1 % (Ohio, Pennsylvania and West Virginia). In the U.S.A. as a whole, the land areas used by mining were only a sixth of those taken for highway construction.

Of the total area disturbed, 59 % was accounted for by excavations, 20 % by waste disposal from surface mining, 13 % by waste disposal from milling, and 5 % by underground mine wastes. The balance of 3 % was subsidence damage.

In 1971 alone land used by the U.S. mineral industries was 83 000 ha (206 000 acres), of which 66 000 ha (163 000 acres) was reclaimed. The distribution between commodities is shown in Table 62, which demonstrates that coal mining actually had an excess of reclamation over disturbance, due to backlog clearing. Metal mining had the poorest performance with only a third of its disturbed acreage being reclaimed.

TABLE 62

LAND UTILISATION AND RECLAMATION IN
THE U.S. MINERAL INDUSTRIES, 1971
(Paone et al.[14])

Commodity	Acres	ha
Metals, utilised	36 400	14 700
Metals, reclaimed	12 600	5 100
Non-metals, utilised	95 100	38 500
Non-metals, reclaimed	53 200	21 500
Fossil fuels, utilised	74 900	30 300
Fossil fuels, reclaimed (excludes oil and gas)	96 900	39 200

Comparisons between Britain and the U.S.A. are difficult because of the different statistical bases. The British total of legally derelict plus disturbed land is about 100 000 ha (250 000 acres), or less than a tenth of the roughly comparable U.S. figure. However, this British figure represents nearly three times the U.S. proportion, about 0·44 %. Taking into account the population density (eleven times greater in Britain), it is possible that the impact of dereliction is some thirty times greater in Britain than the U.S.A.

The importance of derelict mining land relates to more than its acreage. Pollutants emitted from it affect air and water resources while its presence in an area can aggravate the social damage caused by unemployment following closure of the mine. Persons living in the vicinity of derelict areas have been described as having a 'blitz' mentality. It has been estimated that the presence of just 10 % legal dereliction in an area places a major blight upon the interstices forming the other 90 %. Conversely, the clearing

Fig. 65.   A derelict lead/zinc mine in North Wales. Drainage from the adit and erosion of the tailings dumps are serious sources of water pollution, apart from aesthetic considerations.

and 'greening' (by planting) of even small areas can materially assist in the rejuvenation of such localities.

10.1.3. Planned Reclamation

Many countries have recognised that the existence of dereliction can no longer be tolerated, and have attacked the problems in two ways:

(i) Financial incentives have been made available to assist and encourage the reclamation of derelict land. Because ownership of long-abandoned mines is sometimes hard to establish (the term 'orphan land' as used in the U.S.A. is particularly graphic) and the current owner may not be the original mining company, availability of government funding has been essential.

(ii) For new mineral workings, the conditions under which they are allowed to open include provision for restoration or reclamation, thus avoiding the derelict land stage. Often such conditions do not require more than cosmetic work, to allow the mine to 'lie fallow' without causing environmental damage, until a new land use demand emerges. On the other hand, permissions for short-term workings at which it is possible to predict fairly accurately the desirable land use (*e.g.* sand and gravel) sometimes include detailed reclamation requirements.

A combination of these measures should eventually more or less eliminate mining dereliction wherever they are applied, but there is an interim period in which derelict land continues to be formed, as a result of the older, lax permissions already existing. Already many types of mining operation are routinely undertaken with a clear knowledge of the use to which the worked-out land will be put, and many mining methods have been devised to facilitate after-use in a wide variety of circumstances.[41,111,129] It is to be expected that the progressive 'rolling' reclamation and restoration of mines will be of increasing importance in almost every type of mineral working. To a considerable extent therefore the emphasis of this chapter is on the integration of mining and reclamation, although dereliction remains an important current problem.

10.1.4. Surveys

Planning of reclamation to establish some chosen land use requires a detailed knowledge of the circumstances of the mine site, including ownership boundaries, surface and sub-surface hydrology, physical and chemical nature of the tips and other exposed materials, volumes of tips and excavations, the location of hazards such as shafts, faults and unstable wastes such as tailings slurries, and numerous other matters. The most important are usually the types and volumes of the wastes and the excavations; conventional ground surveys can establish volumes, but aerial survey can be quicker and sufficiently accurate for the purpose of computing the feasibility of various earthmoving operations. All surveys require supplementing by the analysis of materials to determine their suitability for revegetation work, if this is to be carried out.

10.2. RECLAMATION OF WASTE TIPS

Although it is sometimes possible to reclaim tips and excavations in one operation, by moving the former into the latter, more frequently the tip has to be reclaimed *in situ*. Many factors require consideration, and those connected with the relationship between regrading and revegetation considerations (in some cases regrading is unwise) are discussed in Section 10.5.3. Regrading is undertaken to lower the height of the tip and decrease slope angles, and Table 63 demonstrates the relative increases in base area, and the volumes moved, for various reductions in height. It is noteworthy that for a conical tip, halving the height doubles the required base area, but involves the movement of only one eighth of the waste volume. Earthworks of this type are usually the most expensive component of mine reclamation (*see* Section 10.5.3), but the main limitation is often the availability and value of land upon which to spread the regraded waste. The landform selected for regrading depends also upon the land use envisaged; although

TABLE 63

TIP REGRADING FACTORS (Arguile[68])

(a) Change in dimensions in creating a lowered cone

Original height, h	Original base area, a	Original volume, V
For $\dfrac{h}{2}$	Base area becomes $2a$	Volume moved $= 0\cdot125\,V$
For $\dfrac{h}{3}$	Base area becomes $2\cdot94a$	Volume moved $= 0\cdot287\,V$
For $\dfrac{h}{4}$	Base area becomes $4a$	Volume moved $= 0\cdot422\,V$

(b) Change in dimensions in flattening cone-shaped tip to plateau

For $\dfrac{h}{2}$	Base area becomes $1\cdot092a$	Volume moved $= 0\cdot125\,V$
For $\dfrac{h}{3}$	Base area becomes $1\cdot25a$	Volume moved $= 0\cdot287\,V$
For $\dfrac{h}{4}$	Base area becomes $1\cdot55a$	Volume moved $= 0\cdot422\,V$

a gently graded cone is often more satisfactory from a landscape point of view, creation of a plateau offers sites for industrial estates, playing fields, etc. In valley areas such as the South Wales coalfield, regrading of tips to plateaux has provided valuable extra areas of level building land.

Tip regrading needs to take account of various other matters:

(i) Pyritic colliery waste tips can contain unsuspected areas of spontaneous combustion, temperatures of up to 800°C being reported, with temperatures of 150–200°C common even at very shallow depths. Such conditions create a hazard for operators. Water spraying can be dangerous because it causes eruptions and it is necessary to dig out the affected area, allow it to cool and regrade in compact, clay-sealed layers.[95]

(ii) The use of some plant, such as face shovels and trucks, can give rise to operator hazards due to working under high faces of unconsolidated waste. Care is needed to limit face heights.

(iii) Excavation of many tips discloses a variety of scrap metal, rubble, concrete, etc., which has been dumped upon it over the years. Such material needs to be separated out and reburied or otherwise disposed of.

(iv) During regrading, saleable material, such as burnt red shales, may be encountered. It may be feasible to achieve significant reduction of tip volume, plus some revenue, by the sale of such material.

(v) During regrading, dust can be a severe local nuisance. Provision may be needed for watering, or halting operations at high-risk times.

Severe, and largely unexamined, reclamation problems are posed by impoundments of tailings from metal, fluorspar or phosphate mining, washing of coal, stone, clays, etc., and dredge tailings from alluvial tin or other mine types. The fine particle sizes contribute to the frequently thixotropic nature of the tails, and thus render them unstable and unsuited to load bearing for many years if not for ever. Non-ferrous metal tails and coal washings may have additional problems of metal toxicity or acidity. A wide range of uses has been proposed for more or less level tailings impoundments, such as playing fields, airstrips, industrial estates, etc., but all are limited either by revegetation difficulties, or else a lack of experience of suitable construction methods for use on tailings.

10.3. RECLAMATION OF EXCAVATIONS

Excavations usually present greater problems of reclamation than tips. During the creation of an open pit or quarry, the valuable overlying top-soils and subsoils can be destroyed, ground water flows are interrupted, and strata which may leach out toxic ions are exposed. These factors, plus the steep, sometimes unstable, slopes which are created, render after-use possibilities limited in many cases. Primarily, the problems relate to the depth of the pit ('shallow' down to a somewhat arbitrary depth of about 30 m (100 ft) and 'deep', below that depth) and to whether or not sufficient volumes of usable waste are generated to enable the pit to be backfilled either during or immediately after working life. Four categories of pit can thus be recognised, each of which has its own reclamation problems, and these are discussed below, together with the fifth case, of underground mines.

10.3.1. Shallow Pits Without Overburden
Shallow pits which have negligible amounts of overburden or other wastes are most commonly found in sand and gravel and smaller quarries working limestone, sandstones, igneous rocks, etc. Reclamation to a large extent depends upon whether or not the pit is flooded.

(a) Flooded Excavations
The reclamation of wet pits poses numerous technical and land-use problems. Wet pits fall into two categories:

(i) 'Permanent' wet pits, typical of excavations for river gravels or in impervious rocks, where the standing water table is always above the pit floor. The water level may nonetheless fluctuate, occasionally to the extent that parts of the working may be periodically dry.

(ii) Intermittently-wet pits, where the water table only rises above the pit floor at certain times of year. This category also includes excavations with poor drainage, or those in very high rainfall areas, which are subject to periodic flooding whenever runoff into the pit exceeds its drainage capacity.

Of these types, the first is more common, but the second creates greater problems of reclamation. In either case, the choice is broadly between filling the pit with some suitable material, or capitalising on the presence of water (*see* Fig. 66). An exact knowledge of fluctuations in the water table is an essential in planning the after-use of either type of pit.[83,84]

Fig. 66.   Reclamation of a wet gravel pit near London. The landscaped pit margins and scattered trees add to the appeal of the site.

Filling a wet pit presents not only the difficulty of finding a bulk fill material, but also of finding one which is environmentally acceptable. If the pit is in hydraulic continuity with groundwater, rivers, etc., then the fill has to be chemically inert and physically acceptable if extensive water pollution is to be avoided. This usually renders a major source of material —raw domestic refuse—inapplicable. Instead, the most suitable wastes for backfilling purposes are rubbles and hard cores, furnace clinker, excavated materials from road construction, building sites, etc., and completely incinerated domestic refuse. In certain areas of Britain pulverised fuel ash (pfa) from coal-fired generating stations is used for fill but such instances are unusual and the growing demand for pfa as a concrete extender renders it likely that its use for fill will decline. In one example, Meering gravel pits near Nottingham, U.K., over 4 million tons of pfa have been used, restoring 54 ha (133 acres) of wet pit. The land use is in fact upgraded.[15]

Unless disposed of with care, inert but fine-particle materials such as pfa, flue dusts, etc., can be drawn into groundwater and interrupt flows. The most unfortunate instances of pollution have arisen from the disposal of raw domestic refuse. In effect the results of this are similar to those of adding any other organic, biodegradable waste to water (see p. 104), namely a rapid deoxygenation of the water (especially in warm weather) and resultant anaerobic, partial decomposition and putrefaction accompanied by water discoloration and the generation of hydrogen sulphide. A further danger with the use of raw domestic refuse is that pathogenic bacteria may enter the groundwater and contaminate potable water supplies.

However, the advantages which would accrue from the development of a means of disposing of raw domestic refuse in wet pits have prompted certain experiments and, today, there are occasional examples of this method. In experiments at Egham, Surrey, U.K., a wet gravel pit was divided into small cells, the size of which permitted rapid filling. Aeration equipment was installed to prevent deoxygenation and odour problems. It has been shown that in such carefully-controlled conditions raw domestic refuse can be safely used.[121]

Tipping into wet pits requires care to avoid accidents when backing vehicles up to the tipping point. It has been found that in wet pits the fill consolidates rapidly and few subsidence problems arise. The tip face should be kept advancing on a broad front, thus avoiding isolated pools of stagnant water (see Fig. 67).

Less selectivity is needed if the excavation is entirely isolated from the groundwater, although anaerobic decomposition should still be avoided. Problems of finding sufficient volumes of suitable fill are increased when deep excavations, rather than shallow pits, are considered, and very deep wet pits are frequently not filled. It is also inevitable that filling of wet pits is a visually offensive procedure and the combination of these factors has resulted in alternative strategies in which the pit is maintained as an open water area.

The presence of a lake provides great opportunities for creative landscaping, and for a variety of leisure and conservation uses. Among the many uses appropriate for wet pits, especially in urban areas, are sailing, rowing, power-boating, water skiing, swimming and fishing.[118] A further

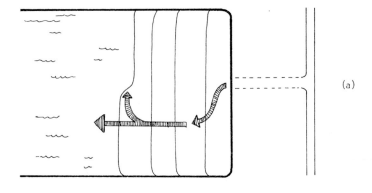

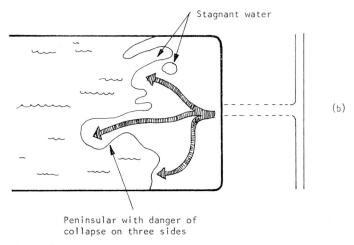

FIG. 67. Methods of filling a wet pit: (a) single face with controlled vehicle routing; (b) uncontrolled, haphazard filling creates safety hazards and stagnant water pools.

important feature of such pits is their value as wildlife sanctuaries and many flooded gravel pits are noted for bird life. Superimposed upon such uses can be the establishment of residential areas which have a particular attraction for persons wishing to take advantage of aquatic pursuits.

The exact work required to render a pit suitable for leisure uses of these sorts varies but usually requires provision of one or more of the following features:[36,39]

(i) Access, by private or public transport, plus car parking areas. Hard standings can often be adapted from former processing plant areas.

(ii) Landscaping is needed if visitors are to be attracted to the site. The work includes grassing of banks and planting of trees tolerant of wet conditions, such as willows and alders. Removal of excess regularity in the shoreline, by selective backfilling with overburden, increases the landscaping possibilities (*see* Fig. 68). Overburden can be used to create islands which will serve as breeding places and refuges for birds.

(iii) Facilities such as boathouses, changehouses, toilets, cafeterias, etc., according to the use envisaged.

(iv) For safety reasons parts of the bank should be regraded to provide a gently-sloping beach, from which swimmers and boats can gain easy access to the water. In large pits, there is a possibility that wave action may undercut and erode banks; regrading prevents this.

(v) Fisheries established in wet pits are not always self-sustaining (unless closely-calculated limits are placed upon the number of fishing permits issued) and provision is therefore needed for regular restocking.

In addition to leisure uses, wet pits may have significant water-storage value because the volume of water which can be stored in an area is greater after excavation has taken place than before. Such a use is not appropriate to pits which contain toxic chemicals leached out from the surrounding rock (*e.g.* open cast coal mines, open pit metal mines) unless the toxic strata can be sealed off, but is highly relevant to sand and gravel pits.

It is possible to envisage that in the future other productive uses might be found for flooded pits in certain circumstances: for example, fish farming, sources of industrial cooling water, or balancing reservoirs for pump-storage electricity generation. No widely applicable productive use seems probable and it will certainly be the case that leisure, recreation and general landscape improvement will be the prime function of wet pits. This however raises a dilemma which is increasingly experienced in areas of intensive wet-pit generation such as the South East of England: the availability of water-filled excavations is greater than the demand for leisure areas. In only very few cases have attempts been made to establish a full-scale leisure industry on a profit-making basis—although isolated notable examples exist[118]—and a flooded pit usually generates only a very low level of revenue. If however the pit is filled, the resultant land has potential for housing or industry—or indeed other leisure uses, such as playing fields—and an immensely increased value, especially in proximity to urban areas. Such a location can be deficient in suitable filling materials. This paradox is becoming of increasing importance.

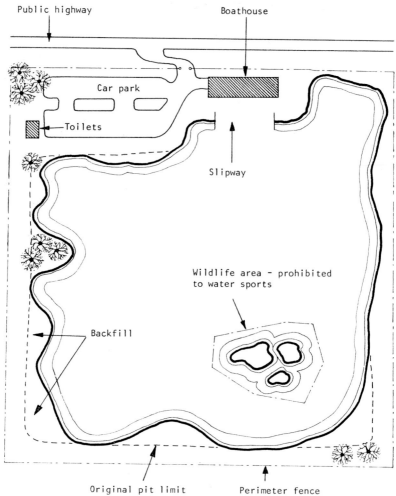

FIG. 68.   Example of after-use of a wet gravel pit.

Quarries subject to intermittent flooding are extremely difficult to turn to productive use, except in the rare cases that a marshland type of habitat might be desired for nature conservation purposes, and filling to a level above the water table is usually beneficial.

(b) Dry Excavations

Excavations which do not become flooded have a variety of after-use possibilities; in some cases these are similar to those of wet pits although problems of implementation differ.

Many shallow pits in stratified deposits can be worked to a rolling reclamation pattern. Provided topsoil is conserved, it can quickly be replaced behind the face and agriculture reestablished, with or without grading up to undisturbed land according to the depth of excavation (*see* Figs. 69–71). Alternatively if an appropriate fill is available the pit can be progressively backfilled and restored.[26,72,74,83,84]

The nature of the fill is often less critical than in the case of wet pits, but precautions against water pollution are nonetheless important.[17,76] Untreated domestic refuse, when rotting, provides a noxious leachate which can permeate porous strata (limestone, chalk, gravels, etc.) and contaminate water supplies. Its use may therefore be prohibited in

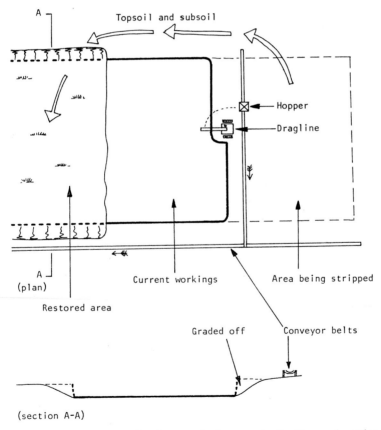

FIG. 69. Plan and section showing a typical sequence of rolling reclamation in a dry shallow pit.

FIG. 70.   Gravel pit reclaimed to agriculture.

susceptible areas. In one case, this problem has been overcome by the installation of an impermeable clay lining in a chalk pit prior to infilling with raw refuse. A drainage layer was provided to intercept leachate and carry it to a sewage works for treatment.

The operation of refuse tips has considerable nuisance potential which can be considerably reduced by controlled tipping methods. Controlled tipping in Britain is essentially the same as the U.S.A.'s sanitary landfill.

FIG. 71.   This shallow magnesium limestone quarry has been returned to grazing by spreading overburden over the floor.

With these methods, the refuse is spread in layers no more than 2·4 m (8 ft) thick (after initial compaction) each layer being covered quickly (after no greater interval than the end of each day's tipping) with inert overburden, soil, etc., in layers at least 23 cm (9 in) thick. The object of this seal is to contain odours, lessen water percolation into the waste, inhibit access to the refuse by vermin, and prevent foraging by scavengers such as seagulls, dogs, etc. Refuse decomposition commences with the generation of heat by bacterial decay action, and progresses until, often after a period of years, an innocuous compost results. Throughout this time, the backfill compacts and settles; with dry tipping of refuse, or any other fill, it is much harder to obtain consolidation of the fill than in the case of wet tipping. Consequently only rough treatment of the tip surface is undertaken at the time of tipping, final improvement work being performed after settling has ceased.

Whether or not the pit is filled, the most appropriate after-use is usually either agriculture, forestry or industrial developments, the latter depending upon the nature of the backfill and its ability to support development.

10.3.2. Shallow Pits with Ample Overburden

In this category are included open cast workings for coal, ironstone, lignites and lesser minerals such as fullers earth, gypsum, some limestones, etc. Coal, lignite and ironstone are by far the most important and in many countries account for by far the greatest proportion of land disturbed by mineral working. These minerals typically occur in relatively thin seams, horizontal or inclined, beneath a high proportion of overburden. The working methods employed are either the basic strip mining technique of exposing the seam with a dragline, and casting the overburden back into the previously worked cut, or variations thereon, notably benching because of increased depth. A wide variety of modifications to basic open cast techniques, for example box cuts and block methods of working, have been devised for particular circumstances.[41,99,110] Working is advanced on a broad front, face lengths of 3–5 km (2–3 miles) being not uncommon. Currently depths of overburden to a maximum of about 35 m (120 ft) can be handled by the largest machinery in a single lift. Although the dragline is common, many shovel and truck operations exist and bucket wheel excavators are used in some instances.

Reclamation, and usually full surface restoration, is feasible in such circumstances because the stripping ratio is high (often 10–15:1 and up to 25:1) and volume of overburden is more or less sufficient to compensate for the excavated mineral when bulking factors are taken into account. Methods of restoration are simple but numerous problems arise, usually as a result of inefficient execution of the works. Restoration is fully integrated with production, and typically comprises the following stages (illustrated in Fig. 72):

FIG. 72. The classic strip mining sequence, at an ironstone mine in Northamptonshire. The dragline is chopping down the boulder clay (right) and casting it back (left) before removing the limestone beds upon which it stands. The railway track (bottom left) is laid on top of the iron ore.

  (i) Stripping of topsoil and subsoil, by scraper, ahead of the working face, and their separate storage. Where possible, it is usually advantageous to avoid the intermediate storage and respread soil directly on the backfilled area.

 (ii) Bare the mineral, casting the overburden back into the previously-worked area. In some cases, it is advantageous to attempt to redeposit the overburden as nearly as possible in its original stratigraphic sequence. The case study in Section 10.6.2 demonstrates this method. This is because rock strata cause problems of cultivation if allowed to intermix with the surface layers, and because it is likely to be easier to re-establish soil water relationships if the strata are returned in their original order. It is also possible that this would benefit groundwater flows and minimise interruption of supplies (Fig. 73), but this aspect has been little studied.[108]

(iii) Remove valuable mineral.

(iv) Regrade the hill and dale formation which results from casting back overburden and respread subsoil and topsoil.

This very simple sequence can easily become unnecessarily complex. At strip coal mines it is common practice to dispose of coal preparation plant refuse in the open cast mine, rather than form a surface tip of highly acid,

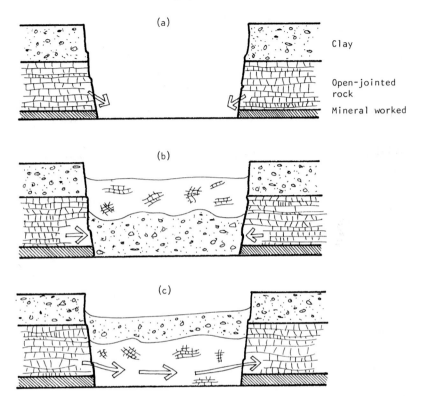

FIG. 73. Possible influence of backfilling upon groundwater movements: (a) during mining, water enters freely; (b) reversing the order of strata during backfilling seals off water flows; (c) re-establishing the correct order of strata permits water flow to be restored.

pyritic material. Unless this waste is properly buried, the restored land may develop 'hot spots' of acidity and consequent vegetation failure. Without careful overburden handling, excessive stoniness can be produced in the surface layers, resulting in the extra cost of stone picking. In many circumstances, controlled overburden replacement can actually upgrade the area in terms of soil quality and increased water resources. The following benefits have been attributed to strip mining in the Appalachian coal region of the U.S.A.:

(i) Creation of level areas in hilly terrain.
(ii) Planned hills and ponds in level country.
(iii) Replacement of old, acid soils by non-acid soils.

(iv) Higher levels of plant nutrients (P, Ca, Mg and K) than in original soils.
(v) Deeper rooting zones.
(vi) Boulder-free soils.

Whether or not these benefits are achieved, depend largely upon the original nature of the area. It is more common for reclaimed land to be, at least temporarily, less productive than before mining began but this can be remedied by intensive maintenance over a period of years. In some cases, benefits to water resources have been achieved:[108,113,114] studies in the U.S.A. showed that undisturbed land supporting a natural vegetation 'consumed' (i.e. took into the vegetation and lost by evapo-transpiration) about 50% of the total precipitation it received. 25% of the input entered the groundwater by infiltration and 25% was lost via run-off. In strip-mined land, consumption was 30%, 60% was infiltration, and 10% run-off. Although the percentage of run-off on disturbed land is often relatively low, the damage it does by erosion is relatively much greater because the mined land is inherently more erodable, as has already been discussed (p. 108). However, the unconsolidated mined land also has an increased capacity to absorb and store water, and can thus contribute to smoothing out storm-water and drought flows in water courses draining from the area. The chemical quality of such drainage remains a matter of concern, however, as indicated by the data in Table 64.

The after-uses of reclaimed strip mine land are usually agriculture or forestry, because the extensive land areas required for economic large-scale

TABLE 64

ALTERATIONS IN WATER QUALITY ARISING FROM COAL MINING IN WEST VIRGINIA (Plass[89])

Parameter		Before mining	After reclamation
pH (units)		6·4–7·3	6·9–7·8
Specific conductance (mmho/cm)		55–100	87–483
Alkalinity	(mg/litre)	10–20	13–33
Bicarbonate	(mg/litre)	11–25	19–38
Zinc	(mg/litre)	0·12–0·18	0·29–0·57
Potassium	(mg/litre)	1·7–2·7	2·5–4·6
Sulphate	(mg/litre)	8–26	17–2·07
Calcium	(mg/litre)	3·4–7·8	7·7–54·5
Magnesium	(mg/litre)	1·8–3·6	3·3–26·6
Iron	(mg/litre)	0·01–5·70	0·01–0·13
Manganese	(mg/litre)	0·01–1·60	0·01–1·50
Aluminium	(mg/litre)	0·01–0·32	0·20–3·60

strip mining render rural or remote sites imperative. Occasionally leisure pursuits are established, particularly in cases where lakes can be created, and are most often found at older sites which had become derelict and subject to encroachment by housing developments.

10.3.3. Deep Pits with no Overburden

Large quarries or open pits with little or no overburden available for backfilling are typical of the crushed rock industry, in limestone or igneous strata. In very occasional instances, some other bulk fill is available—colliery waste has been so utilised in the North of England—but generally backfilling is not possible. In the case of, for example, British carboniferous limestone quarries their locations are usually too remote for filling with domestic refuse to be feasible and in any case water pollution problems can remain.

A wide variety of esoteric uses exists for deep pits—firing ranges, leisure pursuits such as rock climbing, or electricity pylon testing, for example—but the most obvious uses are limited to two possibilities. If the quarry has been worked sub-water table, or if it is poorly drained and becomes flooded it may have great value for water storage purposes. Such use is proposed for the large quarries which may be developed in the Mendip limestones of the U.K. A quarry producing over its life a total of 400 million tonnes of stone, from an excavation about 150 ha (370 acres) in extent would have a water storage capacity of about 120 million m^3 (4200 million ft^3), depending upon the location of the water table. The value of this can be judged from the fact that the adjacent Bristol conurbation has a daily consumption of 300 000 m^3 (10 million ft^3). A variety of leisure activities could be superimposed upon this basic use.

The alternative possibility, which is already commonly applied, is to use the quarry as a nature reserve, or to conserve interesting geological exposures. Long-abandoned quarries often provide excellent refuges for rarer species of plants adapted to rock-ledge habitats, which are threatened in their natural localities by tourist pressure, farming developments, etc. In Britain several such quarries have been scheduled as nature reserves and without doubt it would be possible to create artificially such reserves in quarries for endangered species. It is clear also that the availability of such quarries would exceed the demand.

A major problem with deep workings is that their level areas are usually divided among a succession of benches so that no level space of useful size exists. Proposals for factory development are thus difficult to implement, quite apart from the problems of long-term sidewall stability that may exist. Forestry or agriculture suffer the same drawbacks. Therefore if the excavation is benched down to its limits, cosmetic work for visual reintegration may be the most appropriate use. Such work would require eliminating regularity of the benches (*see* Fig. 74) and revegetation.

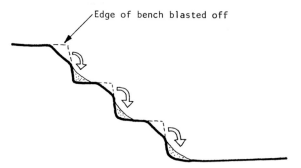

FIG. 74.   Treatment of benches during reclamation.

An alternative strategy is to limit the depth of the quarry and increase its lateral extent. This would provide extensive level floor space for development or agricultural purposes (*see* Fig. 75); indeed the complete removal of anticlinal structures could be envisaged (*see* Fig. 76). Problems of such a solution include the inevitably greater environmental impact during working life and the possible lack of space for lateral development. Nevertheless, this course of action has been adopted in certain circumstances where the benefits of better reclamation possibilities outweigh the artificially low mineral yield per hectare.

10.3.4. Deep Pits with Ample Overburden
In this category are included the major open pit operations with relatively high stripping ratios, notably non-ferrous metals, and the deeper open cast coal mines which are tending to employ open pit techniques. It also

FIG. 75.   Reclamation in a deep (40–50 m) chalk quarry where the extensive, gently graded floor has permitted agriculture to be restored.

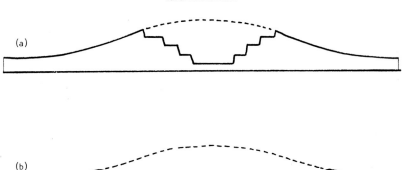

Fig. 76.   Working out of anticlines: (a) conventional method has relatively low environmental impact during operation, but limited after-use potential; (b) complete removal of hill increases nuisance during working life but permits better reclamation.

encompasses operations at which beneficiation of the ore produces large volumes of waste, usually tailings; examples are non-ferrous and certain types of ferrous metal mining. China clay working is a particular example of these circumstances in Britain.

The methods applied in the British open cast coal industry consist of more or less complete restoration of surface topography at the cessation of mining, and mining sequences are designed specifically to enable back-filling.[70,85,86] Details of the techniques are given in the case study (Section 10.6.1).

Although the working life of open cast coal mines is increasing, the average life is still under 7 years, so that changes in restoration philosophies or legislative requirements are relatively quickly reflected in the actual workings, and can often be accommodated without a drastic alteration in the costings upon which the mine was first planned and opened. This is not the case at open pit metal mines which commonly have a life expectancy of several decades. Few, if any, such mines were planned with backfilling in mind and therefore any such requirement would have a most drastic effect upon their financial circumstances.

If an open pit mine were to be backfilled with waste rock and over-burden, there would be a surplus of waste if the overall stripping ratio exceeded about 1·75:1, depending upon the precise pit topography, bulking factors and the compaction achieved during backfilling. If tailings were to be returned to the pit, there would be a surplus of tailings unless the original stripping ratio was at least 0·6:1, depending upon similar factors to those above. No technical problems would arise if solid waste were to be used, but several difficulties may be anticipated if tailings were returned to the pit. If the impounded tailings were unsuited for excavation by truck and shovel—which would probably be a high cost method—a reslurrying system would be needed. Hydraulic monitors are employed for reslurrying with some degree of success, the object being value recovery rather than backfilling. Few methods of relocating tailings have yet proved wholly satisfactory.

Serious environmental disturbance could arise if a tailings backfill were to be employed in unsatisfactory circumstances. There would be the potential for extensive groundwater and surface water pollution unless the open pit were wholly water-tight, which is unlikely. The probable continued instability of the redeposited tailings could result in the backfilled pit having no greater land-use flexibility than the original tailings impoundment. Concurrent backfilling with solid and slurry wastes could perhaps overcome this. However, it is the case that almost no major open pit has ever been reclaimed or restored by backfilling, so that practical experience is seriously lacking.

If the waste accumulations are not used for backfill, landscaping or other treatment is required, and the opportunities for reclamation of the open pit are substantially the same as those of the quarries discussed above (Section 10.3.3). Potable water storage possibilities may be limited by toxicity problems, while the remote location of most major mines further decreases the choice of after-uses.

10.3.5. Underground Mines

Underground excavations present unique reclamation and after-use features. If the mine is operated by a deliberate, controlled subsidence, mining method (notably longwall coal mining), only the main roadways and shafts/drifts remain open at closure. No significant after-use has ever been applied under such conditions, excepting only residual uses such as pumping for adjacent mines. A form of 'rolling restoration' is applied to some collieries, in the sense of backfilling dirt by hydraulic, pneumatic or mechanical stowing, by which some lessening of subsidence is achieved, as well as reductions in the volumes of wastes to be surface tipped.[92-94] Such techniques are only in local use, due to cost considerations, and are considered further in Chapter 12.

After-use possibilities exist mainly in mines worked by pillar and stall, or some variation thereof, under which the excavation remains open and in a more or less stable and permanent condition. In Britain, two important examples are:

(i) Room and pillar mines in the Bath Stone Beds, Wiltshire, have been extensively used for munitions storage, with subsidiary uses for mushroom cultivation, light engineering, and high-security storage. The mines are laterally very extensive (in excess of about 4·5 million

m^2 [50 million ft^2]), the strata are substantially level, about 30 m
(100 ft) below ground surface, with entry by adits. The somewhat
remote location was (and is) well suited to munitions purposes, but
tends to limit other possibilities. One feasible use could be low
temperature storage, for the mines have a very constant ambient
temperature of 11°C, a relative humidity of 90% and good insulating
properties which would enable a refrigeration plant easily to lower,
and maintain, the temperature. As the mines exist at the present,
they incorporate basic facilities such as concrete or asphalt flooring,
minor roof support in places, electric lighting, air conditioning,
ventilating fans, drainage, and conveyer systems for transport.
These were adequate for munitions use.

(ii) A slate mine in North Wales is used for explosives storage. The slate
beds are about 10–12 m (30–40 ft) thick, dip at about 35° and the
levels are interconnected by rope haulages. Rooms are about 12 m
(40 ft) wide on the strike, with pillars 6–12 m (20–40 ft) thick. The
very remote location of this mine is advantageous. Another mine,
also remote, is used for Government high-security storage. It was
originally developed for this purpose during World War II, when
art treasures and the Crown Jewels were stored here, and has
continued in use.

Very much more significant after-uses have been developed in U.S.A.
room and pillar limestone mines.[3–6,125] In comparison with the three or
four small British mines of this type, there are about 100 room and pillar
mines in the U.S.A., most of which are mining in thicknesses of 5–15 m
(15–45 ft), and producing up to 4 million tonnes/a, typically at depths of
less than 100 m (300 ft) although some mines are very much deeper.
Extensive underground space is therefore available. A survey of 47 underground mine operators in 1971 showed that 25% make some positive after-
use of the space they create, the most common uses being warehousing
and civil defence purposes. Detailed examination of the features of this
type of after-use is reserved for the case study in Section 10.6.5. The overall
conditions which govern the utility of underground space can be summed
up as follows:

(i) The mine must be safe and stable, free of water problems (especially
percolation through the roof) and must usually be substantially
level.

(ii) Room dimensions, spacing and orientation should be appropriate
for the desired use.

(iii) Where possible, access should be on the level, for direct entry by
road and rail vehicles. Steep drifts or vertical shaft entries place
severe limitations on use.

(iv) The costs of converting the space to a new use, and subsequently operating it, must be less than the equivalent surface costs. Precise costs vary but, except for very undemanding warehouse use, can be between 25 and 75 % of the equivalent initial surface costs. Creation of underground office space has been found to cost more than on surface. Running costs may be lower underground.

v) The location of the mine must be suitable.

10.4. MISCELLANEOUS RECLAMATION PROBLEMS

Closed mines, and especially long-abandoned derelict mines, pose a variety of reclamation problems other than the major ones of the excavation and tips discussed above.[22,23,25,64] This section reviews a number of these.

10.4.1. Shafts and Adits

Problems of reclaiming mined land which contains shafts and adits are particularly prevalent in old mining fields and fall into two parts: locating the shafts, because, in the last century or earlier, plans were frequently not kept; and the technical problems of rendering shafts safe. As an illustration of the lack of knowledge of where shafts exist, the small East Shropshire Coalfield, U.K., which is being developed as a major new town, contains some 3000 known shafts and, it is estimated, at least the same number again which remain to be found. This scale of the problem is typical of most of the old British coalfields, and shaft depths are typically up to 400–500 m (1200–1500 ft). Locating old shafts relies greatly upon historical research and careful site examination, supplemented by excavation, trenching and short-hole drilling where the indications warrant them. Magnetometer and resistivity surveys have been made without much success.

Treatment of abandoned shafts used to be on a very casual basis; frequently they were left open and fenced, and many such shafts, with their fencing long-since gone, are now safety hazards. Common treatments used to be either a cap or a wooden plug (Fig. 77) but these become dangerous with time. There is no doubt that shafts which are today being capped by the best methods will, in a century or more, again present hazards. Consequently filling is ultimately the most desirable method. Fill should be inert, non-toxic as far as possible, and not liable to decomposition. Demolition rubble, or inert mineral wastes are suitable. Volumes of the fill and the shaft should be regularly compared during operations, to permit the formation of large cavities to be detected. For coal mines, care should be taken against accidents due to gas build-up. The fill should be allowed to consolidate, and be periodically topped up until settling ceases.

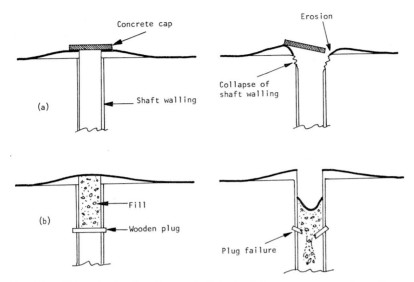

FIG. 77.   Old methods of sealing vertical shafts: (a) concrete cap can be undercut by vandals or erosion, or shaft walling may collapse; (b) wood or steel plug, with superimposed fill, may decay and give rise to sudden failure.

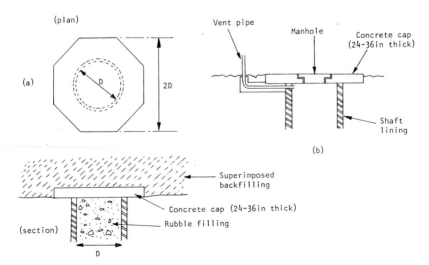

FIG. 78.   Current designs for shaft capping: (a) plan and section of a cap used on open cast coal sites where capped shaft will be filled over; (b) section of a cap plus vent for use at the surface (Arguile[71]).

Figure 78 illustrates current designs of shaft capping and plugs for adits.[71] Even if the shaft is filled, capping can be desirable. All shaft positions should be marked to enable them to be relocated. If any ventilation pipes are installed to prevent build-up of gas pressure in the mine they should not discharge into a confined space; there are cases of children being asphyxiated in these circumstances.

10.4.2. Subsidence Flashes

Subsidence often disrupts water-flow patterns and flooded areas can be created. In some cases the water area may be permanent and of sufficient extent to be a useful landscape and recreation feature. Otherwise, the procedure is typically to pump the flash dry, remove and conserve the mud and soil from it, backfill with some mineral waste to the desired contours, and respread the soil.[71]

10.4.3. Communications

Railways, roads and canals often fall into disuse along with the mines they served. The linear pattern of these features has always rendered constructive after-use difficult. Canal beds can be restored by backfilling, or the waterway can be improved to fulfil a water storage or leisure function. Both are common uses in Britain. Roads can be broken up and the ground returned to its former use, but access to reclaimed sites usually needs to be retained. Railways pose greater difficulties because of the presence of cuttings and embankments, often on a considerable scale. Land severed by the construction of the railways can often be rejoined, either by backfilling cuttings or regrading embankments. Because of the earthworks associated with railways, drainage has always been important and abandonment of drain and culvert maintenance can create flooding problems. Return of the route to a derelict, wilderness state has often been accompanied by vermin problems, so that dereliction is best avoided. Long-distance footpaths are one use to which some old railways are now being adapted.

10.4.4. Buildings

The buildings and fixed plant of mines usually requires demolition and/or salvage. Metals can be salvaged and in the past the derelict shells of buildings, being worthless, were often abandoned. In Britain complete demolition is usually required under the terms of the original planning consent. In certain cases, the growing interest in industrial archaeology renders desirable the preservation of individual items (kilns, engine houses, etc.) or, more rarely, complete installations. Incorporation of such features add aesthetic and historical interest to reclaimed sites as a whole, and are often to be found at reclaimed leisure areas or new light industrial estates on the sites of former mines.

10.5. REVEGETATION

In the overwhelming majority of cases, the reclamation, for whatever purpose, of abandoned mineral workings requires the establishment and maintenance of vegetation on the disturbed land. No other medium can achieve rapid visual reintegration, surface stabilisation, or reductions in air and water pollution, nor offer a wide variety of land use possibilities which can be achieved at acceptable cost.[119] Even though the engineer may be seldom called upon to actually design and execute revegetation schemes—for numerous specialist organisations exist in most countries to carry out such works—the design of the mining programme can materially ease revegetation problems. In this respect revegetation is but another example of the virtues of planning the mine site not merely for its life, but afterwards.

10.5.1. Characteristics of Soils

Most revegetation work requires that plants be grown in mineral wastes and it is helpful to examine the main features of natural soils before drawing comparisons with mine wastes. In natural circumstances plants exist in a soil formed by weathering and biological processes from a parent rock which may or may not underlie it. The precise type of soil which forms depends upon the complex interactions between the parent rock, the climate, and the types of plants and animals which first colonise it. The soil can be divided into the following components:

  (i) inorganic mineral matter;
 (ii) organic matter;
(iii) water;
 (iv) air;
  (v) living organisms;
 (vi) nutrients.

The mineral matter exists in a definite structure for various soil types, defined by the percentages of sand, silt and clay, which have the following sizes:

Coarse sand	2·0–0·2 mm (0·08–0·008 in) particle diameter
Fine sand	0·2–0·02 mm (0·008–0·0008 in)
Silt	0·02–0·002 mm (0·0008–0·00008 in)
Clay	<0·002 mm (0·00008 in)

Different soils are classed on the relative distributions among each particle size; most natural soils are mixtures (called 'loams') of each size. Thus a sandy loam has most of its weight in the coarse sand category, a loam has an almost equal distribution among all four categories, and a clay is predominantly in the silt and clay fractions. The importance of these divisions is their influence on waterholding capacity which is greater in clays because of their colloidal retention properties. Thus plants growing in very sandy soils suffer frequently from water stress and may have specialised water retention systems. Conversely in clays plants may suffer from waterlogging.

Organic matter in natural soils is the dead remains of plants and animals, and has several vital functions in soil. It contributes to the soil structure, sticking together mineral particles into aggregations called crumbs. A soil with a well-developed crumb structure has good aeration, waterholding and drainage characteristics, and therefore provides a good medium for plant growth. Organic matter also has vital chemical functions. It passes through a cycle of biological decomposition to provide inorganic nitrogen nutrient essential for plant growth, and is usually the major nitrogen source, as well as an important source of sulphur and phosphorus. The cation exchange capacity of organic matter enables it to act as a major reservoir for nutrients.

Soil water is the medium of nutrient transport from the soil to the plants, for all soil nutrients enter the roots as a very dilute aqueous solution. In a well-drained soil at 'field capacity', that is, the quantity of water retained in the soil after drainage by gravity, a film of water surrounds the soil particles and infills many of the interstices. It does not however wholly saturate the soil, but permits the exchange of air. The soil atmosphere is also vital for the growth of plant roots; in poorly aerated conditions (*i.e.* waterlogging) root growth of intolerant species is drastically reduced, while deleterious soil chemical changes may also occur in a deoxygenated environment.

Soils contain extensive populations of organisms—bacteria, fungi, algae, protozoa, nematodes, earthworms, insects and burrowing animals—apart from plant roots. Many of these organisms participate in one of the various nutrient cycles, which are responsible for decomposition of organic matter and fixation of atmospheric nitrogen, and are essential for the maintenance of soil fertility.

The nutrients which are regenerated as a result of these processes can be classed as macro- or micronutrients. Of the former, the soil supplies six: phosphorus, potassium, nitrogen, sulphur, calcium and magnesium. Micronutrients are required for growth in only very small quantities—though they are nonetheless vital—and include iron, boron, manganese, copper, zinc, chlorine and molybdenum.

This brief résumé provides two important conclusions:

(i) Mine wastes, even at first impression, clearly fail to meet many of these criteria of a good natural soil.

(ii) However, many natural soils are also deficient in one or more of these 'desirable' characteristics—and often seriously deficient. Nonetheless, either natural evolution has provided plants adapted to difficult circumstances, or cultivation by man alters the soil and permits standardised crop plants to be grown. These principles are the fundamental basis of revegetation of mining lands.

10.5.2. Problems in Revegetating Mining Wastes

In the light of the comments above it is apparent that there are two possible approaches to revegetation work on existing wastes:

(i) Accept the poor soil conditions as they exist and select plants which have a tolerance of the inhibiting factors. This approach might be termed an 'ecological' approach. In natural situations, bare land is colonised by invading (or 'volunteer') species which can withstand the harsh environment. Such plants are termed pioneers. As they become established they alter and ameliorate the environment and permit the growth of secondary species. This process of succession, if allowed to go unchecked, eventually results in a 'climax' community, composed of vegetation perfectly adapted to the prevailing soil and climate. Pioneer species may thus be an economical solution to a revegetation problem, and an ecologically desirable one, but they suffer from limitations, particularly in the case of the more severe soil conditions for which few if any pioneers exist. Moreover, such an approach may not achieve the rapid results so much desired today.

(ii) Alter the soil conditions and render the site suited to the growth of the particular plant species which are required. This is probably the more common approach. If it can be carried out successfully this method enables more rapid results, and permits a more positive choice of after-use. It is precisely the same process as is employed all over the world when bringing deficient natural soils to a better state of cultivation. However, amending the soil conditions also has disadvantages. Considerably increased costs of site preparation are entailed, together with an indefinite requirement for maintenance. The latter arises because artificial alterations of soil properties are seldom other than temporary and, unless repeated and maintained, the vegetation may revert to the poorer-quality natural succession described in the preceding paragraph. Permanence without excessive maintenance may not be obtained unless the plantings have been carefully designed to generate a self-perpetuating, relatively maintenance-free, vegetation cover.

These two alternatives are not of course mutually exclusive and in fact combinations of them are frequently applied.[18,20,21,43,49,50,65,66,67]

The objective has been defined as 'synthesising an ecosystem' or 'synthesising a soil'.

Before considering in detail the methods available for the revegetation of mine wastes it is useful to examine in more detail the various factors which render substrates inimical to plants.[10] Goodman[1] offers a useful classification of type and severity of problem in relation to the type of waste (Table 65). Obviously such a table can only represent approximately the likely problems, and it was in fact drawn up for British circumstances, but is broadly applicable in Western Europe as a whole. In numerous other parts of the world the problems may differ—notably, because the table applies to British mining, it lacks reference to metal tailings—while at any particular site specific problems may assume prominence. Nonetheless this overall view of the difficulties is valuable. Further detail of individual problems, which illustrates the ways in which mining wastes differ from soils, is given below.

(i) Toxicity of Metals and Other Contaminants

The heavy metals are defined as those with a density greater than 5, which includes some 38 elements. In mining situations the metals most generally of concern are copper, cadmium, lead and zinc, although most metals can cause problems in isolated cases. These include boron, iron (Fe^{2+}), manganese (Mn^{2+}), aluminium, chromium, nickel, fluorides, arsenic, mercury and cyanide. Although, as has been mentioned, some metals are essential plant micronutrients—for example, zinc deficiency is a feature of some soils and zinc fertilisers are used in orchards in Florida—the heavy metals are highly toxic to plants when present in excess. There is frequently a very sharp dividing line between the levels required for healthy growth and those which cause toxic symptoms to develop.

Plants are not completely indiscriminate absorbers of soil ions but their capacity for selective uptake is often insufficient to prevent metal ions being taken in. The mechanisms of toxicity are by no means fully understood, and indeed some plants simply concentrate metals in their cell walls without discernible ill effect. In most cases however metals disrupt active metabolic sites within the cells and severely inhibit respiration and some synthetic processes. Sterility is a common effect, as is root stunting. This last is especially unfortunate in that the plant's ability to absorb water and nutrients is thereby diminished. The toxicity of metals in the wastes depends on several factors. Different plant species, and races of the same species, vary in their susceptibility, while the chemical combination in which the metal exists is also important because it controls the availability of the metal to the plant.[53,58]

Although a mine may produce only one metal, most non-ferrous ores contain traces of a variety of minor elements.[12] Certainly in aquatic environments, it is known that different metals may have synergistic

TABLE 65

MAIN ENVIRONMENTAL FACTORS INHIBITING PLANT GROWTH ON THE PRINCIPAL WASTE MATERIALS DERIVED FROM MINERAL EXTRACTION AND OTHER INDUSTRIAL ACTIVITIES (Goodman[1])

Substrate type

Inhibitory factors	'Normal' soil material	Smelter slags and wastes	Chemical wastes	Coal wastes	Pulverised fuel ash	Metal mine wastes	Brick-claypits	Slate and shale wastes	Stripped peatland	Quarry stone pits and wastes	China clay pits and wastes	Sand and gravel working	Ironstone wastes	Domestic refuse
(a) Instability	−	++	−	++	−	+	−	++	−	−	−	−	+	−
(b) Spontaneous combustion	−	+	+	++	−	−	−	−	+	−	+	−	−	++
(c) Unworkably steep slopes	−	+	−	+	−	+	+	+	−	++	++	+	+	+
(d) Inhibitory water regime	−	−	−	+	+	−	+	+	++	+	+	+	+	++
(e) High levels of toxic elements	−	++	++	+	++	++	++	−	+	+	−	−	−	++
(f) Compaction and cementation	−	−	−	+	+	−	+	−	−	+	+	−	+	+
(g) Inhibitory surface temperature regime	+	++	+	++	++	+	++	++	++	++	++	++	++	+
(h) Wind turbulence, wind erosion, 'sand blasting'	+	++	+	++	++	+	+	++	+	+	++	++	++	+
(i) Low nutrient status	−	++	++	++	++	++	++	++	++	++	++	++	++	+
(j) Excessive stoniness, absence of fine soil-forming material	−	++	+	++	−	++	++	++	+	++	++	+	++	−
(k) Broken, uneven surface (no seed bed)	−	++	−	++	−	++	−	++	−	−	−	−	−	−
(l) Sheet and gully erosion	−	++	+	+	−	+	++	++	−	++	−	+	+	+
(m) Absence of soil micro-organisms and soil fauna	−	+	+	+	+	+	+	+	+	+	+	+	+	−

Note. Inhibitory factor very marked (++), present (+) and negligible or absent (−).

properties, *i.e.* their toxicities when in combination are greater than would be suggested by their individual toxicities. Copper and zinc show this phenomenon, which may well manifest itself in mine wastes.

To specify available levels at which particular metals become toxic is therefore difficult. Soluble Al, Cu, Ni, Zn and Pb are commonly toxic to normal plants at concentrations in the range 1–10 ppm. Boron can be toxic at levels above 10 ppm and is very toxic at 200 ppm or more. Mn and Fe toxicity levels are variable, but usually require to be in the range 20–50 ppm before damage may be caused.

Severe growth inhibition of vegetation can be caused by a variety of other contaminants. Soluble salts of the alkali metals may give rise to salinity problems which can persist if the climatic regime does not allow leaching of the material. If the specific conductivity of the waste exceeds 7 mmho, toxicity is to be suspected and may be regarded as severe at 10 mmho. Such conditions are found in such wastes as asbestos and gold tailings, and pulverised fuel ash wastes.

Reclamation by rubbish tipping, or the spontaneous combustion of pyritic tips, may generate H_2S which, as a free gas, is toxic to vegetation at concentrations of about 10 ppm.

(*ii*) *Acidity and Alkalinity*

Most higher plants thrive in soils which differ little from neutrality and, once the range exceeds pH 5–8·5, most species are almost eliminated. Many species have distinct preferences for acid or alkaline soils. Highly acid materials (less than pH 4) are especially inhibiting, both because of the acidity itself and the fact that toxic metal ions become mobile and hence available in acidic conditions. Colliery spoils and metal mine wastes typically contain excess hydrogen ions.[20,42,56,60,77,103,115] When fresh, colliery spoils are often alkaline but within 5 years of tipping, oxidation of pyrite and pyrrhotite generates free sulphuric acid and the pH may be as low as 2. In some cases the acid may leach out gradually and, after about 50–90 years of exposure, conditions may have improved to about pH 6. Figure 79 shows the pattern of such changes. However, other spoils remain extremely acidic and in these cases the increased availability of metal ions such as Al, Mn and Fe may generate metal toxicity problems in colliery waste.

Wastes suffering from high alkalinity are less common, but include asbestos tailings, red muds from aluminium processing and lime-burning rejects.

(*iii*) *Water Balances*

Vegetation cannot be established without an adequate water supply. Excess water can, however, be damaging, as can marked fluctuations from drought to surplus. These are all conditions widespread in mining wastes.[79,109,122]

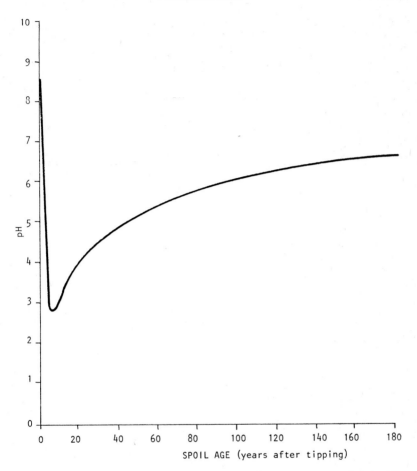

FIG. 79. Acidification cycle in colliery waste after exposure to the air. The magnitude, rate and duration of acidification is very variable, depending upon the quantities and forms of the sulphides present, and external conditions.

Excess water is likely in some clays (particularly in the 'dales' of hill-and-dale topography), flashes caused by subsidence where the ground surface has been lowered beneath the water table, or pits in porous strata (*e.g.* chalk) which have been worked nearly down to the water table. Water table fluctuations determine whether or not a permanent lake is formed, or whether marshland—or even complete drought—occurs at a certain part of the year. Revegetation possibilities are limited in these circumstances.

Without the water-holding capacity of organic matter, vegetation on raw mineral wastes is likely to show the effects of drought. Extremely free-draining materials such as slates, or pit faces in impermeable rocks, are common points for water stress. The presence of adequate water, apart from its intrinsic benefits for vegetation, inter-reacts with other factors, such as soil temperature. It is considered that in harsh, arid regions such as the south-western states of the U.S.A., Australia and South Africa, lack of water is the single most important inhibitor of revegetation attempts.

(iv) Temperature

Extremes of temperature as a result of solar heating by day and radiation cooling at night can be very marked. In severe cases they can kill vegetation. The temperature regime in tips, etc., is influenced by climate, tip slope angle, slope orientation, moisture content and colour of the wastes. A dark, south-facing waste tip may attain temperatures at the surface of the spoil in excess of 50°C in summer. Figure 80 shows examples of such measurements in British colliery waste; it may be noted that a 35° south-facing slope exceeded by nearly 12°C the daily temperature maximum on an adjacent flat area.

The diurnal temperature fluctuations may thus be 40°C or more and it is possible that seedlings, at the point of emergence through the spoil surface, may experience temperatures above their thermal death point only 12 h after suffering frosting at night. The damage so caused is aggravated in drought conditions in the surface, and explains why volunteer plant species often tend to colonise adjacent to boulders and bits of pit props, conveyer belts, etc., dumped with the waste. Moisture is retained closer to the surface by such objects, thus lessening the thermal effects. A reduction in spoil particle size also has this effect. When a good plant cover is established, surface temperature extremes are reduced by 25–30°C.

Although light-coloured wastes absorb less heat than dark wastes, plants growing in them may suffer from the heat reflected upwards to the undersides of their leaves.

(v) Wind Damage

The unnatural topography of many mines can create unusual wind turbulence. The funnelling of wind between tips, buildings, etc., can cause direct mechanical damage to young trees.[127] If the locality is exposed, trees may thrive only on the leeward side of tips and be unable to grow above them (*see* Fig. 81). Turbulence also 'stirs' young trees and, if the waste is angular, the bark and the phloem (the nutrient-conducting tissue beneath the bark) may be abraded and the tree killed. If fine abrasive particles exist, notably but not wholly at tailings dumps, a sand blasting effect is created at critical wind speeds, which can cause complete destruction of the bark and phloem on the windward side.

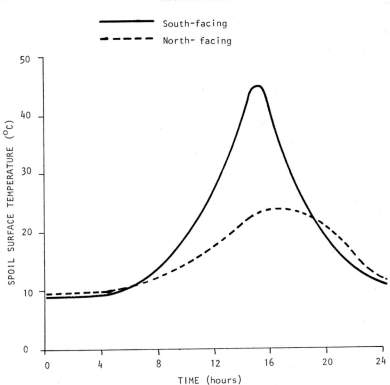

FIG. 80. Diurnal temperature fluctuations at the surface of a colliery waste tip, showing the difference between north- and south-facing slopes.

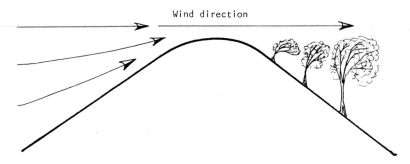

FIG. 81. Wind-blasting effects upon trees.

(vi) Compaction

The commonest cause of compaction is the use of heavy machinery, often during reclamation work. Compacted wastes cannot be penetrated by roots and form a barrier to water movement. Consequently vegetation failures from this cause are widespread. Any waste with a high clay content is liable to become over-compacted or 'puddled' even by light machinery, particularly if it is used in wet weather. A somewhat similar phenomenon in its effects on vegetation is cementation. A waste such as pfa sets like cement because of the formation of hydroxy-silicate complexes. Pyritic

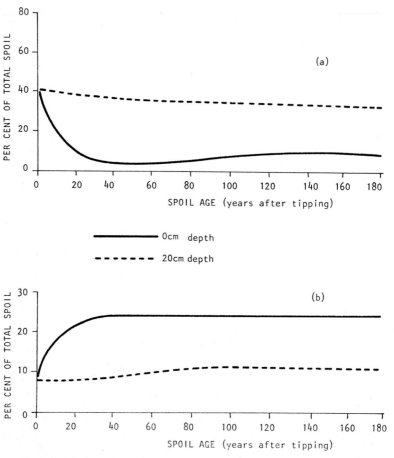

FIG. 82.   Weathering of colliery spoil to produce fine particles, at the surface of a tip and at a depth of 20 cm: (a) decrease in the proportion of large (>9530 μm) particles; (b) increase in smaller particles in the range 2000–1000 μm.

wastes commonly show surface cementation if the pyrite content exceeds about 0·8 %. Additionally, any fine-particle wastes, if wetted, tend to settle out in layers with the finest particles at the surface. On drying-out, an impenetrable crust is often formed.

(vii) Particle Size

Natural soils usually possess a fairly wide gradation of particle size, but it is very common for mine wastes to exhibit an unbalanced distribution. Tailings, for example, are predominantly in the silt and clay size ranges, with about 50 % passing 74 μm. Other wastes, such as slates or resistant igneous rocks, are excessively stoney and, at least at the surface, may be composed wholly of boulders. However, even in such extreme cases fine material usually exists in the depths of the tip. The presence of even as little as 20 % fine material has a dramatic effect upon free establishment and growth.[124] Blocky spoils of this type are more or less impossible to seed to grass, unless regrading can be undertaken to expose finer material to comprise a rooting layer.

Many wastes weather to produce fine particles and an acceptable seed bed. Colliery shales are a good example, and the degradation which occurs is illustrated in Fig. 82.

(viii) Nutrient Deficiencies

Although a large number of elements is required for healthy plant growth, only three are usually lacking in wastes to such an extent that corrective action is required. These are nitrogen, phosphorus and potassium, without which no plant growth will occur. It is important to note that mineral wastes, lacking both soil structure and organic matter, have little or no ability to retain these nutrients. Thus, it is usually essential to refertilise new plantings, for a number of years, to replace the leached-out fertiliser. Organic matter levels increase with time (Fig. 83) and thus nutrient retention capacity also tends to increase.[116]

(ix) Surface Instability

There are many causes of instability of waste surfaces. Wind or water erosion are severe on unvegetated slopes, but the magnitude of the erosion varies markedly. Even in fairly good conditions steep bare slopes can lose in excess of 400 tonnes/ha (160 tons/acre) by water erosion (see Figs. 84 and 85). Any soil or soil-forming material may therefore rapidly be lost. Erosion gullies which have breached a topsoil covering over toxic waste are commonly found. Another cause of instability is surface creep due to the effect of gravity, supplemented by wetting and drying, on the surface layers. Frost heaving may occur (Fig. 86) in which the expansion of water on freezing lifts soil particles at right angles to the slope but gravity causes them to fall vertically. All forms of surface movement can severely damage

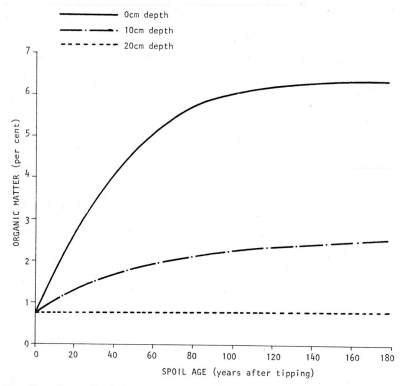

FIG. 83.   Generalised picture of organic matter build-up with time in colliery waste.

vegetation, tearing young roots and opening cracks to the influence of eroding agents.

(x) Soil Flora and Fauna

These are usually absent, with the exception of sulphur-oxidising bacteria, unless topsoil exists among the wastes. It is vital to the objective of establishing a self-perpetuating vegetation cover that a population of soil organisms able to carry out the requisite nutrient recycling is developed.

10.5.3. Earthworks

Before the site can be planted, considerable preparation is usually needed. In essence this comprises the creation of the most favourable seed bed, so that the vegetation has the best initial start that can be obtained. As such, site preparation can include major earthmoving works and landscaping,

Fig. 84.   Massive erosion gullies (1 m wide and 2–3 m deep) in the side of a gold slimes dump in South Africa.

FIG. 85.   Once started, erosion can cause serious damage even on long-restored
sites with shallow slopes, as in this example in South Wales.

undertaken to suit the land for some end use. Specifically in relation to
revegetation an essential preliminary to any resculpting is comprehensive
soil analysis to permit the burial of toxic materials—or prevent them from
being unearthed—and to select for final spreading the most favourable
substrate for growth. The immense variations in composition, both
laterally and vertically, of most wastes make analysis important.[102, 110–112] In certain cases, topsoil is available for final spreading.

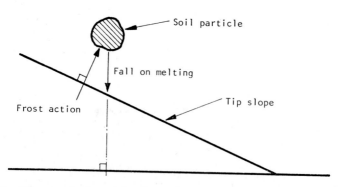

FIG. 86.   Movement of particles down slopes due to frost heaving: expansion
of water on freezing lifts the spoil particle at right angles to the slope, but it falls
vertically on melting.

The benefits of regrading can be summed up as follows:

(i) Removes steep slopes which limit land use potential, and render access for revegetation difficult.
(ii) Toxic wastes can be buried.
(iii) Simplifies seed bed preparation.
(iv) Lessens problems of soil temperature and stability by erosion or slumping.
(v) By necessitating expansion of tip area by spreading, enables stripping of surrounding topsoils which are available for respreading.
(vi) Partial or complete backfilling of pits and quarries may be possible.

There are also disadvantages associated with regrading:

(i) Weathered, relatively innocuous spoil may be lost and fresh, possibly toxic, waste exposed.
(ii) Can be a dangerous operation for operators.
(iii) Requires the taking of extra land, which may be impracticable.
(iv) Earthmoving is usually the largest single cost in reclamation work.
(v) Often creates compaction problems.

The balance of these alternatives differs from site to site, but in most recent cases in Britain regrading is considered desirable. The final contours are determined by the relative requirements of land use, cost of earth moving and availability of space to spread the waste, apart from the needs of revegetation. In dark-coloured materials, where soil temperatures may become damaging, it is sometimes wise to retain the south-facing slope almost at its angle of repose, thus minimising the acreage which will suffer from temperature extremes. Generally speaking, gradients of 1 in 4 to 1 in 10 are selected if land is to be planted with trees, whereas for agricultural purposes, slopes as flat as 1 in 50 are sometimes provided. Generally 1 in 10 is the maximum slope for grazing land, and 1 in 20 the maximum for arable farming. Flat slopes do present drainage problems if the material is relatively impermeable, so that, unless under-drains are to be installed at a later stage, slopes no flatter than about 1 in 25–30 are often selected. These provide a reasonable compromise between erosion damage and flooding. Nonetheless, drainage can be a most difficult problem: graded spoil must be left to consolidate for about 5–7 years before tile or mole drainage can be undertaken. In the interim therefore contour and perimeter ditches may be provided. In the early months after regrading a great deal of silt run-off is to be expected and provision should be made for this.[11,64,101]

The machinery used for regrading includes tractor-scraper units, bulldozers, traxcavators, front-end loaders and, sometimes, face shovels and trucks. Recently equipment has been designed specifically for regrading mine wastes; in the U.S.A. bulldozer units constructed to level hill-and-dale coal stripped land have included the largest such machines in the world, an illustration of the cost of earthmoving and the hoped-for economies of scale.[33,75,100]

The cost of earthmoving as a percentage of total reclamation costs naturally varies considerably. However, it is graphically illustrated by cost figures for colliery waste tip reclamation in County Durham, U.K. Overall costs per acre of tip varied from £2000 to £6000 (£4900–£14 800/ha) in 1973; if account was taken of the acreage of adjacent undisturbed land used for spreading the tips, these variations were ironed out to £1700–£1900/acre (£4200–4700/ha). The cost breakdown (%) was:

Demolition	2·2
Earthworks	62·5
Drainage	20·8
Cultivation and seeding	8·1
Fencing	3·8
Miscellaneous	2·6

Compaction is extremely difficult to avoid and therefore, once the final contours have been established, ripping is needed. This is undertaken by bulldozers or a variety of purpose-built ripping machines. To some extent the depth and spacing varies according to the use, but ripping should never be less than 450 mm (18 in) depth at a spacing of 1 m (3 ft). For large-scale agricultural reclamation ripping depths of 1 m (3 ft) are often used, but the availability of machinery able to undertake such jobs is limited. Ripping can have up to five objectives:

  (i) To break up compaction and permit root penetration.
 (ii) To assist in water infiltration and thus lessen run-off. Ripping should therefore be carried out *along* the contour and not down-dip.
(iii) To bring to the surface all stones in the plough layer large enough to turn the blade of a plough.
(iv) By creating tiny north-facing furrows on the south faces of slopes, providing suitable, sheltered niches for vegetation establishment.
 (v) In the event of further layers, of sub- or top-soil, being spread over the waste, the furrows created by ripping are valuable in keying the new layer to the old, lessening or preventing slippage and helping to establish hydraulic continuity between the layers.

Each successive layer requires rooting and, on flat or nearly flat areas, rooting of one layer is usually done at right angles to the direction used on the previous layer. However, on slopes working along the contours is desirable for each layer. Depth of ripping for layers of deposited material should be slightly less than the thickness of the layer, or else unwanted mixing of layers can occur.

Handling and spreading of topsoil require special attention. If topsoil is stripped and dumped for storage, it usually undergoes marked deterioration and acidification, which is evidenced by the sour, rotten smell obtained when opening up the storage dump. This is caused by anaerobic decay of organic matter in the dump. Acidification can be corrected by lime additions (*see* Section 10.5.6) but it is preferable, where possible, to strip and replace topsoil as one operation and avoid intermediate storage. This also enables economies in handling to be achieved. Retention of topsoil, even if the quantities are small, is a valuable procedure because even as little as 5–10 cm (2–4 in) of topsoil covering is very beneficial to vegetation.[96] Respreading of topsoil should only be done in dry weather when the soil is friable and the machinery used should be selected for minimum compaction. Subsequent cultivation operations will achieve fully adequate consolidation.[72]

The remaining operation before seeding is the final preparation of the seed bed. This is vital to the success of the more demanding works such as re-establishment of arable land.[55, 106] Commonly the land is disc-ploughed and harrowed, but if a high clay content exists more elaborate ploughing or ripping is required to break up the clods. Stone picking is also required if the land is to be ploughed; picking is a costly, manual process which has so far defied effective mechanisation, although some success has been obtained by the use of potato-picking machines.

If there is any likelihood of waterlogging occurring then provision for drainage is required; temporary systems suffice until the area has consolidated to permit the installation of permanent works. Open ditches, placed beside the proposed new field boundaries, roads, etc., are usually the simplest and cheapest method of removing surface run-off. Gradients should be a compromise between those required for self-cleansing and erosion, while 1 in 1 is the steepest desirable slope for the sides of the ditch. Temporary reinforcement, with rubble or rough brickwork, at confluences, etc., is worthwhile. At a later stage in the work (say after about 5 years) permanent pipe or tile drainage may be installed. Clay tile drains are particularly appropriate if an arable use for the land is intended, but may be omitted entirely in areally small or unsophisticated schemes.

The vexed question of whether or not to form terraces on slopes creates much confusion. As a general rule, terraces are only useful in climates characterised by flash floods, and/or on steep slopes, which render control of surface run-off necessary if serious erosion is to be avoided. In addition, if access roads have to be cut across slopes, these function as terraces and need to be designed as such.

Where terraces are decided upon they should be as simple as possible and can usually be created simply by ploughing along the contour. Such a terrace can be perfectly level, to hold water and increase its absorption into the soil, or can be gently graded (1 in 50–80) to carry the water to a main downslope ditch. The size of terrace required must be calculated to safely retain the maximum volume of surface run-off to be expected from the catchment area above it. The great danger of installing terraces is that, if a breach does occur, the whole volume of run-off is concentrated at one point and massive erosion results. For this reason, it is only in extreme conditions that terraces are widely used; in most European and North American situations, simply tilling along the contours is a sufficient control.[104,107]

In addition to the earthworks required during the restoration of surface dumps and derelict land, which is by far the most common operation, one special situation requires mention: cases in which it is desired to establish vegetation on bare rock faces, quarry side walls, etc. Such conditions can be severe and much experimental work is in progress at the present time. The task is two-fold:

(i) To create small rock pockets, screes and crevices in which plants can find suitable niches for growth (*see* Fig. 87).

(ii) To introduce sufficient soil into these niches to enable plants to grow.

FIG. 87.   Natural recolonisation by plants of an abandoned chalk quarry face. Note how the vegetation has established only in crevices and on ledges.

The first operation can be done by carefully-controlled blasting, using small charges to break up and bring down small areas of face. Soil can be introduced either by tipping from above, or by slurrying the soil and pumping it from above or below. If seeds or minced turf are included, soiling and seeding can be carried out simultaneously.

Finally, it should be noted that all fencing, ditching and construction of access roads, etc., are carried out at this stage of the works.

10.5.4. Methods of Planting

The methods suitable for planting depend upon what type of vegetation is being established.[106] Grasses, clovers and other herbs are almost invariably planted as seeds (the alternatives of turf or minced turf are extremely unusual and rarely applicable) whereas trees are seldom if ever planted from seed, but young trees are used instead. The methods used are:

(i) Tree Planting

Trees are usually planted by hand but in exceptional circumstances machinery may be used; even in this case, it may be necessary to follow up by hand to finish the work properly.

Conventionally, trees are obtained from tree nurseries and are 2–3 years old. The most successful trees are those with extensive fibrous root systems, because only the fibrous roots absorb water and nutrients; the bulk of a tree's root system serves only for anchorage. Nursery cultivation methods are designed to produce such stock, by means of transplanting or undercutting.

Trees are sold with a designation indicating the treatment received: a plus (+) sign indicates a transplant; and a letter 'u' indicates undercutting. Thus, the designation 2 + 1 shows that the tree grew for two years before transplanting for a further year in a transplant line, while 1u 1u 1 denotes a tree that has been undercut twice but not transplanted. Some nurseries produce trees grown in containers in a sandy loam which also encourages fibrous root growth. Planting densities vary greatly, from about 2500 to 4350 trees/ha (1000–1800/acre). For young trees such as those referred to, the heights would be no more than 30 cm (12 in) for conifers and 45 cm (18 in) for hardwoods. Planting is extremely simple in these cases: a 'notch' is made with a spade, opened out by moving the spade about, and the tree placed in. Its roots are spread out, and the soil heeled firmly back around it. Under no circumstances should the soil level be either higher or lower than was the case in the nursery; the stem usually shows clearly what this level was.

If 'whips' are planted—older trees about 1–1·5 m (3–4·5 ft) high—a proper planting pit has to be dug, the bottom well broken up, and a fertiliser such as basic slag added before placing the tree and carefully backfilling. As has been discussed in Chapter 3, the use of 'instant' mature trees creates great problems of cost and maintenance: such a tree requires a large pit, well broken up and fertilised, plus stakes and guy ropes and copious watering. The specialised equipment required to transplant a mature tree is illustrated in Fig. 88.

A most important point in connection with nursery stock is that trees do not survive well transplantation to very different surroundings. If every attempt is made to buy stock from nurseries in similar climatic and geographic conditions to the mine site, these problems will be minimised.

(a)   (b)

(c)   (d)

FIG. 88   Use of the Vermeer tree spade for transplanting semi-mature trees. The details of the device are shown in (a), and the process of lifting the tree in (b) and (c). In (d) the root ball of the tree is being protected by baling.

(ii) Grasses and Other Herbs

Hand broadcasting of seeds is slow and expensive. Under good conditions, it takes about 1–1½ h to hand-seed an acre (2·5–4 h/ha), but only 20–25 min (50–60 min/ha) by tractor.[106] Mechanical seeding, the standard agricultural method, falls into two main categories. Ground broadcasting equipment, such as tractor-mounted rotary spreaders, can produce segregation of seed mixtures (due to different seed weights) while the seed is not buried. It may therefore be necessary to follow-up with a rake to bury ('mulch') the seed with a small amount of topsoil. In good conditions, drilling-in seed is advantageous because seed distribution and seeding rate are uniform and the seed is buried to a constant depth. All mechanical methods are limited by the terrain.

Hydroseeding is a relatively recent method which has achieved much popularity.[81] The equipment comprises a truck-mounted tank (about 6800 litres [1500 gal]) with pumps and agitation devices to mix and propel a jet of slurry up to 50 m (180 ft) away from the vehicle. The basic slurry is simply a suspension of seeds and fertiliser in water, with the addition of a mulch such as straw, wood cellulose fibres, sewage sludge, glass fibre, etc. A green dye is also included to permit the operator to see which areas have already been seeded. The technique has the great advantage of not being limited by terrain. If the mulch and seed are sprayed in one operation, up to 60–70% of the seed may be retained and die on the mulch; thus, excessive seeding rates may be needed.

Aerial seeding has been found useful at some U.S.A. strip mines, where 30–50 ha (75–125 acres) have been seeded in an hour. Aerial broadcasting requires windless conditions but seed mixture segregation can be a serious problem.

Seeding rates vary but are commonly at least double the normal agricultural rates of 17–28 kg/ha (15–25 lb/acre). For most purposes, a rate of no more than 45 kg/ha (40 lb/acre) suffices, although some authorities recommend up to 135 kg/ha (120 lb/acre). Seeding rates of this magnitude are standard on British road verges, but are often regarded as excessive.

10.5.5. Surface Stabilisation

Vegetation is the only method of achieving long-term surface stability of bare wastes, but numerous other methods have been tried. None has been found satisfactory on its own, but many have great value in improving conditions sufficiently to enable vegetation to become established.[45,46,90] The subject has received much attention by the U.S. Bureau of Mines and Table 66 summarises some of the conclusions of this work.

Specific surface stabilisation methods are required mainly on fine particle wastes such as tailings, in arid conditions where wind erosion is likely; south-west U.S.A., South Africa and Australia are some of the main areas of application. Two common physical methods of stabilisation

TABLE 66

COST COMPARISON OF STABILISATION METHODS[90]

	Effectiveness	Maintenance	Cost £/ha	(£/ac)
Physical				
Water sprinkling	Fair	Continual	—	
20 cm slag	Good	Moderate	60–180	(24–73)
Straw harrowing	Fair	Moderate	7–13	(3–5)
Bark covering	Good	Moderate	150–170	(61–69)
10 cm gravel and soil	Excellent	Minimal	40–100	(16–40)
30 cm gravel and soil	Excellent	Minimal	120–290	(49–117)
Chemical				
Elastomeric polymer	Good	Moderate	50–130	(20–53)
Lignosulphonate	Good	Moderate	43–100	(17–40)
Vegetative				
10 cm soil plus vegetation	Excellent	Minimal	50–105	(20–42)
30 cm soil plus vegetation	Excellent	Minimal	130–295	(53–119)
Hydroseeding	Excellent	Minimal	34–77	(14–31)
Chemical-vegetative	Excellent	Minimal	17–40	(7–16)
Sewage sludge pellets	Excellent	Minimal	56–250	(23–101)

are covering the waste with soil or overburden, or water spraying.[19] The former is only satisfactory if revegetation is carried out.

A defect of all methods which cover the waste with soil, unless a very thick layer of overburden can be placed, is that toxic ions and acidity may leach upwards and eventually contaminate the soil. Another problem can be that on toxic wastes, plant roots will not penetrate the innocuous surface skin; there is thus no connection between the two layers and it is common for the skin to be eroded, re-exposing the toxic waste. This has been a particular problem on tailings.

Other physical methods are essentially developments of mulching, namely covering the waste with straw, brushwood, refuse, etc. Windbreaks built of reeds or brushwood have been the subject of extensive study in South Africa/Rhodesia (Fig. 89). They are often overwhelmed by drifting sand.[47,48,51,52]

Chemical stabilisation is the reacting of an appropriate reagent with the waste to form an erosion-resistant crust. Chemicals which have been found effective include lignosulphonates, resinous adhesives and bitumen.

10.5.6. Fertilisers and Other Soil Amendments
The nutrient deficiencies of many mineral wastes mean that N, P and K fertilisers usually have to be added, at least during the early years of plant

Fig. 89.   Small enclosures of reed wind-breaks on a South African gold slimes dump.

establishment. Fertilisers are usually supplied in a single compound (or, erroneously, 'complete') inorganic fertiliser. In Britain, the Fertilisers & Feeding Stuffs Act specifies that all fertiliser is sold with a declaration of its content of available N, P and K—always in that order—with the result that the description (say) 10:20:10 denotes a fertiliser containing 10% nitrogen, 20% phosphoric acid (in which form the phosphorus content is always quoted) and 10% potassium.

Although compound fertilisers are widely used for convenience, the individual nutrients can also be used separately. Phosphates are available in four forms:

Single super phosphate	18–20% P_2O_5
Triple super phosphate	47% P_2O_5
Basic slag	6–25% P_2O_5
Ground mineral phosphate	25–35% P_2O_5

The first two of these are relatively soluble, whereas the other two are somewhat insoluble. Insoluble forms are often preferable; because raw mineral wastes have little or no nutrient retention capacity it is often the case that most soluble nutrients are lost by leaching within 4–8 weeks of application. Insoluble forms, and specially formulated slow-release fertilisers have value over long periods of time, their rates of nutrient release being such that the young vegetation has the opportunity to absorb the minerals. Simple super phosphates are now seldom used except when incorporated in compound fertilisers. Basic slag is valuable on acid sites because, for every ton applied, about 500 kg (10 cwt) of lime equivalent is also added. The relationship between phosphate and growth is illustrated in Fig. 90.

Nitrogen addition is a topic over which much disagreement exists. Nitrates and ammonium compounds are very soluble and easily lost from the soil and nitrogen deficiency is an extremely common cause of vegetation failure. A slow-release nitrogen source is also available: organic matter, in the form of sewage sludge, farmyard manure, broiler wastes, domestic or industrial wastes, etc. All of these sources have been utilised and in particular sewage sludge, readily available in urban areas, has proved popular and beneficial.[29,38,59,105,126] Extensive use of it has occurred in the U.S.A. However, it should be used with caution because it sometimes contains toxic levels of heavy metals. Further, the use of organic matter can be seen in perspective by recalling the point made on p. 243, namely that the objective is to recreate a balanced soil, in which the organic matter input derives from the dead residues of the vegetation. Large additions of organic matter encourage the development of a population of micro-organisms which rapidly break it down; organic input from the vegetation is, however, insufficient to sustain this early rate of activity so

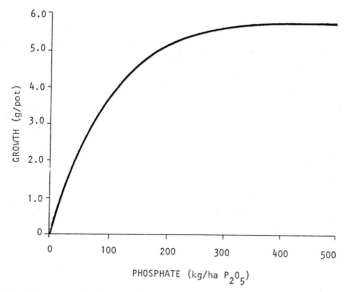

FIG. 90. Response of grass to added phosphate in colliery shale, in experimental conditions (Gemmell[60]).

that, after a year or two, the scheme may suffer serious regression. Indeed, there are good grounds for believing that, unless toxic levels of available metals exist in the spoil (in which case organic matter renders them innocuous by the formation of complex organometallic compounds), the costs of obtaining the large quantities required plus its somewhat dubious effectiveness over long periods make its use superfluous. Even in the case of metal-toxic wastes, breakdown of the organic matter releases the complexed metals.

There exists a third, and vital source of nitrogen which on the grounds of permanence is by far the most desirable. This is the ability of certain plants to fix atmospheric nitrogen. The selection of such plants for the seed mixture is likely to increase the likelihood of long-term success, and they are discussed in Section 10.5.7.

A serious problem with nitrogen-fixing plants is that the symbiotic bacteria which actually undertake the fixation are pH sensitive. The species *Azotobacter* will not survive below pH 6 while *Rhizobium* has a limit of pH 5–5·5. In view of the fact that so many mine wastes are acidic, this is extremely unfortunate, and means that careful neutralising is required. This is by lime or, more usually, ground limestone additions, before and after seeding in many cases, to raise the pH to within the range 6–8.

Calculations of the lime requirement must be made for the specific site conditions. The lime requirement can be calculated in various ways. If defined as the amount of lime needed to raise the plough layer of the soil to pH 7, the calculation requires analysis of the actual percentage base saturation of the soil. Generally the lime requirement is related to pH but not invariably. In colliery spoil, an initial application of 10–15 000 kg/ha (4–6 tons/acre) $CaCO_3$ is sufficient for wastes of pH 3–4 with low acidity potentials but in strongly acidic conditions, requirements are up to 40 000 kg/ha (16 tons/acre). In the worst cases, those with a great acidification potential (*i.e.* high levels of oxidisable pyrite), initial applications of up to 80 000 kg/ha (32 tons/acre) may be required, with re-liming at intervals. In highly pyritic tailings materials, up to 100 000 kg/ha (40 tons/acre) have been employed. Liming of metal-toxic materials has the further advantage of rendering the metals insoluble. In South Africa, very acid gold tailings have not been susceptible to limestone neutralisation and a water spraying method, intended to leach out acids, has been developed with limited success.[52]

Additional methods of reducing or preventing acid formation include floating off pyrite at the milling stage, dosing the waste with antibacterial agents in the hope of preventing bacteria-mediated oxidation from occurring, and coating the pyrite with a chemical blinding agent to prevent the reactants required for oxidation reaching the pyrite. These are all experimental processes at the present.

It is impossible here to specify amendment schedules for every likely substrate. However, the details in Table 67 demonstrate a typical routine in British colliery waste, which is also applicable to similar wastes in North America.

Maintenance of vegetation is as important as initial establishment, but is frequently neglected with the result that the vegetation regresses and fails. As in the initial planting and fertilising, maintenance requirements are highly specific to individual sites. For all except highly toxic sites—at which special maintenance may be required in perpetuity—special maintenance can probably cease after 5–7 years, when the revegetated site can return to the level of care appropriate to comparable types of vegetation on undisturbed lands. Fertilising requirements are determined by analysis of soil conditions plus inspection of any deficiency symptoms exhibited by the vegetation, and amendments are made as required. However, even if all were to appear satisfactory upon analysis, standard practice is to take the precaution of moderate reliming and/or refertilising. In particular, many grass swards benefit from an application of about 75 kg/ha (68 lb/acre) N (60 units/acre) soon after establishment. If however clovers form an important component of the sward, nitrogen applications should be lowered (say by about half) because in conditions where good supplies of soil nitrate are available, the nitrogen-fixing bacteria do not achieve a

TABLE 67

REVEGETATION SCHEDULE ON COLLIERY WASTE

(After Gemmell[60])

Operation	Specification
1. Deep ripping	0·6 m (2 ft) depth at 0·5–1·0 m (18–36 in) centres
Allow winter exposure for saline sites, lime and repeat ripping	
2. Lime spreading (before or after ripping)	Ground limestone at up to 50 000 kg/ha (20 tons/acre) and incorporate into spoil with tine cultivator
3. Phosphate fertilisation	250–1 250 kg/ha (200–1 000 units/acre) of triple super phosphate, and incorporate with tine harrow
4. Prepare seed bed	Disc harrow
5. NPK fertilisation	50–125 kg/ha N (40–100 units/acre) 50–125 kg/ha P and 50–100 kg/ha K (40–80 units/acre)
6. Seed with grass	On shales, increase seed rate by 50%
7. Harrow and roll to incorporate seed and fertiliser	

good rate of fixation. It should not be forgotten that trees also require fertiliser applications, at a rate comparable to grassland, although selectively applied around each individual. Low-nitrogen fertilisers for trees help to limit weed growth and prevent smothering.

Refertilising is by no means the only maintenance needed. All grasses benefit from carefully controlled grazing, which encourages the development of side branches ('tillers') and hence a denser sward. Cattle or sheep are commonly employed but, because newly established grassland tends to be abnormally rich as a result of the high rates of fertilising, the animals tend to overgraze, injuring both themselves and the grass. In particular grazing should never be permitted on grassland on non-ferrous metal mine wastes unless the absence of toxic metals has been conclusively demonstrated. The fatal results of this have been discovered on many occasions. Mowing can perform the same function with greater precision but also more expense.

Maintenance of trees is generally less complex. Apart from fertilisers, the only necessity is to prevent grass or weeds from smothering the young trees. This can be achieved either by grass cutting or selective weed killers applied around each tree.

Other maintenance may include attention to earthwork failures, such as slumping, drainage, etc. On acid wastes, so-called 'hot-spots' develop and must be heavily limed and reseeded. Particularly in urban areas, the whole scheme must be protected from vandalism.

On acid colliery spoils and metallic wastes, where vegetation has been established in the raw wastes, the best course is to leave a soil structure to develop over many decades with attention as required. If however the waste was not toxic but merely infertile, or if subsoil and topsoil coverings have been used, it can be beneficial after about 3 years to plough up the grass, refertilise, relime and reseed. A cereal crop can be grown for a year before reseeding. This has the effect of returning a large amount of organic matter to the soil, breaking up and aerating the surface. However, it would be wise to attempt a trial area first and do the bulk of the work in the following season, unless there are already good reasons to expect success.

10.5.7. Selection of Species

To attempt to list the plant species which have been successfully employed in the investigation of mineral workings would be impossible. The following general points can however be made:

(i) A high proportion of successful species has been capable of fixing atmospheric nitrogen.
(ii) No species or combination of species has been invariably successful, even in the same medium.
(iii) Virtually every successful species was already in routine use in agriculture, horticulture or forestry.
(iv) No amount of maintenance will enable basically unsuitable species to survive in a particular environment.
(v) Exotic 'garden' species are less suitable than unspectacular pioneer species.

From this it can be concluded that, provided the initial choice of species is sensible, success or failure hinges upon the maintenance and care it receives. This section attempts to define the principles upon which this selection can be made.

The most important general principle of plant selection is simple observation. Around any mine site will be found numerous plants growing in undisturbed, natural situations. Also, old dumps and tips in the locality are likely (unless exceedingly toxic) also to show good natural revegetation by volunteer species.[106] Such inspections give valuable clues to the species likely to succeed in the conditions prevailing—the more valuable because, unlike literature surveys (which are, though secondary, also essential), the natural flora is an expression of the prevailing soil and climate. In addition, the selected species must be appropriate to the use envisaged.

If agricultural use is intended, a permanent pasture mixture is needed, containing mainly perennial ryegrass (*Lolium perenne*), timothy (*Phleum pratense*) and wild white clover (*Trifolium repens*). The timothy can be replaced by various other grasses. In very good topsoils, a mixture of 6 lb ryegrass and 3 lb white clover per acre is adequate for agriculture. In Britain, the standard Ministry of Transport mixture for road verges is:

	lb	kg
S 23 perennial ryegrass	60	27·2
S 59 red fescue (*Festuca rubra*)	20	9·1
Smooth-stalked meadow grass (*Poa pratensis*)	10	4·5
Crested dogstail (*Cynosurus cristata*)	12	5·4
S 100 White clover	10	4·5
	112 lb	50·7

This is sown at about 1 lb/90 yd^2 (1 kg/200 m^2) during March–April or August–September, with spring sowing being preferable. The S numbers refer to strains tested at the Aberystwyth plant breeding station and do not apply outside Britain.

Numerous variations in this specification can be made. Other grasses commonly used include *Agrostis stolonifera*, bent grass (*A. tenuis*), and cocksfoot (*Dactylis glomerata*). If grasses are sown around young trees (which is beneficial to their growth), species need to be selected for slow or low growth, to avoid smothering the trees. Suitable species include *Agrostis tenuis*, *Festuca rubra* and sheeps fescue (*F. ovina*). Whatever the use intended for the sward, the inclusion of nitrogen-fixing species is desirable, at about 10–15% of the seed mixture by weight. Among the successful species in Britain are birds foot trefoil (*Lotus corniculatus*), and the clovers *Trifolium repens*, *T. pratense* (red clover) and *T. hybridum*.

In Pennsylvania, extensive study has enabled five seed mixtures, in combination with a single site preparation schedule, to be recommended, on the basis of spoil acidity. These are given below:

	lb/acre	kg/ha
(i) *For spoil pH of at least* 6·0		
Alfalfa (*Medicago sativa*)	12	13·2
used with one of the following grasses:		
Orchardgrass (*Dactylis glomerata*)	5	5·6
Smooth brome (*Bromus inermis*)	10	11·8
Reed canarygrass (*Phalaris arundinacea*)	10	11·8
Tall fescue (*Festuca arundinacea*)	10	11·8
Tall oatgrass (*Arrhenatherum elatius*)	15	16·7
(ii) *For spoil pH of at least* 5·5		
White clover (*Trifolium repens*)		
or ladino clover (*T. repens* var.)	1	1·1
Red clover (*T. pratense*)	2	2·2
Kentucky bluegrass (*Poa pratensis*)	6	6·7
Timothy (*Phleum pratense*)	4	4·5

		lb/acre	kg/ha
(iii)	*For spoil pH of at least* 5·0		
	Crown vetch (*Coronilla varia*)	10	11·2
	used with one of the following grasses:		
	Tall fescue (*Festuca arundinacea*)	15	16·7
	Tall oatgrass (*Arrhenatherum elatius*)	20	22·4
	Perennial ryegrass (*Lolium perenne*)	15	16·7
(iv)	*For spoil pH of at least* 5·0		
	Birdsfoot trefoil (*Lotus corniculatus*)	8	8·9
	used with one of the following grasses:		
	Timothy (*Phleum pratense*)	3	3·3
	Smooth brome (*Bromus inermis*)	8	8·9
	Orchard grass (*Dactylis glomerata*)	4	4·5
	Reed canary grass (*Phalaris arundinacea*)	8	8·9
	Tall fescue (*Festuca arundinacea*)	8	8·9
	Tall oatgrass (*Arrhenatherum elatius*)	15	16·7
(v)	*For spoil pH of at least* 4·5		
	and altitudes not exceeding 1200 *ft*		
	Sericea lespedeza	10	11·2
	used with one of the following grasses:		
	Switchgrass (*Panicum virgatum*)	15	16·7
	Reed canarygrass (*Phalaris arundinacea*)	15	16·7
	Tall fescue (*Festuca arundinacea*)	14	16·3
	Tall oatgrass (*Arrhenatherum elatius*)	20	22·4
	Timothy (*Phleum pratense*)	8	8·9

Among the many trees which have proved valuable, the best-suited to acid soils have been the alders (*Alnus incana* and *A. glutinosa*) and the birches (*Betula verrucosa*). Birches *B. pendula* and *B. pubescens* have also been useful on mine wastes, as have the larches *Larix decidua* and *L. leptolepis*, the black pine (*Pinus nigra* ssp *laricio*), poplars (*Populus robusta* and *P. tremula*), the locust (*Robinia pseudoacacia*) and a variety of willows (*Salix*). Several shrubs have also succeeded, such as sea buckthorn (*Hippophae rhamnoides*) and gorse (*Ulex europaeus*), which are also nitrogen fixing. Because nitrogen fixation depends upon bacteria living in nodules on the roots, and the wastes are usually sterile, it is necessary to innoculate the seed prior to sowing.

Grass and clover mixtures are employed to obtain a variety of desirable properties in a single sward—rapid establishment, palatability for stock, nitrogen fixation, etc.—and exactly the same applies to trees, which are seldom planted as pure stands unless intended as an economic forestry crop.[11] Many of the species considered to be ultimately desirable—for example, ash (*Fraxinus excelsior*), sycamore (*Acer pseudoplatanus*), beech (*Fagus sylvatica*)—are slow growing and require to be planted among 'nurse' crops which are better able to grow on wastes. Nurse trees include the alders, willows, black locust or contorta pine. Nurse species provide protection and, provided they are not allowed to overwhelm the ultimately desired tree, materially assist in its establishment. Additionally, where trees are being used for visual purposes, rapidity of growth is often important. Nurse-type species are thus applicable in these cases.

The above discussion refers to all mine wastes except those which contain high toxic levels of heavy metals. This type of site offers the possibility of selecting species which have a definite tolerance of high levels of particular metals.[44,53,56,61,119] Research into, and exploitation of, the phenomenon of tolerance has been carried out at the University of Liverpool, by Bradshaw and co-workers.[57,58,80,106] Metal tolerance has been a recognised occurrence for several centuries and much use has been made of it in the exploration for orebodies. Now formalised as the sciences of geobotanical and biogeochemical prospecting, such exploration depended upon the observation that certain plant species indicated the presence of metals, or that some species contained high metal levels by concentration from the soil and yet were able to grow. Clearly such species—or, rather, particular individuals within a species population— are able for some reason to withstand metal levels toxic to fellow members of the species. The development and multiplication of such genetically adapted plants has proceeded to the point that three populations have been isolated: one of *Festuca rubra*, for calcareous lead/zinc wastes; and two of *Agrostis tenuis*, one for acidic lead/zinc wastes and the other for copper waste. Such plant material offers the possibility of growing vegetation in wastes containing up to 140 000 ppm combined lead and zinc although either is toxic to normal non-tolerant plants at less than 500 ppm. The only fertiliser treatment required has been shown in tests to be 2·4 tonnes/ ha (20 cwt/acre) of slow release fertiliser, or standard 15:15:15 at 880 kg/ha (7 cwt/acre); the tolerant plants are also able to withstand low nutrient conditions. In 1972, costs of using this method were estimated at up to £247/ha (£100/acre) whereas conventional methods involving burying the waste could cost up to £2470/ha (£1000/acre).

10.5.8. Integration of Revegetation with Mine Working

The numerous problems of revegetating mined land which have been outlined above can in many cases be avoided by appropriate alterations in the mining technique. Although some long-term operations, notably hard-rock quarries and open pit metal mines, present severe problems of revegetation, in numerous other types of mining at least some degree of rolling restoration can be achieved. To a large extent the revegetation problems of U.S.A. coal strip mines, for example, are aggravated by environmentally damaging working methods. Haphazard backfilling of the workings, inadequate topsoil conservation and delays in regrading the backfill mean that acid generation can occur in wastes at the surface, severely hindering vegetation establishment, and cause the land to be removed from circulation—and hence economically productive use—longer than is necessary. Increasingly legislation refuses to permit practices which are technically redundant and there can be little doubt that rolling reclamation will become a much more common feature of mineral operations. With this in view, the case studies in Section 10.6 have been selected to demonstrate the advanced techniques which are in use.

10.6. CASE STUDIES

The case studies below illustrate both integrated and post hoc rehabilitation. The creation of true derelict land in Britain, for example, is virtually absent in the case of open cast coal and sand and gravel winning, both of which sectors have great experience in progressive restoration of their workings. Even in the case of deep-mined coal, historically one of the worst offenders, increasing use is made of progressive restoration techniques for surface tips. There is therefore much evidence that integration of rehabilitation with mineral working will become a more common feature than hitherto, and the emphasis of these case studies is thus upon this aspect. Although much backlog dereliction remains to be cleared, the contribution from mineral working should grow steadily less. This does not however imply that the integration of working and rehabilitation is seen as sufficient. Especially in urban areas there is much interest in the place of mineral working within an overall land use strategy. This is leading to the choice of mine sites being made, not only on the basis of minimum nuisance during operating life, but with a view to the exhausted quarry providing the marina or landfill site which the locality requires in any event.

10.6.1. Open Cast Coal in Britain

British open cast coal mining effectively commenced in 1941 as an emergency war-time measure. Frequently little, if any, restoration was undertaken. Although intended as a short-term expedient, the use of open cast methods persisted after World War II because of coal shortages, and has continued thereafter. Production peaked at 14 million tonnes in 1958, since when it has fluctuated between 6 and 10 million tonnes/a. Open cast mining has survived partly on cost grounds—average costs per tonne before interest charges being 15–20% below deep mined coal—and partly because colliery closures have rendered certain grades of coal in short supply. Open cast production has been the only way to make up such shortfalls.

In 1951 a code of land restoration was agreed. Experience gained since that time now permits total restoration of all sites being worked. Between

1942 and 1975, a total of 62 000 ha (153 000 acres) of land has been taken for open cast coal production, of which 47 000 ha (117 000 acres) have been either restored to their previous use, or to an agreed new use. 15 000 ha (36 000 acres) remain in use, or in the process of rehabilitation. An additional feature since 1964 has been the design of open cast mines in such a way that adjacent derelict land is also reclaimed, the overall result being a gain in amenity as a result of mining. Some 650 ha (1600 acres) of derelict land have so far been reclaimed. Details of the costs incurred in open cast coal mining are given in Table 68, which illustrates the relatively small increase in restoration costs despite the higher standards being achieved. As a percentage of total costs, restoration has increased only from 4 to 5%.[85]

TABLE 68

NATIONAL COAL BOARD OPEN CAST EXECUTIVE COSTS OF WORKING AND RESTORATION (Arguile[85])

Component	Cost £p/tonne saleable			Change over 21 years (%)
	1952	1970	1974	
Prospecting and boring	0·11	0·11	0·8	−27
Production	1·84	3·11	4·65	+153
Haulage to disposal points	0·19	0·22	0·27	+42
Preparation, handling and stocking	0·14	0·66	0·75	+436
Distribution and selling	0·02	0·04	0·05	+233
Restoration and compensation	0·10	0·19	0·24	+140
Administration	0·09	0·20	0·34	+278
Restoration as per cent of total costs	4	4	5	—

The methods and scale of working, and hence the problems of restoration, have steadily increased. This is illustrated by the equipment in use: in 1944 excavators for overburden removal were predominantly in the range 0·5 m^3 (0·6 yd^3) or below, and the largest size was 4 m^3 (5 yd^3). By 1973 the largest size was in the 23–30 m^3 (30–40 yd^3) range, with most being 4–7 m^3 (5–9 yd^3). The size of scrapers has undergone a similar evolution, from a maximum of 14 m^3 (18 yd^3) in 1944 to in excess of 18 m^3 (24 yd^3) in 1973.[8,68,70,86,117]

Originally open cast workings were for single (rarely two) seams at their outcrops and mining was limited to overburden depths of about 12 m (40 ft) maximum, giving a stripping ratio of up to 5:1. This was essentially comparable to contour strip mining in the U.S.A. today. As equipment capabilities and demand increased, overburden depths have become progressively greater; by 1948, depths of 30 m (100 ft) were found with stripping ratios of 12:1. Today 75 m (250 ft) depths are not uncommon, and one site is intended to reach 230 m (700 ft). Stripping ratios of up to 25:1 are now acceptable in favourable circumstances, although difficult geological conditions may cause the maximum to be about 12:1. Multi-seam working, with 4–6 seams commonly worked in a single pit adds greatly to the complexity of mining and subsequent restoration. Indeed some Welsh sites comprise 15 seams.

The simple, classical, single dragline operation is now seldom found, and it is usually necessary to bench and use a combination of dragline, and face shovel plus trucks. Consequently there is an increasing need to double-handle overburden. Examples of the types of situation encountered can be found at Abercrave, South Wales[86] and at several Northumberland sites where it has been found advantageous to work from the deeper to the shallower parts of the site, rather than the more conventional reverse. This entails forming, by scrapers, shovels and trucks, a box cut along the high wall. The initial expense is balanced by the subsequent economies in trucking overburden downhill, and by free drainage. Generally, a seam dip no greater than about 1 in 9 is needed if such methods are not to entail excessive initial excavation. In the Abercrave example (Fig. 91) the stripping dragline is supplemented by another which stands on the cast-back spoil and rehandles it, prior to regrading. This technique is sometimes termed 'haymaking'. Rehandling is also frequently required when temporary tips are formed, as when working by the box-cut method.

The standard method of working includes backfilling and at only one mine (Westfield, Scotland), is this not intended. In this case an extensive overburden tip has been built and progressively regraded and returned to farmland. Overburden crushing to minus 25 cm (10 in) was required, overburden transport being by 1·2 m (48 in) conveyer belt. At the tip, the waste was graded to a maximum slope of 1 in 8, soiled and planted.

To enable extensive excavations to be fully restored, advanced planning of overburden disposal is required, to minimise double-handling. The 1951 Code of Practice referred to above included provision for the separate stripping, storing and replacement of topsoil (ideally 30 cm [12 in]) and subsoil (60 cm, 24 in) and their replacement upon graded overburden which was free, to a depth of at least 1 m (3 ft), of rocks, shales and blue clays.

Stripping of topsoil and subsoil may be progressive or, more rarely, a single operation. Frequently there is a lack of topsoil and in such cases other suitable, soil-forming, subsoil material is retained instead. In a few cases the topsoil and subsoil are infested by rushes and are not saved but buried. Glacial drift materials can be substituted. Stripping operations are usually carried out by scrapers and the materials are stored in banks

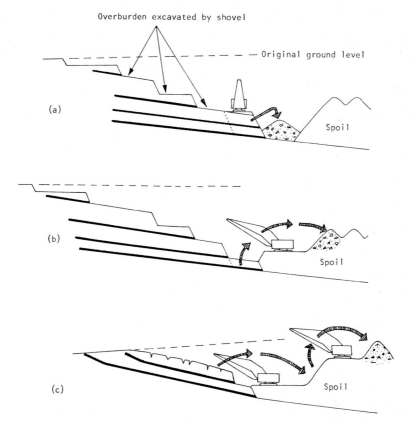

Overburden excavated by shovel

Original ground level

(a)

Spoil

(b)

Spoil

(c)

Spoil

FIG. 91.   Working methods at Abercrave open cast coal site: (a) working from
deep to shallow part of site, upper overburden is removed by shovels and lower
overburden by dragline; (b) dragline then stands on cast-back overburden to
take out lowest overburden; (c) sequence on steep sites showing dragline used
for chopping down, with second dragline rehandling overburden (Whincup[86]).

around the site perimeter, the locations being chosen with a view to the
stockpiles acting as noise baffles and visual screens. The stockpiles are
grassed to prevent erosion and weed incursion.

Coaling and backfilling then takes place, followed by final grading.
The contours selected may be those originally existing, or may be adapted
to permit some new use. In the case of sites in steep terrain, a highwall cliff
is sometimes retained. All grading is planned to take account of subsequent
settlement, which is of particular importance from the point of view of
draining the land.

The contoured overburden is then ripped in two directions at right angles, to a depth of about 45 cm (18 in). This assists in 'keying on' the subsoil and breaks up compacted areas. Subsoil is then lifted and spread, usually in two 45 cm (18 in) layers; these are likewise rooted. Finally the topsoil is spread by scraper, and ripped to remove compaction. These stages are undertaken only in dry weather and require very close supervision to prevent unwanted intermixing of layers.

At this point, the site is handed over to the Ministry of Agriculture which, for the subsequent five years, specifies the restoration methods for the surface. For agricultural re-use these fall into two categories:

(i) Replacement of the fixtures of the farm. Formation of ditches is an early essential since surplus water accumulations are common and damaging if not checked. New hedges are installed, usually beside ditches, or else fences and walls are built, depending on the locality. Water supplies, access roads, etc., are also installed as required. Under-drainage, mainly tile drains, is not installed until about the fourth year, when surface consolidation is less. Drains are placed at 15 m (17 yd) intervals at depths of about 1 m (3 ft).

(ii) Planting, cultivation, manuring and controlled cropping. The precise course followed varies markedly from region to region, depending upon the depth and nature of the soil layers, their degrees of compaction, the climate, and the prevailing type of farming. Grassing is the usual first step, with seeding in July–August after the application of 7·5 tonnes/ha (3 tons/acre) ground limestone and 650 kg/ha (5 cwt/acre) compound fertiliser. Seed mixtures vary, but a common type includes perennial ryegrass, cocksfoot, timothy and various clovers. In the second, third and fourth years additional fertiliser applications are made and in the fifth year reseeding is carried out. The grass crop is either mown or grazed, with care to avoid damage to the sward. In certain areas woodlands are planted and in such cases standard forestry practices are applied. Regrading is not usually completed to smooth contours for forestry purposes, because a hill-and-dale topography is satisfactory for the purpose.

The costs of such schemes vary considerably. In 1974 the average costs of topographic restoration were as follows:

Stripping and stacking separately
topsoil and subsoil   £725–£775/acre (£1790–£1915/ha)
Excavation of overburden by
dragline   6–10p/yd^3 (8–13p/m^3)

Backfilling of overburden, including
transport and consolidation   7·5–12·5p/yd^3 (10–16p/m^3)
Replacement of topsoil and
subsoil, plus rooting, etc.   £775–£875/acre (£1915–£2160/ha)

The agricultural treatment costs in 1971 averaged £230/acre (£570/ha) as under:

	acre	ha
Cultivation, manuring and cropping over 5 years	£80	£197
Permanent drainage	£75	£185
Ditches, fences, hedges, etc.	£60	£148
Water supplies to fields	£15	£37
Total	£230/acre	£567/ha

This figure had, by 1974, increased to £350–£375/acre (£865–£925/ha).

Agriculture and forestry are the most common after-uses for the restored land but, in recent years, certain other uses have become important, in particular recreational use. A number of country parks are now being created as a result of open cast coal working. Druridge Bay, Northumberland open cast coal site has been the subject of a scheme to develop a seaside country park. The scheme will cover some 75 ha (190 acres) and regrading will include provision of an 18 ha (45 acre) lake for sailing, waterskiing, swimming, etc. Similar schemes are in progress in Derbyshire and Warwickshire.

In all these cases, the immense costs associated with the provision of leisure areas would have been unlikely to be borne by the local authorities or other developers. However, the planned, profitable working of the coal, within the context of a restoration scheme agreed as part of an overall land-use strategy permits such amenities to be provided at acceptable cost and, as such, carries the concept of restoration beyond the simple idea of clearing up abandoned mines.

10.6.2. Dunbar Limestone Quarry

The Dunbar quarry and cement works, located 50 km (30 miles) from Edinburgh, Scotland, was opened by the Associated Portland Cement Manufacturers Ltd, in 1963. The mixed limestone-shale quarry is in geologically unusual circumstances which have permitted the development of a progressive restoration scheme almost unique in this mineral. Details of the succession are:

Strata	Thickness (ft)	(m)
Topsoil	1·5	0·5
Overburden	15	5
Limestone and shale	19	6
Upper limestone	15	5
Fireclay and limestone	13	4
Sandstone and shale	17	5
Shale	6	2
Lower limestone	22	7
Total	108·5	34·5

The quarrying problem was to devise a method which would:

(a) allow the extraction of the upper and lower limestone beds, plus a
    portion of the intervening shales, for delivery to the crusher as
    feedstock for cement manufacture (dry process);
(b) allow the ordered deposition of the remaining strata and the
    progressive restoration of the worked-out area to the high grade
    agricultural use which it enjoyed before quarrying.

These objectives have been achieved by the use of a strip mining method.
An idealised diagram of the method is given in Fig. 92; since 1963 there
have been certain alterations as a result of experience, but the basic scheme
is essentially as described below. The features of this method include
working the stone in a slot, 600 m (2000 ft) long and 30 m (90 ft) wide,
advancing downdip at 5°. The 30 m (90 ft) width has latterly been increased
to allow greater working space. Of particular note is a purpose-built
overburden transporter, the jib of which bridges the working slot and
enables overburden to be cast behind ready for spreading. In brief the
working method is as follows:

(i) Topsoil (about 45 cm [18 in]) is removed by bulldozer and stockpiled
    in advance of the cut. It is reloaded and spread on the backfilled
    land once per year, usually between April and June, depending on
    the weather. This double handling enables land in advance of the
    face to be left in cultivation for the maximum length of time. The
    use of scrapers to transfer topsoil is not considered feasible because
    of compaction and 'puddling' problems.
(ii) All overburden above the upper limestone is removed by a 2 m³
    (2½ yd³) dragline loading to the hopper of the transporter. This has
    a 1 m (36 in) conveyor belt and a boom length of 80 m (240 ft);
    allowing for the width of the hopper, maximum transport distance
    is 100 m (315 ft).

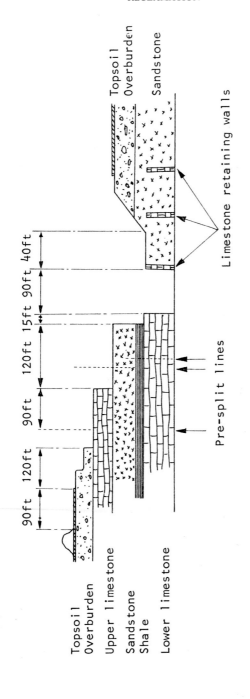

FIG. 92. Section through Dunbar limestone/shale quarry showing the original method of working.

(iii) The overburden is deposited on sandstone already cast back, and spread by bulldozer as a continuous operation.

(iv) The upper limestone is then removed. Because the dimensions of the whole layout are critical—particularly with respect to the relatively unmanoeuvrable transporter—precision blasting based on presplitting was selected.

(v) The whole of the strata down to the top of the upper limestone are taken out as one lift. Presplitting is used to break the sandstone. A 2 m^3 (3 yd^3) walking dragline (chosen because it stands upon the broken sandstone and its low bearing pressure, 4·5 tonnes/m^2 (0·4 ton/ft^2), is thus advantageous) casts the sandstone behind the slot. The shale which is required in the process (10 % of the limestone tonnage) is removed by a traxcavator, the surplus being cast back with the sandstone.

(vi) The lower limestone thus exposed is excavated, leaving a wall of stone about 60 cm (2 ft) thick to act as a retaining wall for the sandstone waste.

(vii) The topsoil stockpiles are returned via the overburden transporter, spread with a bulldozer and rooted in two directions. After stone picking, the land is returned to cultivation.

10.6.3. Ironstone Working in the Midlands

The winning of the Northampton Sand and Marlstone iron ores in the Midland counties of Britain recommenced in modern times in about 1850, and today exemplifies classical strip-mining open cast techniques. Early workings were by hand, under very shallow cover (less than 6 m (20 ft)), and these were fully restored and are now scarcely detectable. In 1896–7 steam navvies and overburden transporters were introduced and the onset of mechanised working, which enabled greater depths of overburden to be economically handled, also markedly increased the problems of restoration. This coincided in the incorporation in quarrying leases of clauses requiring only partial restoration—or payment in lieu—rather than the full restoration clauses of early leases, a recognition of the practical problems of restoration of deep workings. By 1938 increasing concern was expressed at the despoilation which was accumulating and in 1951 a restoration fund was begun. Since that time, virtually complete restoration has been carried out upon current workings and some backlog work has been tackled.[9,27,28,78]

Current workings, in Northamptonshire and Lincolnshire, produced 8 800 000 tonnes in 1972–3. The methods used are typified by the Northampton sand iron ore workings in the Corby area. The stratigraphic succession is:

Stratum	Thickness (ft)	(m)
Upper estuarine clays	10–40	3–12
Lincolnshire limestone	up to 130	40
Lower estuarine clays	up to 35	10
Northampton sand ironstone	up to 25	8

Ironstone is worked to a maximum depth of about 35 m (120 ft), facilitated by the development of large walking draglines with bucket sizes up to 27 m^3 (35 yd^3); one such machine can deal with the full depth of overburden, within the following sequence of operations (Fig. 93):

(i) Topsoil is stripped by scraper and either stored or directly respread upon a worked-out area.

(ii) The dragline stands upon the limestone beds. It chops down the overlying clays and deposits them on top of the limestone overburden which was dumped in the previous cut.

(iii) The limestone and lower clays, previously fractured by blasting, are removed and deposited in the worked-out part of the current cut. Thus, in the backfilled area, the original geological sequence is maintained.

(iv) The ironstone is removed by a face shovel.

Frequently some double-handling of overburden is carried out by a smaller dragline on the backfilled area, which pulls back overburden to maintain working space in the floor of the cut, and also roughly levels the

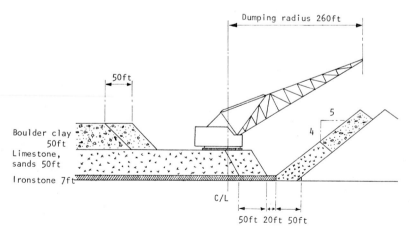

FIG. 93. General sequence of strip mining ironstone in Britain (Pearson[8]).

hill-and-dale formation. Final grading is by bulldozers, prior to respreading of topsoil. Seeding can then take place with rapidity, often within 2–3 weeks.

For various reasons topsoil is not always available and it is necessary to cultivate the estuarine clays or chalky boulder clays. Rooting and stone-picking has not been practised, with the result that ploughing can be difficult or impossible. For planting, an initial seeding of 35 kg/ha (30 lb/acre) timothy, meadow fescue and white clover has been found satisfactory, with a heavy 750–1250 kg/ha (6–10 cwt/acre) preliminary dressing of a compound fertiliser plus basic slag. Refertilising is carried out each subsequent spring. Field boundaries, drainage, etc., are also installed.

Much of the earlier workings was left as hill-and-dale and afforested, and this is occasionally practised for current workings. In damp, heavy clays the only successful tree has been alder (*Alus glutinosa*) but drier areas have been successfully planted with larch (*Larix europaea*), sycamore (*Acer pseudoplatanus*), Scots pine (*Pinus sylvestris*) and others. In a few cases restoration of the land for the use of heavy industry, housing or leisure has been appropriate.

The circumstances under which British stratified ironstones are exploited has rendered applicable a restoration fund, the only one of its type in Britain. The Ironstone Restoration Fund was established under the Minerals Working Act, 1951, to provide money for the after-treatment of worked-out ironstone areas. Since inception, payments into the fund have increased; at the present time they stand at 3p per imperial ton of ore payable by the ironstone producer, and 0·3125p payable by Central Government. In the case of leasehold minerals the lessee can recover from the lessor up to half his payments, provided that this sum does not exceed one-third of the royalty payable under the lease, and subject to a maximum of 0·46875p. The Fund may make payments under the following conditions:

(a) To the producer, reimbursing the cost of complying with planning conditions in excess of £1000/ha (£410/acre). Up to this limit the producer bears the whole cost.

(b) Lessors may recover the costs of new buildings, drainage, hedges, etc., necessitated by the working of the land.

(c) Occupiers of restored land may obtain contributions towards the cost of fertilisers and other measures needed to bring the land back in full fertility.

(d) Local authorities may obtain recompense for restoring backlog land.

By 1961 local authorities had spent £200 000 in regrading 800 ha (2000 acres) of old hill-and-dale, where topsoil is naturally absent. Ironstone producers have restored current workings, mainly with topsoil, at a rate of 160 ha (400 acres)/year at a cost of £2 600 000. However, there has been a marked escalation in the cost of restoration, from £620/ha (£250/acre) in 1951 to £2470/ha (£1000/acre) today. Government grants in respect of restoration of fertility have amounted to £700 000.

10.6.4. Taconite Tailings Revegetation

Dickinson[2] describes the revegetation of taconite tailings produced by the Erie Mining Co. at its mines in the Mesabi Range, Northern Minnesota. Tailings were deposited over some 800 ha (2000 acres) in dams constructed from the coarse fraction (mainly the 26·6% + 65 Mesh component). Revegetation of the outer faces of the dams was considered desirable to minimise erosion and studies began in 1948.

Analysis determined that although there were no toxic constituents the material was extremely infertile. The coarse material was deposited at 2:1 slopes and was therefore subject to temperature extremes. Such a gradient rendered farm machinery unusable and hydroseeding was employed. Initial work applied in one operation 15 lb seed, 500 lb fertiliser and 700 lb of wood pulp mulch per acre (17 560 and 780 kg/ha) but was not wholly successful because of wind erosion. Separate applications of fertiliser, seed and mulch were found better. Infertility was countered by the use of 1000 lb of 11–55–0 fertiliser per acre (1120 kg/ha). Seeding rates were also increased, to 27–33 kg/ha (25–30 lb/acre). In all cases it has been found best to lightly cover the seed, which is done by means of a chain pulled along the slope. Because fertiliser penetration was found to be inadequate, a spiked flexible 'roller' termed a Klodbuster, hauled by two bulldozers (one at the foot of the berm and the other at the top) was used to break up the surface and work the fertiliser into the rooting layer.

Climatic and soil conditions naturally influenced the species selected. Average annual precipitation was 680 mm (27 inches) of which a quarter fell as snow which lay for about four months of the year. Half the precipitation fell during the growing season, and droughts were rare. Consequently, watering was not required, and species choice was wide. Also the tailings were slightly alkaline, pH 7·5–8·4, which eased conditions. Two seed mixtures have been developed, each including a nitrogen-fixing legume. The grasses were chosen because of their ability to form a dense ground cover. The mixtures were:

1. Intermediate Wheat Grass		16 lb	18 kg
Perennial Rye Grass (*Lolium perenne*)		3 lb	3 kg
*Alfalfa (*Medicago sativa*)		10 lb	11 kg
Total		29 lb/acre	32 kg/ha

2. Red Top (*Agrostis alba*)	6 lb	7 kg
Lolium perenne	6 lb	7 kg
*Birdsfoot Trefoil (*Lotus corniculatus*)	10 lb	11 kg
Total	22 lb/acre	25 kg/ha

The seed of each legume (indicated above with an asterisk) is inoculated with the appropriate bacteria before planting. Hay or straw mulch is now used, at a rate of 5 tonnes/ha (2 tons/acre).

10.6.5. Re-use of Underground Stone Mines

The opportunities for re-use presented by underground stone mining are exemplified by the limestone mines of Kansas City, U.S.A.[3−6] The city is a major interchange centre for east–west and north–south traffic in agricultural products and manufactured goods, which has had an important bearing upon after-use possibilities.

Geologically the outcropping rocks around the city are Upper Carboniferous and contain two important limestone beds: the Bethany Falls Limestone and the Argentine Limestone, each being about 7 m (25 ft) thick. The former underlies Kansas City at about 40 m (130 ft) depth, and has a negligible dip of about 1 in 1000. An impervious clay stratum overlies the limestone and prevents water seepage. Extensive quarries were worked in the Bethany Falls Limestone from the late nineteenth century, but increasing overburden forced working to proceed underground prior to the 1920s, since when some 12 million m^2 (130 million ft^2) of underground excavation have been produced. Room and pillar methods were employed from the start, at first to a somewhat haphazard layout but later upon a regular plan. In the older workings, extraction was carried to excess so that subsidence damage at surface is common. Currently a 'typical' mine has 6 m (20 ft) square pillars with 12 m (40 ft) wide rooms, with headrooms of 4–5 m (12–18 ft) (some stone is left to form a good roof). A survey in 1969 suggested that 12 of the companies then mining stone were creating 0·5 million m^2 (5 million ft^2) of space annually.

Of the total of 12 million m^2 (130 million ft^2) of underground space within 40 km (25 miles) of the city centre, about 1·3 million m^2 (15 million ft^2) are in use, 2·75 million m^2 (30 million ft^2) available for use, and much of the balance is unusable because of bad mine design or geological problems. After-use began in 1951 with automobile storage, and other uses such as warehousing and deep freezes were developed. At about 12 sites, 2000 people are employed underground. Nearly 90% of the space is used for warehousing, the balance being divided between offices and factories. Ten per cent of the U.S.A.'s freezer storage space is installed in Kansas City mines, which also house some 15% of the city's total warehouse space.

Mine operators develop their underground space and let it to tenants. Among the modifications are erection of partitions to isolate different rental blocks from each other, installation of services, road and rail access, and various special facilities. Rental costs are some 40% below those of the equivalent surface facilities, while maintenance and overheads are up to 20% cheaper. Underground temperatures are fairly constant at 13–15°C so that heating or cooling can be economically achieved. In the case of cold storage, freezer rooms cooled to -22°C take some 2 years to achieve stability at this temperature. However, it takes months to warm up to 0°C so that damage to frozen goods in the event of a power failure is improbable. Savings on heating, cooling and humidity control can range from 40 to 90% of the equivalent surface costs.

Mining methods have been adapted to render the mines more suitable for re-use. Currently a headroom of 4 m (12–13 ft) is maintained because this coincides with a roof bedding plane and enables an almost maintenance-free ceiling to be obtained. Roof bolts are not usually required. Changes in blasting techniques have been made, but the main alteration has been in pillar spacing to a regular pattern. A reduction in percentage recovery, from 90% by the old methods to 80% today, has occurred. In all cases the mines are profitable on the sale of crushed rock, income from after-use being an added bonus. Overall the importance of after-use can be judged from the fact that five times as many people work in secondary underground uses as in the mining phase, at a wages bill four times greater than paid by the mining industry.

Mine developments are now routinely carried out in the expectation of a triple incoming cash flow. Because surface damage is avoided revenue can be derived from continuing surface tenancies; immediate income is obtained from the sale of stone; and mining creates, after about 5 years, sufficient underground space to permit the first underground lettings to take place. Working the mine so that parts of it can be vacated rapidly for after-use enables a rental income to be obtained quickly enough to be of significance in the initial mine costings.

REFERENCES

1. Goodman, G. T. (1974). Ecology and the problems of rehabilitating wastes from mineral extraction, *Proc. R. Soc. Lond.*, **A339**, 373–87.
2. Dickinson, S. (1972). Experiments in propagating plant cover at tailing basins, *Mining Congress Journal*, **58**(10), 21–6.
3. Dunn, J. R. (1971). The creation and use of underground space by the crushed stone industry, in *5th Annual Convention of the National Crushed Stone Association*, Bal Harbour, Florida, 9 pp.
4. Stauffer, T. (1974). Use of underground space in Greater Kansas City, *Tunnelling Technology*, No. 8, 1–7.

5. Industry tries living in caves (1961). *Business Week*, 166–9.
6. Yearich, B. (1971). After-use of limestone mines, *Quarry Managers' Journal*, **55**(12), 413–16.
7. Brent-Jones, E. (1971). Methods and costs of land restoration, *Quarry Managers' Journal*, **55**(10), 341–54.
8. Pearson, R. M. (1970). British ironstone mining, *Mining Engineer*, **130**(121), 51–60.
9. Precision quarrying and land restoration at Dunbar (1968). *Quarry Managers' Journal*, **52**(7), 243–50.
10. Ranwell, D. S. (1967). *Sub-Committee Report on Landscape Improvement Advice and Research*, British Ecological Society, London, 8 pp.
11. Manchee, J. S. (1974). Some problems and solutions to the establishment of vegetation in mineral wastes, in seminar *Restoration and the Quarrying Industry*, Northumberland & Durham Roadstone & Concrete Training Association, Newcastle, 16 pp.
12. Hill, J. C. R. and Nothard, W. F. (1973). The Rhodesian approach to the vegetating of slimes dams, *J. Sth African Inst. Min. Metall.*, **74**(5), 197–208.
13. Davis, G. (Ed.) (1971). *A Guide for Revegetating Bituminous Strip-Mine Spoils in Pennsylvania*, Research Committee on Coal Mine Spoil Revegetation, Pennsylvania.
14. Paone, J., Morning, J. L. and Giorgetti, L. (1974). *Land Utilization and Reclamation in the Mining Industry*, 1930–71, U.S. Bureau of Mines, IC 8642, Washington, D.C.
15. A gravel operation that 'creates' farmland (1969). *Cement, Lime & Gravel*, **44**(5), 129–33.
16. Whitt, D. M. (1970). After the mining, useful land, *Mining Congress Journal*, **56**(5), 26–9.
17. Crosby, J. G. and Renold, J. (1974). *Where to Put Solid Waste*, Local Government Operational Research Unit, Manchester.
18. Murray, D. R. and Zahary, G. (1973). *An Approach to Revegetation of Mine Wastes*, Internal Report 73/68, Department of Energy, Mines & Resources, Mines Branch, Ottawa.
19. Murray, D. R. (1971). *Factors Affecting Revegetation on Uranium Mine Tailings in the Elliot Lake Area, Part I—Physical Conditions*, Internal Report 71/104ID, Department of Energy, Mines & Resources, Mines Branch, Ottawa.
20. Murray, D. R. (1972). *Factors Affecting Revegetation of Uranium Tailings in the Elliot Lake Area, Part II—Chemical Conditions*, Internal Report 72/55, Department of Energy, Mines & Resources, Mines Branch, Ottawa.
21. LeRoy, J-C. and Keller, H. (1972). How to reclaim mined areas, tailings ponds, and dumps into valuable land, *World Mining*, **25**(1), 34–41.
22. University of Newcastle upon Tyne (1971). *Landscape Reclamation, Vol. 1*, IPC Business Press, London.
23. University of Newcastle upon Tyne (1972). *Landscape Reclamation, Vol. 2*, IPC Business Press, London.
24. Pancholy, S. K., Elroy, L. R. and Turner, J. A. (1975). Soil factors preventing revegetation of a denuded area near an abandoned zinc smelter in Oklahoma, *J. appl. Ecol.*, **12**(1), 337–42.

25. Corbett, B. O. and Lord, J. A. (1971). Rehabilitation of derelict land, Lletty Shenkin tips, Cwmbach, Aberdare, *Civil Engineering & Public Works Review*, **66**(182), 965–7.
26. Bauer, A. M. (1966). How to make more than holes in the ground, *Landscape Architecture*, **56**(2), 115–19.
27. Cowan, R. J. and Dean, R. (1975). Rehabilitation of ironstone workings, in conference *Minerals Extraction and the Environment*, Royal Institution of Chartered Surveyors, Nottingham, 10 pp.
28. Jones, T. W. (1964). Restoration of ironstone workings in the Midlands of Great Britain, in symposium *Opencast Mining, Quarrying and Alluvial Mining*, Institution of Mining & Metallurgy, London, 17 pp.
29. Hinesly, T. D., Jones, J. L. and Sosewitz, B. (1972). Use of waste treatment plant solids for mined land reclamation, *Mining Congress Journal*, **58**(9), 66–73.
30. Gemmell, R. P. (2 November 1973). Reclamation of chemically polluted sites, *Surveyor*, **142**(4247), 36–8.
31. Persons, H. C. (1958). Rewards of land rehabilitation, *Rock Products*, **61**(4), 62–5.
32. Chesson, M. W. (1966). Second careers for Florida phosphates fields, *Landscape Architecture*, **56**(2), 129–31.
33. Maumee pioneers draglines, land reclamation projects (1949). *Mechanization*, **13**, 136–43.
34. Hogg, J. L. E. (1971). Mined land reclamation in British Columbia, *Forestry Chronicle*, **47**(6), 335–8.
35. Grenfell, A. P. (1909). Recent progress in afforestation, *Quarterly Journal of Forestry*, **3**, 26–31.
36. Maneval, D. R. (1975). Reclamation land for recreational development, *Coal Mining & Processing*, **12**(4), 84–6.
37. Sawyer, L. E. (1946). Reclamation and conservation of stripped-over lands, *Mining Congress Journal*, **32**, 26–36.
38. Hortenstine, C. C. and Rothwell, D. F. (1972). Use of municipal compost in reclamation of phosphate-mining sand tailings, *J. Environ. Quality*, **1**(4), 415–18.
39. Nickeson, F. H. (1969). Hanna Coal's reclaimed mine lands, *Coal Mining & Processing*, **6**(10), 48–51.
40. Riley, C. V. (1973). Design criteria of mined land reclamation, *Mining Engineering*, **25**(3), 41–5.
41. Bauer, H. J. (1971). Recultivation and renewal of a balanced landscape in a lignite mining area of the Rhineland, *Geoforum*, **8**, 31–41.
42. Berg, W. A. (1970). How to promote plants in mine wastes, *Mining Engineering*, **25**(11), 67–8.
43. Ludeke, K. L., Day, A. D., Stith, L. S. and Stroehlein, J. L. (1974). Pima studies tailings soil makeup as preliminary to successful revegetation, *Engineering & Mining Journal*, **175**(7), 72–4.
44. Goodman, G. T. (1969). The revegetation of derelict land containing toxic heavy metals, in symposium *Metals and Ecology*, Swedish Natural Science Research Council, pp. 3–16.

45. Dean, K. C. and Havens, R. (1970). Stabilization of mineral wastes from processing plants, in *Proc. 2nd Mineral Waste Utilization Symposium* (ed. M. A. Schwartz), I.I.T. Research Institute, Chicago, pp. 205–13.

46. Dean, K. C. and Havens, R. (1972). Reclamation of mineral milling wastes, in *Proc. 3rd Mineral Waste Utilization Symposium* (ed. M. A. Schwartz), I.I.T. Research Institute, Chicago, pp. 139–42.

47. Chenik, D. (1960). The promotion of a vegetative cover on mine slimes dams and sand dumps, *J. Sth African Inst. Min. Metall.*, **60**, 525–55.

48. Chenik, D. (1963). Addendum to Ref. 47, *J. Sth African Inst. Min. Metall.*, **63**, 212–53.

49. Ludeke, K. L. (1973). Vegetative stabilisation of copper mine tailing disposal berms of Pima Mining Company, in *Tailing Disposal Today* (ed. C. L. Aplin and G. O. Argall), Miller Freeman, San Francisco, pp. 377–408.

50. Leroy, J.-C. (1973). How to establish and maintain growth on tailings in Canada—cold winters and short growing seasons, *ibid.*, pp. 411–17.

51. James, A. L. (1966). Stabilizing mine dumps with vegetation, *Endeavour*, **25**, 154–7.

52. James, A. L. and Mrost, M. (1965). Control of acidity of tailings dams and dumps as a precursor to stabilization by vegetation, *J. Sth African Inst. Min. Metall.*, **65**, 488–95.

53. Cole, M. M. (1973). Discussion on Ref. 57, *Trans. Instn Min. Metall.*, **82**, A147–A153.

54. Grandt, A. F. and Lang, A. L. (1958). Reclaiming Illinois strip coal lands with legumes and grasses, *Bull. Ill. Agric. Expt. Sta.*, No. 628.

55. Currier, W. F. (1973). Basic principles of seed planting, in *Research & Applied Technology Symposium on Mined-Land Reclamation*, National Coal Association, Pittsburgh, pp. 225–32.

56. Gemmell, R. P. (1973). Revegetation of toxic sites, *Landscape Design*, No. 101, 28–32.

57. Smith, R. A. H. and Bradshaw, A. D. (1972). Stabilization of toxic mine wastes by the use of tolerant plant populations, *Trans. Instn Min. Metall.*, **81**, A230–A237.

58. Antonovics, J., Bradshaw, A. D. and Turner, R. G. (1971). Heavy metal tolerance in plants, *Adv. ecol. Res.*, 7, 1–85.

59. Chumbley, C. G. (1971). *Permissible Levels of Toxic Metals in Sewage Used on Agricultural Land*, Agricultural Development & Advisory Service paper No. 10, Ministry of Agriculture, Fisheries & Food, London.

60. Gemmell, R. P. (6 July 1973). Colliery shale revegetation techniques, *Surveyor*, **142**(4230), 27–9.

61. Down, C. G. (1975). Problems in vegetating metal-toxic mining wastes, in *Minerals and the Environment* (ed. M. J. Jones), Institution of Mining & Metallurgy, London, pp. 395–408.

62. Peterson, H. B. and Monk, R. (1967). *Vegetation and Metal Toxicity in Relation to Mine and Mill Wastes*, Circular 148, Utah Agricultural Experimental Station, Logan.

63. Funk, D. T. (1962). *A Revised Bibliography of Strip-mine Reclamation*, Miscellaneous Release No. 35, U.S. Department of Agriculture, Forest Service, Columbus, Ohio.

64. Kerry, F. E. (1974). The problems of derelict and despoiled land and its treatment, in seminar *Restoration and the Quarrying Industry*, Northumberland & Durham Roadstone & Concrete Training Association, Newcastle, 9 pp.

65. Ludeke, K. L. (1973). Soil properties of materials in copper mine tailings dykes, *Mining Congress Journal*, **59**(8), 30–7.

66. Beatty, R. A. (1966). The inert becomes 'ert', *Landscape Architecture*, **56**(2), 125–8.

67. Murray, D. R. (1971). *Vegetation of Mine Waste Embankments in Canada*, Internal Report MR71/31ID, Department of Energy, Mines & Resources, Mines Branch, Ottawa.

68. Arguile, R. T. (1972). Reclamation: five industrial sites in the East Midlands, *J. Institution of Municipal Engineers*, **98**(6), 150–6.

69. *Surface Mining and our Environment* (1967). Department of the Interior, Washington, D.C.

70. Davison, D. J. (1971). Restoration and reclamation of opencast sites, *Colliery Guardian, Annual Review*, **219**, 94–102.

71. Arguile, R. T. (1974). *Protection of the Environment*, Monograph No. 21, Institution of Municipal Engineers, London.

72. Batey, T. (Ed.) (1972). *The Restoration of Sand & Gravel Workings*, Agricultural Development & Advisory Service, Ministry of Agriculture, Fisheries & Food, London.

73. Oxenham, J. R. (1966). *Reclaiming Derelict Land*, Faber & Faber, London.

74. Stearn, E. W. (1971). Guidelines to successful reclamation, *Rock Products*, **70**(6), 64–72.

75. Developing big-scale reclamation systems (1969). *Coal Mining & Processing*, **6**(3), 40–2.

76. Skitt, J. (1972). *Disposal of Refuse and Other Waste*, Charles Knight, London, pp. 26–52.

77. Plass, W. T. (1974). Revegetating surface-mined land, *Mining Congress Journal*, **60**(4), 53–8.

78. Edmond, T. W. (1963). Some aspects of iron ore quarrying in Great Britain, *Mining Engineer*, **122**(34), 806–25.

79. Reclamation under semi-arid conditions (1970). *Coal Mining & Processing*, **7**(11), 34–7.

80. Smith, R. A. H. and Bradshaw, A. D. (1970). Reclamation of toxic metalliferous wastes using tolerant populations of grass, *Nature*, **227**, 376–7.

81. Hydroseed for efficient reclamation (1967). *Coal Mining & Processing*, **4**(4), 74–8.

82. Mitchell, B. A. (1957). Malayan tin tailings—prospects of rehabilitation, *Malay Forester*, **20**, 181–6.

83. Johnson, C. (1966). *Practical Operating Procedures for Progressive Rehabilitation of Sand and Gravel Sites*, University of Illinois, Urbana.

84. Bauer, A. M. (1970). *A Guide to Site Development and Rehabilitation of Pits and Quarries*, Report No. 33, Department of Mines, Ontario.

85. Arguile, R. T. (1975). Opencast coal mining in Britain—the first 32 years, *Colliery Guardian*, **223**(2), 46–52.
86. Whincup, G. T. (1970). Some aspects of opencast coal mining in South Wales, *Proc. South Wales Institute of Engineers*, **86**(1), 15–34.
87. *Derelict Land and its Reclamation* (1954). Ministry of Housing & Local Government, London.
88. *Report of the Committee on Planning Control over Mineral Working* (1976). Department of the Environment, London.
89. Plass, W. T. (1975). Changes in water chemistry resulting from surface-mining of coal on four West Virginia watersheds, in *3rd Symposium on Surface Mining & Reclamation*, National Coal Association, Louisville, Kentucky, pp. 152–69.
90. Dean, K. C. and Havens, R. (1973). Comparative costs and methods of stabilization of tailings, in *Tailing Disposal Today* (ed. C. L. Aplin and G. O. Argall), Miller Freeman, San Francisco, pp. 450–74.
91. Williams, R. E. (1975). *Waste Production and Disposal in Mining, Milling and Metallurgical Industries*, Miller Freeman, San Francisco, pp. 406–43.
92. Fairhurst, C. (1974). European practice in underground stowing of waste from active coal mines, in *1st Symposium on Mine and Preparation Plant Refuse Disposal*, National Coal Association, Louisville, Kentucky, pp. 145–6.
93. Atwood, G. (1974). The technical and economic feasibility of underground disposal systems, *ibid.*, pp. 147–60.
94. Poundstone, W. (1974). Problems in underground disposal in active mines, *ibid.*, p. 164.
95. Magnuson, M. O. and Baker, E. C. (1974). State of the art in extinguishing refuse pile fires, *ibid.*, pp. 165–82.
96. Brundage, R. S. (1974). Depth of soil covering refuse *v.* quality of vegetation, *ibid.*, pp. 183–5.
97. Capp, J. P. and Gillmore, D. W. (1974). Fly ash from coal burning power-plants: an aid in revegetating coal mine refuse and spoil banks, *ibid.*, pp. 200–11.
98. Sorrell, S. T. (1974). Establishing vegetation on acidic coal refuse materials without use of topsoil cover, *ibid.*, pp. 228–36.
99. Saperstein, L. W. and Secor, E. S. (1973). Improved reclamation potential with the block method of contour stripping, in *1st Research and Applied Technology Symposium on Mined-Land Reclamation*, National Coal Association, Pittsburgh, pp. 1–14.
100. Howland, J. W. (1973). New tools and techniques for reclaiming land, *ibid.*, pp. 42–67.
101. Allen, N. (1973). Experimental multiple seam mining and reclamation on steep mountain slopes, *ibid.*, pp. 98–104.
102. Grube, W. E., Smith, R. W., Singh, R. N. and Sobek, A. A. (1973). Characterisation of coal overburden materials and mine-soils in advance of surface mining, *ibid.*, pp. 134–52.
103. Sutton, P. (1973). Establishment of vegetation on toxic coal mine spoils, *ibid.*, pp. 153–8.

104. Riley, C. V. (1973). Furrow grading—key to successful reclamation, *ibid.*, pp. 159–77.

105. Peterson, J. R. and Gschwind, J. (1973). Amelioration of coal mine spoils with digested sewage sludge, *ibid.*, pp. 187–96.

106. Ruffner, J. D. (1973). Perfecting use of new plant materials for special reclamation problems, *ibid.*, pp. 233–42.

107. Jones, J. N., Armiger, W. H. and Hungate, G. C. (1973). Seed ledges improve stabilization of outer slopes on mine spoil, *ibid.*, pp. 250–8.

108. Thames, J. L., Crompton, E. J. and Patten, R. T. (1974). Hydrologic study of a reclaimed surface mined area on the Black Mesa, in *2nd Research and Applied Technology Symposium on Mined-Land Reclamation*, Louisville, Kentucky, pp. 106–16.

109. Watilquist, B. T., Dressler, R. L. and Sowards, W. (1975). Mined-land revegetation without supplemental irrigation in the arid Southwest, in *3rd Symposium on Surface Mining & Reclamation*, Vol. 1, Louisville, Kentucky, pp. 29–39.

110. Stamm, G. G. (1975). Preplanning for coal production and reclamation of mined lands, *ibid.*, pp. 62–9.

111. Gould, W. L., Miyamoto, S. and Rai, D. (1975). Characterizing overburden materials before surface mining in the Fruitland formation of Northwestern, New Mexico, *ibid.*, pp. 80–94.

112. Fisser, H. G. and Reis, R. E. (1975). Pre-disturbance ecological studies improve and define potential for surface mine reclamation, *ibid.*, pp. 128–34.

113. Pennington, D. (1975). Relationship of ground-water movement and stripmine reclamation, *ibid.*, pp. 170–8.

114. Connell, J. F., Plass, W. T., Contractor, D. N. and Shanholtz, V. O. (1975). Water quality models for a contour mined water-shed, *ibid.*, pp. 179–99.

115. Vogel, W. G. (1975). Requirements and use of fertilizer, lime and mulch for vegetating acid mine spoils, in *3rd Symposium on Surface Mining & Reclamation*, Vol. 2, National Coal Association, Louisville, Kentucky, pp. 152–70.

116. Caspall, F. C. (1975). Soil development on surface mine spoils in Western Illinois, *ibid.*, pp. 221–8.

117. Davies, I. V., Brook, C. and Arguile, R. T. (1975). Opencast coal mining: working, restoration and reclamation, in *Minerals and the Environment* (ed. M. H. Jones), Institution of Mining and Metallurgy, London, pp. 313–32.

118. Hartwright, T. U. (1975). Development of gravel-pit lakes for leisure purposes, *ibid.*, pp. 333–40.

119. Jeffrey, D. W., Maybury, M. and Levinge, D. (1975). Ecological approach to mining waste revegetation, *ibid.*, pp. 371–86.

120. Blesing, N. V., Lackey, J. A. and Spry, A. H. (1975). Rehabilitation of an abandoned mine site, *ibid.*, pp. 341–61.

121. *Pit and Quarry Textbook* (1967). Sand & Gravel Association, London, pp. 180–98.

122. Bach, D. A. (1973). Use of drip irrigation for vegetating mine waste areas, in *Tailing Disposal Today* (ed. C. L. Aplin and G. O. Argall), Miller Freeman, San Francisco, pp. 563–70.

123. Blunden, J. R., Down, C. G. and Stocks, J. (1974). The economic utilization of quarry and mine wastes for amenity purposes in Britain, in 4th *Mineral Waste Utilization Symposium* (ed. E. Aleshin), I.I.T. Research Institute, Chicago, pp. 255–64.
124. Down, C. G. (1974). The relationship between colliery-waste particle sizes and plant growth, *Environmental Conservation*, **1**(4), 281–4.
125. Stocks, J. (1974). Underground stone mining, *Quarry Management & Products*, **1**(4), 145–52.
126. Chicago reclaiming strip mines with sewage sludge (1972). *Civil Engineering —ASCE*, **42**(9), 98–102.
127. Douglas, J. S. (1969). Rehabilitation of mined areas, *Mining Magazine*, **120**(2), 106–13.
128. Jones, W. G. (1974). Reclamation today in Pennsylvania, *Coal Mining & Processing*, **11**(6), 33–58.
129. Zube, E. (1966). A new technology for taconite badlands, *Landscape Architecture*, **56**(2), 136–40.
130. Lord Zuckerman, Viscount Arbuthnot, Kidson, C., Nicholson, E. M., Warner, Sir F. and Langland, Sir J. (1972). *Report of the Commission on Mining and the Environment*, London, 92 pp.

11
Waste Utilisation

11.1. INTRODUCTION

Interest in possible ways of utilising the waste products of mining and mineral processing has been apparent since the 1920s but has come into increasing prominence since about 1960. This has coincided with the questioning of the concept of 'waste' in mining and other spheres of life. The topic of waste utilisation is clearly also linked with the realisation that mineral resources are finite, the dramatic increase in the scale of waste production, and awareness of the environmental hazards which arise from waste disposal. These hazards—land use conflict, air and water pollution, rehabilitation, etc.—are examined in other chapters of this book in terms of dealing with the environmental problems on the assumption that the wastes remain *in situ*. This is certainly the most realistic approach in the immediate and medium-term future. It does however conceal the possibility of making some positive use of the waste material. If such uses can be found they could offer two important benefits:

(i) Reduction of the environmental problems of conventional disposal.
(ii) Conservation of resources, by partly replacing natural materials. This could also reduce overall environmental impact by lessening the need to open new mines.

It must be emphasised that disadvantages may also occur in certain circumstances. For example, the excavation for re-use of the materials in an old waste tip which has become naturally revegetated and assimilated creates fresh nuisance and poses again the problem of rehabilitating the site. It has also been the case that some plants erected for processing waste products have been the source of far greater nuisance than that experienced at conventional plants using natural materials—so much so, in fact, that some installations have been forced to cease operation. Partly such experiences have arisen from the experimental nature of the processes employed, and hence a reluctance to expend capital on pollution control, but variation and contamination of the waste feed stock has also been found to create serious problems with the control of, particularly, aerial emissions. It is therefore important to balance the relative advantages and drawbacks before embarking upon a waste utilisation scheme.

There are three ways by which mineral wastes may be utilised. These are:

(i) In a more or less unaltered state, for undemanding applications such as bulk fill.
(ii) In an altered condition. This may range from simple crushing and sizing (for example, crushed slate waste used as an inert filler/extender in plastic), to major processing, such as cleaning and sintering of colliery wastes to produce lightweight aggregates.
(iii) As a source of values that were not fully recovered during the original processing, or which were not sought at that time. Reworking of old copper and lead tailings for unrecovered metals is an example of the former case, while the recovery of fluorspar from lead wastes is an example of a new use being discovered for a material which was useless when originally mined.

Any or all of these uses may be appropriate for any particular waste. For example, before using colliery waste for bulk fill, or as a feedstock for lightweight aggregate production, it is standard practice to separate out the residual coal which is then available for sale. Indeed, it is not uncommon to rework colliery wastes for small coal even if rehabilitation is the only anticipated end for the waste.

Whatever use is being considered for the waste, the examination normally comprises the following stages:

(i) locate, quantify and typify the wastes;
(ii) determine the existing uses, if any;
(iii) examine the technical problems of using the waste in a manufacturing process, determine the cost of the products, and their ability to meet specifications;
(iv) investigate the techniques and economics of value recovery;
(v) for the products or values obtained determine the regional demand which may exist for them, and the economics of satisfying that demand, particularly in view of competition from established materials.

11.2. QUANTITIES AND TYPES OF WASTE

Information on the quantities and locations of wastes currently produced, estimates of backlog stocks, and forward projections, is limited. Tables 69 and 70 provide some information for the U.K. and U.S.A., but caution is needed in interpreting these figures for they are not strictly comparable. Overburden and waste rock are excluded in most cases; British open cast coal production entails the annual generation of some 300 million tonnes

TABLE 69
MINERAL WASTE RESOURCES IN THE U.K. (Gutt et al.[41])

Waste	Annual production of waste (million tonnes)	Backlog of waste (million tonnes)
Colliery shales	40–50	3000
Tin tailings	0·75	—
China clay (quartz, mica, overburden)	22	280
Slate waste	1·2	300
Red muds	0·1	—
Fluorspar tailings	0·23	—

of such wastes but because these are backfilled into the pit they are not classed as wastes for the present purpose. The figure of 40–50 million tonnes referred to in Table 69 includes only process wastes plus development wastes from underground collieries, i.e. wastes which are tipped. On the other hand, wastes from metal mining listed in Table 70 comprise only the tailings (with some exceptions); if for example copper overburden and waste rock were also included, the annual production of wastes from that source would be around 800 million tonnes.

TABLE 70
MINERAL WASTE RESOURCES IN THE U.S.A.
(Miller and Collins[42])

Waste	Annual production of waste (million tonnes)	Backlog of waste (million tonnes)
Alumina red/brown mud	5–6	50
Phosphates slimes	20	400
Anthracite coal refuse	10	1000
Bituminous coal refuse	100	2000
Asbestos tailings	1	10
Copper tailings	200	8000
Feldspar tailings	0·25–0·5	5
Gold mining waste	5–10	100
Iron ore tailings	20–25	800
Lead tailings	10–20	200
Taconite tailings	150–200	4000
Zinc tailings and smelter waste	10–20	200

Moreover, neither table includes wastes from the winning of many non-metallic minerals, such as limestone, granites, sandstones, etc. Considerable, but unquantified, tonnages of overburden, waste rock, etc., are produced from these sources, much of which has to be tipped.

11.3. EXISTING USES AND POSSIBILITIES FOR UPGRADING

The following sections briefly review the position with regard to some of the main types of waste.

11.3.1. Colliery Waste

Colliery wastes consist mainly of sedimentary mudstones, siltstones, shales, seat earths, sandstones and limestones, together with varying proportions of unrecovered coal or carbonaceous matter. The major waste source is usually coal preparation—hand picking, washing, dense medium separation, froth flotation and jigging. There is also run-of-mine dirt, from drivages, faults, etc., which goes direct for tipping. Some material, mainly washery coal fines, is kept as a slurry and impounded, but most waste can be surface tipped. The mineral content of the spoil varies, but is commonly quartz, mica, clays (illite and kaolinite), pyrite and carbonates of calcium, magnesium and iron.[1,2,5]

Changes after deposition are numerous. Weathering reduces the particle size and, after a period of years, lessens the sulphate levels, or spontaneous combustion may occur. Burnt spoils ('red dog') have low levels of combustible matter and other constituents which are decomposed by heat, and tend to be physically and chemically more stable than raw spoils. A comparison between the two types is given in Table 71. Very frequently a tip contains a mixture of both burnt and unburnt spoils. However, better tipping methods are almost eliminating spontaneous combustion of tips, so that the resources of burnt spoil are likely to grow proportionately less. A survey of spoil owned by the National Coal Board in Britain in 1966 showed that there were 1184 million tonnes of unburnt spoil, 628 million tonnes of burning spoil and 316 million tonnes of burnt spoil. Since that time, and allowing for privately owned old tips, some 900 million tonnes should be added to these figures.[41]

About 50 million tonnes of coal waste is surface tipped each year in Britain and the total positive utilisation is about 7 million tonnes. About 80–90% of this is used for bulk fill, most of which is backlog waste. For many years burnt spoil has been in great demand at road embankments and similar civil engineering works.[24] Attempts to increase the utilisation of unburnt spoil are also showing success. Research into methods of tipping has shown that if the spoil is compacted, the percentage of voids can be reduced to the point that access of air is minimal and spontaneous

TABLE 71

COMPARISON OF UNBURNT AND BURNT COLLIERY SPOILS
(Gutt *et al.*[41] and Gutt[43])

	Burnt spoils (% by weight)	Unburnt spoils (% by weight)
SiO_2	45·4–60·2	42·9–51·9
Al_2O_3	21·2–31·3	19·4–21·5
Fe_2O_3	3·9–13·4	4·7–6·1
TiO_2	0·17–0·24	0·90–1·03
CaO	0·36–6·3	0·66–1·30
MgO	0·82–2·88	1·21–1·49
Na_2O	0·20–0·65	0·12–0·44
K_2O	2·06–3·45	0·79–3·00
SO_3	0·10–4·66	0·35–0·49
S	0·01–0·10	Trace–0·02
Loss on ignition	1·9–6·3	16·13–24·12

combustion highly improbable. Proper compaction also eliminates water infiltration and hence acidic drainage. In certain respects, unburnt material is proving superior to burnt spoil, notably in its lower sulphate levels (and hence smaller likelihood of corrosion of concrete structures) and ease of compaction.[34]

Burnt anthracite spoil has been proved to be a good anti-skid material in Pennsylvania for use on highways in winter.[16,17] Some 30 million tonnes of the material exist to supply an annual consumption of about 1·5 million tonnes.

The earliest uses of coal wastes for the manufacture of building products were as raw materials in brick manufacture. This was standard practice in many British coalfields for a century or more. The process normally consisted of grinding the shales to a powder, pressing into moulds and then low-temperature firing in kilns. Because the waste often had a high coal content firing was cheap and required little fuel. However, the quality of the bricks was often very poor, with a high reject rate, and in some cases bricks needed to be rendered because they were not weatherproof. Nonetheless, brick manufacture is still an outlet for about 600 000 tonnes/a of colliery wastes although the quantity is declining.

Cement manufacture is a minor use of shales in Britain (130 000 tonnes/a), although at only one cement works is this usage regular. It is found that the coal content is so variable that it has to be removed.

A minor but interesting use of lignite wastes in North Dakota, U.S.A., is as soil conditioners, employing a naturally oxidised lignitic material

'Leonardite' as an artificial humus. Similar trials, many successful, have been made throughout the world.

A possible major future use of coal wastes is in the sintered form as lightweight aggregate.[32] Large numbers of studies have been made on the technical and economic possibilities of manufacturing lightweight aggregates from coal shales. The process depends upon whether or not the shales will 'bloat' at high temperatures. Bloating is an expansion process which requires the simultaneous development of a glassy phase over a wide range of temperature, plus the generation of gases (SO_2, SO_3, CO_2 and O_2) by mineral decomposition, in sufficient quantities and at the appropriate temperatures. Unless a glass phase of suitable viscosity occurs at the same time and temperature as gas evolution, a cellular structure will not be formed. Either the gas escapes, or the viscosity is too low to retain the gas, or the cellular structure which forms is too open for the aggregate to have adequate strength.

The stages of the reaction in clays and shales are as follows:

(i) up to 200°C, drying and removal of free water;
(ii) 200–480°C, dehydration of absorbed water;
(iii) 480–700°C, dehydration of water in chemical combination;
(iv) 500–1000°C, oxidation of iron, sulphur, carbon, etc.;
(v) 500–1000°C, dissociation and reduction of sulphates, carbonates, oxides, etc.;
(vi) 900–1300°C, glass formation;
(vii) 1150–1320°C, pyroplastic stage at which bloating occurs;
(viii) 1300–1500°C, melting occurs with loss of cellular structure.

The mineralogy of coal spoils is complex and there is no agreement as to which minerals evolve the gases necessary for bloating. Carbonates, sulphides and ferric oxide have been implicated, but the probability is that several minerals are the gas sources. A particular problem is that the presence of coal in large quantities hinders the process; after glass formation oxygen cannot reach the coal particles which remain and cause the material to be friable. Burning off the coal before firing the waste more or less eliminates bloating and it is therefore necessary either to ensure complete coal combustion during firing (*e.g.* by passing air through the bed) or, as has been tried successfully, undertake a dense medium separation stage as a preliminary treatment.

In Britain there are several lightweight aggregate plants in commercial production with a total capacity of about 1 million tonnes/a of spoil. The plant at Gartshore, Scotland (200 000 tonnes/a) uses anthracite waste from an abandoned mine. The waste is crushed to 7·5 cm (3 in), followed by a dense medium separation to remove coal and thus assist in controlling the quality of the feedstock. Washed waste is further crushed, blended with returned fines, pelletised at 10 mm ($\frac{3}{8}$ in) and fed to a moving sinter bed. Sinter cake is produced and crushed to size. Much of the output is utilised in a lightweight concrete-block plant. The economic success of the plant does not depend upon selling the floated anthracite.

A development of this use is a plant opened in 1975 at Snowdon colliery, Kent, with an output of 500 000 tonnes/a.[36] A similar plant is planned in South Wales.[35] The Snowdon plant, located in an area of severe aggregate shortage, will at first take existing tipped waste for its feed, but it is hoped that it will eventually become an 'on-line' extension of waste production at the colliery.

Another plant, in Derbyshire, has an output of about 500 000 tonnes/a and uses a feed of 50 % colliery shale and 50 % natural clays. This is because the plant has no coal separation facilities and it is necessary to dilute the raw waste with clay to reduce the proportion of coal.

Other possible manufactured uses for colliery spoil include dense aggregates. Experimental work on both untreated and heat treated wastes is in progress. Experiments have also been done on the manufacture of skid-resistant roadstone. Burnt spoil alone has been turned into a road-stone with a PSV of 70, but only on a laboratory scale. In the U.S.S.R., sized tailings are utilised as a combustible additive in brick manufacture, some 78 000 tonnes being so used in 1969.

In view of the massive backlog stocks of colliery wastes, some 6000 million tonnes in the U.K. and the U.S.A. alone, it is scarcely conceivable that any uses on a sufficiently large scale that would remove these quantities could be found. The major use, as bulk fill, does however avoid the environmental damage of existing tips, reclaim some usable fuel and obviate the excavation of borrow pits. There is likely to be scope for an increase in the production of lightweight aggregates, while developments in economic dense aggregate production could extend the markets still further. Certainly recent British work concluded that colliery spoil was the only waste material that could make a significant impact upon requirements for natural aggregates.

11.3.2. China Clay Waste

Winning of china clay in Britain entails the production of several distinct waste types: coarse quartz sand; fine clay and mica; overburden; and internal waste rock (stent). On average the production of 1 tonne of china clay produces 9 tonnes of waste, mainly comprising:

3·7 tonnes coarse sand
2 tonnes overburden
2 tonnes stent
0·9 tonnes micaceous residue

Details of the composition of the process wastes (sand and mica) are given in Table 72: the overburden and stent are not utilised. About 10 million tonnes/a of coarse sand is produced, composed of quartz (60–80%), feldspar (1–15%), tourmaline (2–10%) and mica (0·5–15%). Some 2 million tonnes/a of micaceous wastes is produced.

1 million tonnes/a of coarse sand is utilised as bulk fill and for the manufacture of building materials, in the latter case using the current production of sand rather than the backlog. The products manufactured are concrete blocks and calcium silicate bricks. Screening the coarse sand enables four grades to be obtained, which supply most of the local markets.[2]

TABLE 72

PROPERTIES OF CHINA CLAY WASTES
(Gutt et al.[41])

	Coarse sand (% by weight)	Micaceous residue (% by weight)
SiO_2	75–90	50·3
Al_2O_3	5–15	32·7
Fe_2O_3	0·5–1·2	2·37
TiO_2	0·05–0·15	0·11
CaO	0·05–0·5	0·07
K_2O	1·0–7·5	5·3
Na_2O	0·02–0·75	0·24
MgO	0·05–0·5	0·35
Loss on ignition	1–2	8·3

Many studies have been undertaken with a view to creating new outlets. For concrete aggregate, the presence of mica renders the sand relatively unsuitable. The white colour of the waste may enable its beneficial use in undemanding applications for white concrete. Other minor uses have been studied.

11.3.3. Slate Wastes

In Britain slate production entails an overall proportion of waste: saleable product of about 20:1. The waste comprises chert, igneous rocks, small quantities of overburden, and slate fragments which form the bulk of the waste. The mineralogical composition of slate is mainly chlorite, sericite, quartz, haematite and rutile, the chemical composition being largely SiO_2 (55%), Al_2O_3 (18%) and FeO (5–6%).

The flaky texture of slate renders its use as aggregate difficult. Currently the main uses are as a crushed and sized inert filler for use as a bitumen extender, and in rubber, paint, linoleum and plastic products. Slate granules are sold as a surfacing material for roofing felt. Other than these, bulk fill is the main outlet.[2,37,38]

A use which has proved technically practicable but economically unsatisfactory is for expanded aggregate production. Tests have shown that most British slates expand at 1150°C and some increase in thickness by up to eight times. A 150 000 tonnes/a rotary kiln plant was put into production in North Wales, but is now closed.[44]

11.3.4. Aluminium Red Mud Waste

Conversion of bauxite to alumina by the Bayer process yields a red mud waste which can be sub-divided into black sands (8–18% by weight of the total waste) and fine red mud. A combination Bayer/sinter process yields a brown mud of dicalcium silicate. The total waste is about 50% of the ore by weight. There are marked variations in the waste depending upon the source of the bauxite (*see* Table 73). The muds are discharged as slurries containing about 20% solids and have very slow settling times sometimes not exceeding 50% solids after many years. This hampers utilisation.

TABLE 73

COMPOSITION OF RED MUD WASTES

(Pincus,[3] and Gutt et al.[41])

Per cent by weight	U.S.A. ore	Jamaican ore	Brown mud	Ghanaian ore
Al_2O_3	26·5	3–5	6·4	16
Fe_2O_3	10·7	75–80	6·1	53
SiO_2	22·9	—	23·3	6
CaO	8·1	—	46·6	—
Na_2O	11·8	—	4·1	5
TiO_2	3·3	—	3·0	8
SO_3	2·8	—	0·5	—
Loss on ignition	12·9	—	7·3	9

There are no regular or large scale uses of red muds at present, but studies have included uses such as thermal insulation, controlled porosity materials, concrete additives, acid soil conditioner (the muds are about pH 12), and pigments.[39] An important process is to use bauxite as a source of iron oxides, which are extracted from high-iron ores, but this represents a modification of bauxite processing methods rather than a beneficial use of the red mud tailings.

11.3.5. Phosphate Wastes

In Florida, the U.S.A.'s main source of mined phosphates, the wastes from processing comprise silica sand tailings and a high-phosphate slimes fraction.[45] The composition of the slimes is given in Table 74. The slimes are no more than 5% solids and, although settling to about 15% solids is rapid, subsequent settling is extremely slow (no more than 30% solids after a decade).

TABLE 74

MINERALOGICAL COMPOSITION OF
PHOSPHATE SLIMES
(Cox[4])

Mineral	Per cent by weight
Calcium fluorapatite	20–25
Quartz	30–35
Montmorillonite	20–25
Attapulgite	5–10
Wavellite	4–6
Feldspar	2–3
Zircon, rutile, etc.	2–3
Dolomite	1–2
Other	0–1

Because about 40% of the mined phosphate is lost into the slimes, there is considerable interest in value recovery. The only alternative uses which have been considered are lightweight aggregate manufacture, bricks and pipes.[11,21] Only the first has so far shown any promise, with good quality aggregates being produced by pelletising dried slimes and firing in a rotary kiln at 1050–1100°C. Both the aggregate, and lightweight concrete incorporating it, met A.S.T.M. specifications. Bricks and pipes suffered from excess porosity and bloating.

11.3.6. Iron Ore Wastes

Waste production from iron ore mining is a particular problem in the U.S.A.'s Lake Superior District. Originally the ores worked were so-called 'natural' ores, brown and red haematite of high-grade which had been geologically altered from the original taconite, a low grade iron ore. Working of natural ores produced glacial overburden wastes and the ores were shipped raw. Depletion of these resources has led to the exploitation of the taconite ores which require beneficiation. The wastes from this process are slurries of 15–20% fine-particle solids. Chemically and mineralogically they are complex and variable, commonly including magnetite, haematite, goethite, quartz, chert, siderite and a wide variety of iron silicates. Combined waste production of the natural and taconite operations is now in excess of 200 million tonnes/a. Because of the considerable residual iron content (up to 20% Fe) of the taconite tailings, interest has centred on value recovery.

Other uses on a small scale have included fill, base-course aggregate, de-icing material and bituminous concrete aggregate. None has proved viable on any significant scale.[23,29]

In Quebec, iron mine tailings have been considered as a source for brick manufacture. After grinding to 50% minus 74 μm, a binding agent (1% calcium lignosulphonate) was added, the bricks pressed and fired at 1100–1200°C. A satisfactory product has been obtained.[6,10,22]

11.3.7. Non-ferrous Metal Tailings

The exploitation of low-grade metal deposits has resulted in a marked increase in the problems of tailings disposal. In particular, the quantities are very much greater, and the particle size very much smaller, than was the case even two decades ago. The magnitude of the disposal problem is illustrated by the fact that copper tailings alone comprise half of the total backlog wastes in the U.S.A. This also indicates the improbability of any waste utilisation processes making a significant impact upon the problem as a whole: the tonnage of tailings produced at any single major copper mine in the U.S.A. is equal to the tonnage of material excavated for, say, brick manufacture in the entire country. Nonetheless the existence of large stockpiles of accessible, finely-ground material has prompted extensive examination of numerous possible uses.[7,8,10,11,18,20,22]

The application of siliceous copper tailings to brick manufacture has been studied. Standard brick clays have been found able to sustain dilution by up to 50% tailings without affecting the structural properties of the product. Minor problems include alteration in brick colour, a greater sensitivity to firing temperature, and the need to separate out pyrite from the tailings before use. This is primarily because of the air pollution problems which would otherwise result.

Use of tailings as a substitute for natural sand in glass manufacture has been experimented with, an important difficulty being the presence of iron minerals which impart undesired colour to the glass. Trials on the manufacture of crystallised glass have been successful.

Work on the utilisation of Californian gold tailings illustrates some of the problems encountered.[12] The use envisaged was the manufacture of calcium silicate bricks and Table 75 compares the criteria normally employed in selecting natural sands for this use, with the ranges of properties found in gold tailings. No tailings sample analysed met every criterion. However, when test bricks were manufactured, it was discovered that the materials selection criteria in use in the industry did not entirely relate to

TABLE 75

COMPARISON OF NATURAL SANDS AND GOLD TAILINGS FOR MANUFACTURE OF CALCIUM SILICATE BRICKS

(Mindess and Richards[12])

	Natural sand criteria	Range in selected gold tailings
% particles >4·8 mm	0	0–37·2
% particles <150 μm	10–15 max	3·9–97·9
% clay and silt	10 max	0·6–48·3
Total soluble salts %	1 max	0·05–0·82
Total alkali %	3 max	0·59–12·08
Quartz %	50 min	28·0–95·0
Feldspar %	15 max	4·0–50·0

the quality of the tailings bricks produced; some tailings which had originally been rejected as wholly unsuitable did in fact make acceptable bricks. This is indeed a general truth in waste utilisation: marked differences between the waste and traditional natural materials may not be important (if the process is suitably altered) and may even indicate the limited factual basis for some of the standards hitherto applied.

A number of studies have been carried out on lead/zinc wastes in South-West Wisconsin,[15] where the main wastes are dolomitic limestone and iron pyrites; the latter was not in demand as such, and the only possible products from the former were raw dolomite, dead-burned dolomite, and refractory magnesia. Minor utilisation of raw dolomite for fill and railroad ballast was the only usage undertaken, although a sulphuric acid plant had once operated using the pyrites. Whatever usage was developed, it was thought essential to separate the pyrite from the dolomite, for which flotation was found most satisfactory. Studies however failed to discover an economic outlet for the wastes.

With the exception of minor uses for bulk fill, the most widespread positive use for tailings is as hydraulic backfilling for support purposes in underground mines. This well-known use normally incorporates a cycloning stage to remove a proportion of undesirable fines. Selection of material for backfill is still to some extent a rule-of-thumb procedure.

11.4. VALUE RECOVERY

There are large numbers of examples of the recovery of saleable minerals from the waste heaps produced by former mining. The circumstances under which this occurs may include any or all of the following:

(i) Technological improvements in processing methods permit the recovery of minerals previously rejected as unprocessable.

(ii) Changes in mineral prices which increase the value of the residual mineral content to the point at which reworking becomes viable. This is particularly found with non-ferrous metal wastes.[13]

(iii) Recovery of minerals for which no use existed at the time of the original processing.

These three factors do of course imply an interval of several decades in most cases between the original deposition of the waste and its reworking.

A typical example of value recovery is the Kadina, South Australia, copper recovery plant.[46] Between 1860 and 1923, when the copper mines in the area were operating, some 12 million tonnes of ore were produced. In 1969 the waste dumps were examined and found to contain $1-1\frac{1}{2}\%$ copper; after laboratory tests had been carried out, a small (180 tonnes/day) plant was constructed. This was later increased to 300 tonnes/day and most recently to 700 tonnes/day. Two different waste types existed, one of which required grinding to liberate the copper mineral. Excavation of these two waste types is co-ordinated to give a mixed feed of 0.6% copper overall, which will ensure the simultaneous exhaustion of both wastes. After grinding in a ball mill and combining with unground waste, the feed is floated in a bank of 10 roughing cells, after which it passes to scavenging, cleaning and recleaning banks. The concentrate grades $20-25\%$ copper, and overall about 9000 tonnes of copper in concentrates will be produced.

Coal recovery from waste heaps is likewise a long established process in both the U.K. and the U.S.A.[30,33] In Pennsylvania, anthracite has been recovered from tips by washing since 1894. Commonly, dense medium separation is employed for coal recovery. Another possibility is the recent development of combustion plants able to burn fuels with up to 40% ash and a power station (250 000 kW) for this purpose has been mooted in Pennsylvania, which would consume some 65 million tonnes of high-anthracite wastes. In Britain a self-sustaining fluidised-bed process has been developed to utilise coal tailings containing 50% moisture. Fuel is provided by the residual coal content of the waste, and the resultant ash may have potential in the manufacture of building products, as does pulverised fuel ash from coal burning utilities at the present time.

An example of the environmental problems associated with reworking is provided by fluorspar production in Derbyshire, U.K. Here numerous fluorspar dumps exist, the product of lead mining over several centuries; fluorspar was a waste product until about 1890. Of the U.K.'s annual output of about 200 000 tonnes of fluorspar a significant proportion comes from reworking of dumps. The excavation and removal of old tips which had become part of the rural scene has caused some public complaint, and exemplifies a situation in which the disadvantages of recovery could be considered as outweighing the advantages.

Cases of value recovery are also found in the wastes of low value minerals. Reprocessing by screening and/or washing of old waste tips or scalpings stockpiles can sometimes recover significant quantities of saleable clean stone.[40]

A most important aspect of value recovery is that, although it possesses obvious benefits in terms of resource conservation, in most cases it does not significantly reduce the waste disposal problem. In very few cases is more than a marginal reduction in waste volume obtained as a result of resource recovery and in the case of metal recovery effectively no reduction is achieved. Although the toxicity of wastes may be thus lessened, and the opportunity of applying modern, less environmentally-damaging disposal methods may arise, reprocessing may, as when grinding coarse wastes, result in a waste which is more difficult to handle than was originally the case.

11.5. DEMAND, TRANSPORT AND ECONOMICS

The results of the extensive studies undertaken on waste utilisation can be summarised as follows:

(i) For the two bulk wastes—coal shales and metalliferous tailings—which are of the greatest environmental concern, it is technically feasible to use them as bulk fill in many situations, and it is usually environmentally advantageous (*e.g.* by obviating the excavation of borrow pits) so to do.

(ii) Many different wastes appear to have great technical potential for use in the manufacture of building products or lightweight aggregate.

Despite this technical suitability for many different purposes, positive waste utilisation is negligible at the present. This relates either to the cost of producing the saleable product, the extra cost incurred in using it over the cost of the equivalent natural product, the lack of local demand, the cost of transporting wastes to areas of demand where they have then to compete with locally excavated natural materials, and to problems in meeting existing specifications.

The cost of manufacturing building materials (such as bricks) from wastes often compares favourably with the use of equivalent virgin materials. Certain extra costs may be incurred by virtue of the variability of the feedstock, which is likely to be greater than in many natural resources, but this may well be more than offset by the low cost of purchase of waste land and waste minerals relative to virgin deposits.

It does not appear that manufactured lightweight aggregate can be produced at prices enabling it to be sold in direct competition with natural products. Experiments upon china clay wastes for this purpose indicate a price differential of at least four times the cost of natural resources. Sintering processes have the particular problem of being fuel-intensive and consequently sensitive to price fluctuations in this commodity. Existing lightweight aggregate plants are most successful in areas lacking natural resources so that transport costs provide some protection for the manufactured product.

In most cases, there seems to be no extra cost incurred as a result of using a waste or manufactured product assuming that the delivered price is competitive. Small increases in cost are likely in cases such as the use of colliery waste for highway construction, which may entail extra compaction or the use of sulphate-resistant concrete.

Lack of local demand for the waste is a crucial factor in the current lack of exploitation of waste resources. In Britain, the reserves of slate and china clay wastes are in relatively unpopulated areas; so too in the case of many U.S. base metal mines. It is almost universally found that no significant local markets either exist, or can be created, to absorb any large proportion of the available wastes.

This therefore indicates the examination of markets remote from the site of the waste. In such cases the high cost of transporting low value waste products is a major inhibiting factor in a free market.[2,26] Studies undertaken in Britain on the possibility of selling china clay sands up to 400 km (250 miles) away from their source indicated that delivered costs would be virtually double those of locally produced sands. The magnitude of the differential will vary according to the efficiency of the transport mode and the circumstances under which the waste is available.

An important question is that a full cost-benefit analysis of long-distance waste transport and its alternatives is lacking. That the matter is not as simple as it appears can be gathered from considering proposals made at the time when the abortive third airport for London was planned.[26] The airport would have required 400 million tonnes of fill material, which could either have been locally-dredged material costing £0·40/tonne or else mining wastes. Colliery waste could have been delivered for a total cost of £1·50/tonne over distances of 320–480 km (200–300 miles), and at the sources a total of 47 km² (18 square miles) of waste tip would have been removed, at a cost equivalent to £98 000/ha (£40 000/acre). This was at least ten times the cost of rehabilitating the waste *in situ* and, although the calculation does not take into account the benefits of avoiding a massive dredging operation, it nonetheless becomes doubtful that the cost incurred in long distance transport of the waste would have resulted in benefits that could not have been obtained by more modest expenditure.

This example clearly demonstrates the inability of mineral wastes to compete with conventional materials unless some form of subsidy is available. Recent British practice has in fact been to subsidise the use of waste materials on new road contracts. The system is one of dual tendering, in which the contractor submits two bids, one based on the use of his preferred material for bulk fill and the other based upon waste utilisation. This permits an informed examination of the cost of waste utilisation, and many new contracts have been let and executed in which colliery wastes have been the specified fill. In one case, in excess of 2 million tonnes of waste were utilised.

The final problem is in connection with compliance with building standards. Very frequently mining wastes or products therefrom do not wholly comply with official standards drawn up for natural materials. There is considerable scope for adjustment of standards, their relaxation where appropriate, or the derivation of separate standards for wastes.

REFERENCES

1. Gibson, J. (1976). Making use of colliery waste materials, *Recycling and Waste Disposal*, 38–42.
2. Verney, R. B. (1976). *Aggregates: the Way Ahead* (Report of the Advisory Committee on Aggregates), H.M.S.O., London.
3. Pincus, A. G. (1968). Wastes from processing of aluminium ore, in *Proc. 1st Mineral Waste Utilization Symposium* (ed. M. A. Schwartz), I.I.T. Research Institute, Chicago, Illinois, pp. 40–9.
4. Cox, J. L. (1968). Phosphate wastes, *ibid.*, pp. 50–62.
5. Sullivan, G. D. (1968). Coal wastes, *ibid.*, pp. 62–5.
6. Fine, M. M. and Heising, L. F. (1968). Iron ore waste-occurrence, beneficiation and utilization, *ibid.*, pp. 67–72.
7. Bingham, E. R. (1968). Waste utilization in the copper industry, *ibid.*, pp. 73–8.
8. Dean, K. C. (1968). Utilization of mine, mill and smelter wastes, *ibid.*, pp. 138–42.
9. McNay, L. M. (1970). Mining and milling waste disposal problems—where are we today? in *Proc. 2nd Mineral Waste Utilization Symposium* (ed. M. A. Schwartz), I.I.T. Research Institute, Chicago, Illinois, pp. 125–30.
10. Nakamura, H. H., Aleshin, E. and Schwartz, M. A. (1970). Utilization of copper, lead, zinc and iron ore tailings, *ibid.*, pp. 139–48.
11. Cutler, I. B. (1970). Ceramic materials from mineral wastes, *ibid.*, pp. 149–54.
12. Mindess, S. and Richards, C. W. (1970). Criteria for the selection of mineral wastes for use in the manufacture of calcium silicate building materials, *ibid.*, pp. 155–66.
13. Kennedy, A. D. (1970). Recovery of copper from Michigan Stamp Sands, *ibid.*, pp. 167–76.
14. Stoops, R. F. and Redeker, I. H. (1970). North Carolina feldspar tailings utilization, *ibid.*, pp. 177–80.

15. Heins, R. W. and Geiger, G. H. (1970). Potential utilization of mine waste tailings in the Upper Mississippi Valley lead-zinc mining district, *ibid.*, pp. 181–94.
16. Spicer, T. S. and Luckie, P. T. (1970). Operation anthracite refuse, *ibid.*, pp. 195–204.
17. Charmbury, H. B. and Maneval, D. R. (1972). The utilization of incinerated anthracite mine refuse as anti-skid highway material, in *Proc. 3rd Mineral Waste Utilization Symposium* (ed. M. A. Schwartz), I.I.T. Research Institute, Chicago, Illinois, pp. 123–8.
18. Dean, K. C. (1972). Reclamation of mineral milling wastes, *ibid.*, pp. 139–42.
19. Faddick, R. R. (1972). A data bank on the transport of mineral slurries in pipelines, *ibid.*, pp. 143–52.
20. Pettibone, H. C. and Kealy, C. D. (1972). Engineering properties and utilization examples of mine tailings, *ibid.*, pp. 161–70.
21. Vasan, S. (1972). Utilization of Florida phosphate slimes, *ibid.*, pp. 171–8.
22. Collings, R. K., Winer, A. A., Feasby, D. G. and Zoldners, N. G. (1974). Mineral waste utilization studies, in *Proc. 4th Mineral Waste Utilization Symposium* (ed. E. Aleshin), I.I.T. Research Institute, Chicago, Illinois, pp. 2–12.
23. Emery, J. J. and Kim, C. S. (1974). Trends in the utilization of wastes for highway construction, *ibid.*, pp. 23–32.
24. Maneval, D. R. (1974). Utilization of coal refuse for highway base or sub-base material, *ibid.*, pp. 222–7.
25. Toyabe, Y. and Matsumoto, G. (1974). Manufacturing ceramic goods out of mining wastes, *ibid.*, pp. 240–4.
26. Blunden, J. R., Down, C. G. and Stocks, J. (1974). The economic utilization of quarry and mine wastes for amenity purposes in Britain, *ibid.*, pp. 255–64.
27. Bean, J. J. (1973). Tailing as an orebody, in *Tailing Disposal Today* (ed. C. L. Aplin and G. O. Argall), Miller Freeman, San Francisco, pp. 606–14.
28. Sims, W. N. (1973). Remining of tailings by hydraulicking and other methods, *ibid.*, pp. 615–33.
29. Emery, J. J. (1975). Use of mining and metallurgical wastes in construction, in *Minerals and the Environment* (ed. M. J. Jones), Institution of Mining & Metallurgy, London, pp. 261–72.
30. Platt, J. and Hellewell, E. G. (1975). Coal-tip recovery, *ibid.*, pp. 273–90.
31. Building Research Establishment (1975). *Utilization of Industrial By-Products and Waste Materials*, Library Bibliography 255, B.R.E., Garston.
32. Edwards, J. A. (1962). Coal refuse for building materials, *Colliery Guardian*, **205**, 340–6.
33. Coal reclamation from waste heaps (1969). *Mining & Minerals Engineering*, **5**(4), 41–2.
34. Sherwood, P. T. and Ryley, M. D. (1970). The effect of sulphates in colliery shale on its use for roadmaking, *Surveyor*, **136**(4083), 37–44.
35. Coal shale reclamation (1973). *Materials Reclamation Weekly*, **123**(20), 27.
36. £2 million lightweight aggregate plant for Kent colliery (1973). *Cement Lime & Gravel*, **48**(9), 196.
37. Trauffer, W. E. (1967). Vermont light aggregate's new plant serves New England area, *Pit & Quarry*, **59**(8), 104–8.

38. Coleman, E. H. and Nixon, P. J. (1974). *A Survey of Possible Sources in Wales of Raw Materials for the Manufacture of a Lightweight Expanded Slate Aggregate*, Building Research Establishment, publication CP 78/74, B.R.E., Garston.
39. Fursman, O. C. and Mauser, J. E. (1970). *Utilization of Red Mud Residues from Alumina Production*, U.S. Bureau of Mines, Report of Investigations 7454, Washington.
40. Concrete aggregate produced from quarry waste (1973). *Cement Lime & Gravel*, **48**(4), 82.
41. Gutt, W., Nixon, P. J., Smith, M. A., Harrison, W. H. and Russell, A. D. (1974). *A Survey of the Locations, Disposal and Prospective Uses of the Major Industrial By-products and Waste Materials*, Building Research Establishment, publication CP 19/74, B.R.E., Garston.
42. Miller, R. H. and Collins, R. J. (1974). Waste materials as potential replacements for highway aggregates, in *Proc. 4th Mineral Waste Utilization Symposium* (ed. E. Aleshin), I.I.T. Research Institute, Chicago, Illinois, pp. 50–61.
43. Gutt, W. (1974). *The Use of By-products in Concrete*, Building Research Establishment, publication CP 53/74, B.R.E., Garston.
44. Harrison, W. H. (1974). *Synthetic Aggregate Sources and Resources*, Building Research Establishment, publication 100/74, B.R.E., Garston.
45. *The Florida Phosphate Slimes Problem* (1975). U.S. Bureau of Mines, Information Circular 8668, Washington, D.C.
46. Copper from mine tailings: re-treating a million tons of waste (1971). *Australian Mining*, December, 64–5.

12

Subsidence

12.1. INTRODUCTION

The removal of material from the earth's crust by underground mining creates an obvious potential for ground movement and consequential deformation of the surface. The circumstances under which this may arise vary widely, the main parameters being:

(a) The geometry of the mineral deposit—this can range from thin stratified flat seams to steeply dipping irregular veins or lenses, and massive orebodies large in all three dimensions.
(b) The method of mining—there may be partial or total extraction, with or without artificial support, and caving of the roof or hanging wall may be undesired or deliberately induced.
(c) The nature of the mineral deposit and the overlying strata—there is wide variety in the physical characteristics, hydrology, geology, depth of cover and other factors pertinent to ground behaviour.

Despite significant advances in the science of rock mechanics in the last two decades, analysis and prediction of stress and strain in large rock masses remains formidably complex because of such factors as anistropy, lack of homogeneity, and the presence of geological discontinuities. These problems have so far prevented the development of a unified phenomenological theory capable of predicting satisfactorily ground movement and surface subsidence in the wide range of mining situations described above.

The environmental importance of subsidence is related to three main factors:

(i) the surface land area affected;
(ii) the nature of the land uses within the affected areas;
(iii) the type and magnitude of ground movement.

A large majority of the investigations into surface subsidence so far undertaken have concentrated upon areally extensive mining in countries of high population density. Underground extraction of seam deposits, such as coal, normally requires that large areas are undermined if significant tonnages are to be produced. Particularly in Western Europe, this type of mining has for many years co-existed with important surface land uses. There has thus been strong pressure to devise techniques to predict and minimise surface subsidence damage, especially since collapse of the roof and overlying strata behind the working faces is normally an integral part of the longwall mining systems most often adopted. Other types of mining have seldom presented the same urgency to understand and control the subsidence mechanism. Many mining methods commonly used in steeply dipping and irregular deposits require that the structural integrity of the hanging wall is preserved, as far as possible, by the use of natural or artificial support. For those methods which rely upon caving of the hanging wall, it is often assumed that major surface disruption is inevitable and that the affected surface zone must be cleared of installations and effectively left derelict. There have been very few attempts to predict or measure subsidence for these types of mining method.

The discussion in this chapter is necessarily oriented towards stratified deposits and particularly the European coalfields, since most of the available data pertains to this type of mining.

12.2. PREDICTING SURFACE DEFORMATION

Mining one or more seams most commonly produces a relatively continuous surface deformation which can be measured in terms of vertical and horizontal displacements. Various calculation techniques are used to predict these displacements, founded on extensive investigation in European coalfields. These are principally based on the complete extraction of stratified deposits in situations where the depth and area of excavation are large in relation to the seam thickness. Most techniques currently in widespread use rely upon an empirical approach, although there are continuing research efforts to base prediction upon theoretical or phenomenological (*i.e.* behavioural) considerations.

12.2.1. Empirical Methods
It is observed that surface deformation over a single extraction area in a flat seam has the following characteristics:

(a) A subsidence trough is formed by the vertical displacements of surface points and this normally extends beyond the limits of the mined area.

(b) Horizontal displacements occur with magnitudes and directions approximately proportional to the slope of the subsidence profiles.

(c) If the extraction is geometrically regular in shape, the distribution of vertical and horizontal displacements is approximately symmetrical about the centre.

Two concepts are fundamental to most empirical calculation techniques. The first is that vertical displacement has a maximum possible value (S_{max}) for a particular excavation and this 'full subsidence' is determined from:

$$S_{max} = am \tag{1}$$

where m is the seam thickness and a is the 'subsidence factor', which varies with local conditions.

The second concept is that full subsidence only develops if a sufficient area of the seam, relative to the depth of mining, is extracted. The 'critical area' is defined as the extraction area which produces full subsidence at one surface point only. Areas that produce no full subsidence are termed sub-critical, and those which produce full subsidence at more than one point are super-critical.

Figure 94 shows the typical profile of a subsidence trough with the vertical and horizontal displacements of surface points. The angle of draw (or limit angle) is the angle of inclination, measured from a vertical axis, from the edge of the mine workings to the point of zero subsidence. It is a function of seam dip and local geology and in Britain is commonly in the range $25°–35°$. In practice an arbitrary small value of vertical displacement is often taken as the limit of subsidence.

Figure 95 shows typical distributions of strain, horizontal and vertical displacements for critical, super-critical and sub-critical extraction areas. Maximum tensile strain normally occurs approximately vertically over the edges of the extracted panel and horizontal displacement is zero where full subsidence has developed.

An empirical relationship in common use is that the maximum subsidence over a sub-critical area depends upon the size of that area in relation to its depth. Provided the length of the extracted area is at least 1.4 times the width, it is found in practice that the two-dimensional case can be considered adequate. In this case the relationship states that, for sub-critical widths, extraction areas having the same width-to-depth ratio produce the same maximum subsidence, other parameters being equal. Figure 96 shows a typical graph used for prediction by the National Coal Board[2] based on the above principle. The graph assumes that there are no zones of special support and that the average panel width is used where panel sides are not parallel. An accuracy of $\pm 10\%$ is claimed for this method in the majority of cases in British coalfields.

The greatest possible subsidence S_{max} is found to occur at width-to-depth ratios exceeding 1.2, where the subsidence factor a is approximately 0.9 for full caving. Solid stowing reduces a to about $0.4–0.5$.

It is generally assumed that a law of superposition applies to surface displacements caused by different extraction areas. This states that the displacements which result from more than one extraction area are the sum of the displacements which would occur if each area were mined alone.

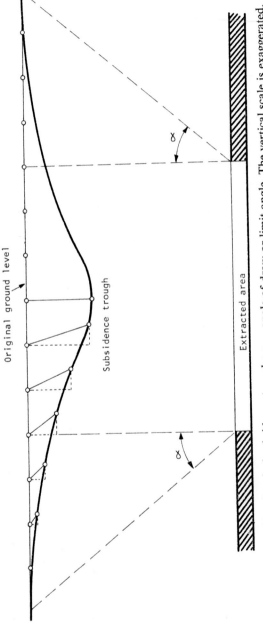

Fig. 94. Profile of typical subsidence trough; α = angle of draw or limit angle. The vertical scale is exaggerated.

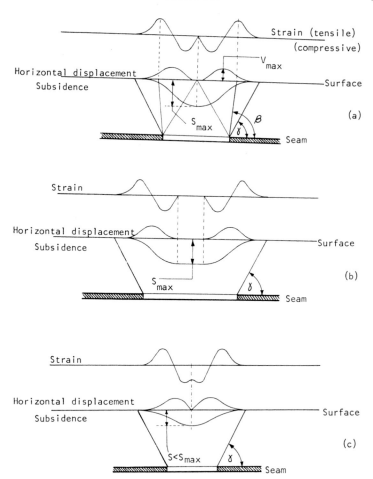

FIG. 95.   Critical (a), super-critical (b) and sub-critical (c) mining depths.

Despite the fact that each subsidence process must obviously change the state of the affected rock mass, in practice the principle of super-position is often found to be accurate within acceptable limits.

A second principle used is the equivalence of extracted areas. Figure 97 illustrates this principle for the two-dimensional case. Equivalence implies that the same percentage of full subsidence would be developed at point A by an extraction width bounded by two straight lines radiating from A. Thus in the example shown extraction width W_1 at depth H_1 and W_2 at H_2 would each cause the same percentage of full subsidence at A. The technique illustrated in Fig. 96 is a special case of this general principle.

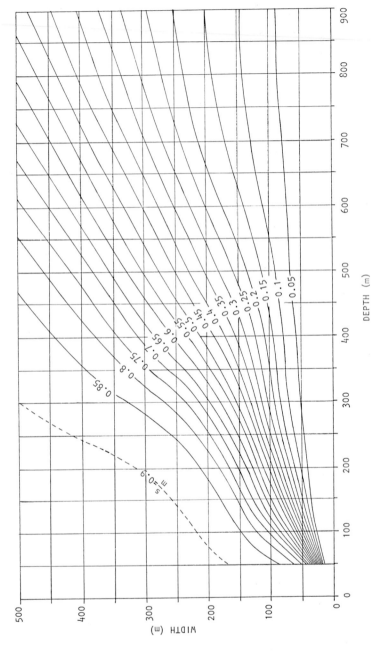

Fig. 96.   Relationship of subsidence to width and depth.

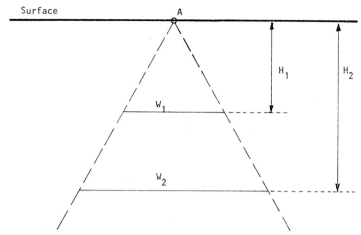

Surface

A

H_1

H_2

W_1

W_2

FIG. 97. Equivalent sub-critical areas.

In the three-dimensional case, cones would replace the straight lines. This principle is approximately true provided there are no major changes in strata with increasing depth.

The subsidence factor a used in eqn. (1) is determined by measuring the full subsidence over super-critical areas. Variations in a may be attributable to the type of packing (if any) used behind the working face, and additionally significant variations are found from one coalfield to another, indicating the influence of particular roof strata. Brauner[1] cites values of a determined in a number of major coal mining countries and these are summarised in Table 76.

Calculation methods which follow the principles discussed above generally predict the final subsidence of horizontal formations and may be extended to other factors such as horizontal displacements and dipping strata. Two general methods are in use and involve functions termed *profile* or *influence* functions.

Profile functions express mathematically the distribution of displacements over two-dimensional critical extraction areas. For super-critical widths the central trough has a constant subsidence of S_{max} and for sub-critical widths the profile is determined from the critical case using empirical relationships. The method is normally restricted to rectangular extraction areas.

Influence functions apply the principle of super-position. The extraction area is considered as an infinite number of infinitesimal elements and likewise the subsidence trough is regarded as a composition of infinitesimal troughs produced by the extraction elements. The subsidence of any surface point is the sum of the individual subsidences due to each extraction

TABLE 76

TYPICAL VALUES OF SUBSIDENCE FACTOR a
(after Brauner[1])

Location and method of packing	Subsidence factor, a
Britain: roof caving or strip-packing	0·90
solid stowing	0·45
Germany (Ruhr): roof caving	0·90
pneumatic stowing	0·45
other solid stowing	0·50
France (Pas de Calais): roof caving	0·85–0·90
pneumatic stowing	0·45–0·55
hydraulic stowing	0·25–0·35
Upper Silesia: roof caving	0·70
hydraulic stowing	0·12
U.S.S.R.: roof caving	0·60–0·90
Pennsylvania	0·50–0·60

element. This method imposes no restrictions on the geometric shape of the excavation area.

Brauner[1] provides an excellent review of the various profile and influence functions in common use.

The time required for full subsidence to develop depends upon several factors, of which the principal are:

(a) The time to extract the critical area—this depends upon the rate of advance and the depth of working;

(b) The nature of the strata—stiff or thick-bedded strata require longer time than compliant or thin-bedded strata;

(c) Depth of extraction—subsidence begins later and lasts longer at greater depths.

(d) Stowing—ground movement takes place over longer periods of time when stowing than when caving.

(e) Previous mining activity—strata already deformed by earlier working respond more quickly than virgin ground.

Extensive observations in the coalfields of Britain and Northern France have shown that about 95% of final subsidence occurs by the time the critical area has been extracted and further movement takes place over the ensuing 6 months. These time factors only apply to trough subsidence. In other cases, such as room-and-pillar mining in strong rock, there may be negligible surface subsidence for many years (often in excess of a century), until pillar deterioration results in a sudden collapse and discontinuous deformation (Fig. 98).

FIG. 98. This major collapse over a room-and-pillar limestone mine in Derbyshire lowered the ground surface by about 10 m and exemplifies discontinuous deformation.

The great advantage of empirical methods of subsidence calculation is that they provide a practical means of predicting surface subsidence and hence assessing the potential for damage. In countries such as Britain the methods in use have been developed painstakingly from a consideration of large amounts of data and have proved to be of great use in conditions where trough subsidence predominates. However, the procedures are based mainly on surface survey data without reference to topographical and geological factors or the properties of the rock mass. The predictions do not require an understanding of the subsidence mechanism and do not establish a relationship between the type and amount of subsidence and mining conditions. The techniques are thus of almost no value beyond the range of mining conditions for which they were developed.

12.2.2. Physical Model Studies
Quantitative simulation of subsidence by the use of physical models has seldom been undertaken because of the difficulties of devising three-dimensional models which accurately reflect the mechanical behaviour of actual rock masses. Most experiments have therefore concentrated upon qualitative investigations. Problems such as the justification of the principle of super-position or the effects of previous workings can be studied with the use of models.

12.2.3. Theoretical Studies

Theoretical techniques consist essentially of replacing the rock mass with an idealised material which deforms in accordance with the principles of continuum mechanics. Most of the traditional mathematical models of solids have been applied to the problems of subsidence. These include isotropic and anisotropic elasticity, viscoelastic and viscoplastic behaviour.

The complex mathematical details of these studies are beyond the scope of this book. Up to the present time this abstract model approach has had little practical significance. The great potential advantage of the method is that it can lead to a much deeper understanding of subsidence mechanisms than is achieved by empirical techniques. The current disadvantages are that application to actual problems requires simplifying assumptions which seem to preclude realistic analytical solutions.

12.2.4. Conclusions

Quantitative prediction of subsidence presently is confined almost entirely to flat dipping stratified deposits using empirical techniques developed in the Western European coalfields. There is an inadequate understanding of the subsidence mechanism and hence an inability to deal quantitatively with ground movements associated with the wide range of conditions encountered in underground mining. The young science of rock mechanics is being extensively applied in research to understand and predict the behaviour of large scale rock masses. This effort should yield a steadily growing knowledge of rock behaviour and hence an increasing ability to cope with subsidence prediction as a general, rather than specific, case.

12.3. SUBSIDENCE DAMAGE

Knowledge of damage occasioned by subsidence is mainly based on stratified deposits, although other types of mining have been the subject of some study.[38] Surface deformations may conveniently be classified in two categories; discontinuous and continuous.

12.3.1. Discontinuous Deformation

Major fracture of the ground surface can cause severe damage to buildings and installations and render land unsafe and unfit for any positive use. Three main types of discontinuous fracture are identifiable and all can vary in magnitude from millimetres to metres. These types are illustrated in Figs. 99 and 100.

It is not possible to predict the magnitude of discontinuous deformations. However, some qualitative assessment of the likelihood of this type of subsidence can be made. Discontinuous fractures are not commonly associated with mining stratified deposits, except when very close to

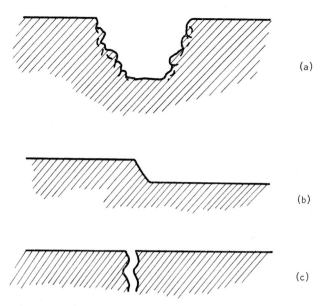

Fɪɢ. 99.   The main types of discontinuous surface fracture: (a) cave-in; (b) step;
(c) crack.

surface. They are more likely to occur when extracting vein or irregular
deposits and are often associated with natural geological discontinuities
such as faults. Fracture is most likely to occur in an area of high tensile
strain and is usually exacerbated by further mining in the same region.
Major cave-in is normally associated with excavations of large vertical
dimensions in relation to the depth of mining.

12.3.2. Continuous Deformation

Continuous surface deformation is usually in the form of a surface trough.
The major components are classified as vertical displacement, horizontal
displacement, slope of the trough, curvature of the trough and strain.

Uniform displacement, whether horizontal or vertical, seldom causes
major damage. Problems are more often, although not always, occasioned
by differential displacement. The main types of damage are:

(a) Vertical displacement—flows of surface and groundwater can be
affected and, in the extreme case, low-lying land may become
flooded. Subsidence 'flashes' have been a common feature in older
British coalfields. Interference with groundwater flow may be
aggravated by internal deformation or fracture within the rock mass.

FIG. 100. Tension cracks caused by mining activity; note also the collapse of the wall.

(b) Slope—the gradients of road, railways and public utilities such as water, gas and sewage pipes may be changed with consequent disruption. Tilting of structures occurs and this can be particularly serious for chimneys, tall buildings and machines whose vertical or horizontal alignment is important. Maximum slope is usually of the order of 0·2–2% but can occasionally be up to ten times the maximum value of this range.

(c) Vertical curvature—differential vertical movements cause distortion due to shear strain and to bending. The radius of curvature is normally in the range 1000–20 000 m.

(d) Strain—both tensile and compressive strains occur at different locations on the subsidence trough. Most observed damage is caused by strains, and installations affected include all types of building, pipes, walls, bridges and railways. Compression zones feature crushing, spalling, buckling and shear fractures. Tensile zones cause cracking and rupture of pipes and cables. Strain is often used directly or indirectly to classify subsidence damage. Table 77 uses such a classification based on the change of length of a structure, which depends upon horizontal strain and the original length of the building.

TABLE 77

NATIONAL COAL BOARD CLASSIFICATION
OF SUBSIDENCE DAMAGE[2]

Change of length of structure	Class of damage	Typical damage
Up to 0·03 m	Very slight or negligible	Hair cracks in plaster
0·03–0·06 m	Slight	Slight fractures showing inside the building. Doors and windows may stick slightly
0·06–0·12 m	Appreciable	Slight fractures showing outside the building. Service pipes may fracture
0·12–0·18 m	Severe	Open fractures allowing weather into structure. Door frames distorted. Floors sloping and walls leaning or bulging
More than 0·18 m	Very severe	Partial or complete rebuilding required. Roof and floor beams need shoring. Severe slopes on floors. Severe buckling and bulging of roof and walls

In assessing subsidence damage there may be confusion with 'pseudo-mining' damage which can be similar in effect but is not caused by mining. Foundation settlement due to the weight of the building or plaster cracks caused by bad construction techniques are common examples of 'pseudo-mining' damage. Changes in the level of the groundwater table, which may be caused by mine pumping, can cause soil shrinkage and settlement of buildings. In localities where damage due to mining subsidence might be expected, it can be difficult to determine unequivocally the cause of such damage.

12.4. CONTROL OF SUBSIDENCE DAMAGE

The alleviation of subsidence damage may be undertaken either by precautionary measures on surface to protect installations or by modification of the mining method so as to minimise deformation of the surface.[1,2,6]

12.4.1. Structural Precautions

The location of new installations is an important control measure. It is advisable to avoid areas of natural geological discontinuity, such as the outcrop of a fault, because of the higher probability of discontinuous deformation. Likewise the hanging wall of a steeply dipping orebody, particularly if caved or unsupported, should be avoided. Long buildings suffer least strain damage if the major lateral axis is oriented parallel to the lines of equal subsidence.

New buildings erected in subsidence areas are least affected if the design is either completely flexible or completely rigid. This is because a rigid structure can resist the forces transmitted by surface deformation and a flexible design permits the building to adapt itself to the deformation without losing its inherent strength. New large buildings may be designed to resist horizontal movement by the incorporation of a slippery membrane which allows the superstructure to slide over the foundation. Superstructures may be flexible to adjust to ground curvature or rigid, in which case the building can if necessary be jacked up to restore it to the horizontal. These techniques are useful for large and costly buildings but may not be economically justifiable for large numbers of small units, such as individual houses on an estate, where it may be cheaper to repair those units which are damaged than to incorporate special arrangements in every building.

If no special stiffening is used, wherever possible flexible constructional materials are preferable. Thus lime mortar should be used in brickwork, fibre-board preferred to plaster, arches of brick or stone avoided, and macadam selected for drives, roads and parking lots.

Pipelines for water, sewage, etc., can be protected by the choice of flexible pipes and the use of flexible joints. Pipelines buried in concrete are vulnerable to high local stresses caused by deformation and failure of the concrete. Subsidence can alter the gradient of the surface and hence it is preferable to lay pipes with an exaggerated gradient to ensure that the direction of flow does not reverse.

Large buildings may be designed as a series of independent units with gaps left between adjacent units. These gaps permit the building to change in length, under compression or tensile stress, without damage, and also allow for curvature.

The precautions described above are applicable to new installations in areas of subsidence risk. Some protection may be afforded to existing structures lacking special design elements, provided the cost of protection can be justified. Resistance to tensile strain is improved by the judicious use of supports such as tie rods, although these must be used with caution as they can lead to unnecessary damage by concentrating strain at particular points in the structure. The effects of compression may be reduced by removing part of rigid elements to provide space for compression. Slots may be cut in floors or paved areas, although this is difficult for some materials such as concrete. Sections may be removed from walls or panels and, in extreme cases, complete houses may be removed from long terraces. Telescopic and flexible joints inserted in pipelines reduce the risk of damage.

Buildings of special architectural merit, such as churches, may warrant extensive and costly protection measures including the removal and storage of stained glass windows and other features particularly susceptible to damage. The effects of horizontal strain, particularly compression, are reduced by trenching around buildings. Trenches should be as close as possible to the outside of the structure and extend just below the foundations. The trench should be filled with a compressible material such as boiler clinker.

12.4.2. Underground Precautions

Surface areas can be protected by leaving a safety pillar of mineral of sufficient dimensions or by controlled mining which ensures that allowable deformations are not exceeded.

The size of safety pillars has often been determined by using a constant 'angle of protection' often equated with the angle of draw (limit angle), as illustrated in Fig. 101. This is now commonly regarded as unacceptable because most structures are only susceptible to differential movements and unnecessary sterilisation of mineral may occur. This loss of mineral increases with depth of working. Furthermore, adequate protection may not be ensured, particularly in multi-seam mining, and other problems such as high stress concentrations underground can result. In workings where subsidence effects are predictable with reasonable accuracy, a better approach is to assess the allowable surface deformations and to undermine as far as possible without exceeding these deformations, subject to a reasonable safety factor. This method is particularly applicable to deep seams where a large loss of mineral may thus be avoided. In shallower workings, the loss of mineral is less severe and the potential surface deformations are more damaging and it may be economical to leave a relatively large safety pillar. Nowadays, it is unusual to leave pillars except under the most fragile of structures—ancient monuments, railway viaducts, reservoirs, etc.—where any ground movement might cause unacceptable damage or danger.

A number of mining techniques are in use to minimise surface damage.

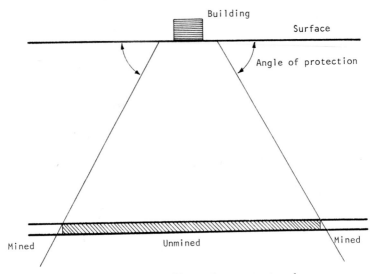

FIG. 101.   Safety pillar under constant angle.

In seam mining it may be possible to avoid adverse configurations which produce maximum differential movements. These are illustrated in Fig. 102.

An alternative to leaving a completely unmined safety pillar is to mine only part of the deposit, leaving the remainder *in situ* to reduce surface deformation. Mining systems of this type are usually termed room and pillar or bord and pillar. The percentage extraction and individual layout can vary widely according to the degree of protection required, the strength of the mineral and similar factors. Brauner[1] cites data from the European coalfields which indicate a preference for relatively long narrow panels separated by permanent pillars, the width of both panels and pillars being of the order of 20–30% of the depth of mining. With extraction ranging from 40 to 70%, observed subsidences are 3–20% of seam thickness. In less compressible strata, such as limestone, only very slight trough subsidence normally occurs although bad layout or an excessive extraction ratio can lead to failure and discontinuous deformation.

Filling or stowing of the excavated area reduces the potential for subsidence damage. In stratified deposits all components of surface deformation are directly proportional to the full subsidence and hence also to the subsidence factor. Table 76 indicates the effect of various stowing methods upon the subsidence factor. At best, a combination of partial extraction and hydraulic stowing can reduce subsidence to negligible proportions. This may however be a costly method of minimising surface

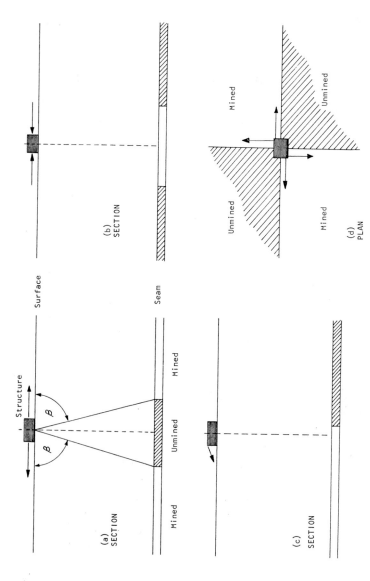

Fig. 102. Unfavourable extraction layouts: β = angle of break and arrows show directions of strain. In (a) the surface structure is placed under maximum extension and in (b) under maximum compression; (c) gives maximum slope, while (d) produces maximum distortion.

damage. Filling of worked-out areas is an integral part of some mining methods, particularly for steeply dipping metalliferous orebodies. This undoubtedly reduces subsidence potential and is of special value since unpredictable discontinuous deformation is characteristic of this type of mining.

The rate of advance of working faces affects surface damage risk. Rapid mining ensures that any surface point is only subjected to the adverse situations indicated in Fig. 102 for a minimum of time. In longwall mining of coal under British trunk roads, for example, three-shift, 7-days per week mining is employed during the critical period, thus avoiding the intermittent stresses arising from faces which stand for one shift in three and at weekends.

In theory it should be possible to arrange the layout of a series of adjacent underground workings so that there are no differential movements on surface. This could be effected by a mining sequence which causes individual positive and negative deformations to cancel each other. In practice this is not possible, but ground movements can be reduced using this type of approach which is usually termed 'harmonic mining'. The principle is illustrated in Fig. 103 for working two seams and can equally be applied to adjacent workings in a single seam. In practice, it is operationally inconvenient, and may be costly, to arrange the juxtaposition of workings in this manner and hence it is only likely to be undertaken where subsidence damage is a special problem.

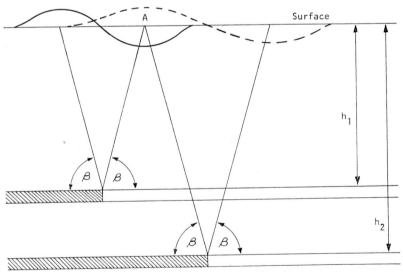

FIG. 103. Strain-reducing configuration of working in two seams. ———: strain curve, upper seam; – – –: strain curve, lower seam.

12.5. CONCLUSIONS

Most of the available knowledge relates to trough subsidence over stratified deposits, particularly coal, worked by longwall methods. The prediction and control techniques in common use are based largely upon empirical investigation and measurement. The science of rock mechanics has enabled some progress on a phenomenological approach with an understanding of the behaviour of the rock mass overlying an excavation. However, this method still yields mainly qualitative rather than quantitative assessment of subsidence phenomena. Continuing research, both empirical and theoretical, will undoubtedly aid the mining engineer in the dual objectives of maximising the percentage extraction of mineral whilst minimising surface disruption.

12.6. CASE STUDIES

12.6.1. Case Study A

King et al.[57] discuss an interesting example of preventive measures taken on the surface to minimise damage to a large nineteenth century hall and its associated buildings.

Precalculation indicated a probably maximum subsidence of 1·50 m (4·9 ft) from an extracted coal seam height of 1·68 m (5·5 ft), with anticipated maximum extension of 2·0 mm/m (0·002 in/in), compression 3·5 mm/m (0·0035 in/in) and slope 15·8 mm/m (0·0158 in/in). Faulted ground underlying part of the area was expected to cause some complication of the subsidence trough profile, although 'step' failures were thought unlikely.

Detailed examination of buildings within the complex showed a probability of severe damage to most structures, because they were neither of sufficient flexibility to sustain the predicted movements nor of adequate rigidity to withstand the induced stresses.

There was no economically feasible underground method of reducing surface movements and hence it was necessary to devise a scheme of surface works to reduce structural damage. The measures taken are illustrated in Fig. 104. In essence they comprise:

(a) Cutting connecting corridors to minimise the effects of lengthening and shortening of the site caused by extensional and compressional strain.

(b) Completely dividing the long Marketing Block into three by cutting two breaches suitably equipped with flexible weatherproofing arrangements.

(c) Digging trenches around the main buildings which connected up with the corridor cuttings and which were filled with compressible boiler clinker ash.

(d) Installing flexible expansion joints in pipes and freeing their passage through walls and doors.

(e) Securing plaster ceilings in old buildings and providing protection against falling pieces.

(f) Taping large glass windows to avoid the risk of flying glass caused by failure in compression.

(g) Providing support for the boot lintels on the side elevations of the new block.

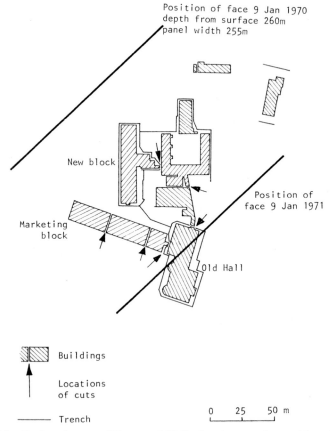

FIG. 104.   Surface layout of Eastwood Hall, showing number and locations of wall cuts, and position of trenches (after King *et al.*[57]).

These protective measures reduced the actual damage to 'slight' to 'appreciable' compared with the 'very severe' classification which would otherwise have resulted. The cost of these preventive works, together with reinstatement and repair costs, was estimated at £13 000 to £15 500 compared with £20 000 for remedial works if no protection had been attempted. Furthermore, temporary inconvenience to those using the buildings was much reduced.

12.6.2. Case Study B

The development of the new town of Peterlee in northeast England provides a classic instance of the successful co-ordination of surface building and underground coal mining activity.[55,56] Peterlee was designated in 1948, with a population target of 30 000, in an area of extensive past and ongoing coal extraction from five seams with a total thickness of 6 m (20 ft).

A very thorough collaborative subsidence research programme was undertaken by the National Coal Board and the Peterlee Development Corporation. Detailed studies provided local subsidence data and facilitated the preparation of the Corporation's Master Plan Report. Surface development was phased to lag behind mining in order to avoid major subsidence effects and the studies also provided valuable information to assist in the design of surface structures. Buildings were constructed to a flexible layout with the plan area of any structure reduced to a minimum and division into units practised where necessary.

A wide range of special construction techniques was used in the development of Peterlee, and underground faces were phased so that tensional effects from one were partially balanced by compressional effects from another. The success of the measures adopted is illustrated by the extraction of 500 000 tons of coal from a 1·2 m (48 in) seam in an area underlying 274 houses. Despite a subsidence wave of 0·9 m (36 in) vertical movement only minor surface damage was reported.

By 1970 Peterlee had a population exceeding 20 000 with 20 factories employing 3000 workers. This major development had been achieved without an unacceptable incidence of surface damage and without sterilising large reserves of coal.

REFERENCES

1. Brauner, G. (1973). *Subsidence Due to Underground Mining* (in Two Parts), U.S. Bureau of Mines Information Circulars IC 8571 and IC 8572, U.S. Department of the Interior.
2. *Subsidence Engineer's Handbook* (1975). National Coal Board, London.
3. Orchard, R. J. (1969). The control of ground movement in undersea working, *Mining Engineer*, **128**(101), 259–73.

4. Berry, D. S. (1963). Ground movement considered as an elastic phenomenon, *Mining Engineer*, **123**(37), 28–41.
5. Salamon, M. G. D. (1963). Elastic analyses of displacements and stresses induced by the mining of reef deposits (in 4 parts), in *Symposium on Rock Mechanics and Strata Control in Mines*, South African Institution of Mining and Metallurgy, Johannesberg.
6. Voight, B. and Pariseau, W. (1970). State of the predictive art in subsidence engineering, *Journal of the Soil Mechanics and Foundations Division, Proceedings of the American Society of Civil Engineers*, **96** (SM2), 721–50.
7. Rankilor, P. R. (1971). The construction of a photoelastic model simulating mining subsidence phenomena, *International Journal of Rock Mechanics and Mining Sciences*, **8**(5), 433–44.
8. Ropski, S. T. and Lama, R. D. (1973). Subsidence in the near-vicinity of a longwall face, *International Journal of Rock Mechanics and Mining Sciences*, **10**(2), 105–18.
9. Ishijima, Y. and Toshiro, I. (1973). The simulation to analyse surface subsidence using a three dimensional finite element method, in *Symposium on Subsidence in Mines*, Australian Institution of Mining and Metallurgy, Wollongong University College.
10. Wardell, K. (1957). The minimisation of surface damage by special arrangement of underground workings, in *Proceedings of the European Congress on Ground Movement*, Leeds, pp. 13–20.
11. Denkhaus, H. G. (1964). Critical review of strata movement theories and their application to practical problems, *J. Sth African Inst. Min. Metall.*, **64**(8), 310–32.
12. Cochran, W. (1971). *Mine Subsidence. Extent and Cost of Control in a Selected Area*, U.S. Bureau of Mines Information Circular IC 8507, U.S. Department of the Interior.
13. Panek, L. A. (1970). Methods and equipment for measuring subsidence, in *Third Symposium on Salt*, Northern Ohio Geological Society, Cleveland, Ohio, pp. 321–38.
14. Obert, L. and Duvall, W. I. (1967). *Rock Mechanics and the Design of Structures in Rock*, John Wiley, New York.
15. Litwiniszyn, J. (1957). The theories and model research of movement of ground masses, in *Proceedings of European Congress on Ground Movement*, University of Leeds, pp. 202–9.
16. Litwiniszyn, J. (1964). On certain linear and non-linear strata theoretical models, in *Proceedings Fourth International Conference on Strata Control and Rock Mechanics*, New York.
17. Sweet, A. L. and Bogdanoff, J. L. (1965). Stochastic model for predicting subsidence, *Journal of the Engineering Mechanics Division, American Society of Civil Engineers*, **91** (EM2, Paper 4292), 22–45.
18. Sweet, A. L. (1965). Validity of a stochastic model for predicting subsidence, *Journal of the Engineering Mechanics Division, American Society of Civil Engineers*, **91** (EM6, Paper 4573), 111–28.
19. *Principles of Subsidence Engineering* (1963). Information Bulletin 63/240, National Coal Board Production Department, London.

20. Marr, J. E. (1959). The estimation of mining subsidence, *Colliery Guardian*, **198**(5116), 343–52.
21. Deere, D. U. (1961). Subsidence due to mining—a case history from the Gulf Coast region of Texas, in *Proceedings Fourth Symposium on Rock Mechanics*, Bulletin Mineral Industries Experiment Station, Pennsylvania State University, pp. 59–64.
22. Nishida, T. (1962). *Mining Subsidence*, Report No. 32, Research Institute of Science and Industry, Kyushu University, pp. 1–74.
23. Whetton, J. and King, J. (1957). Mechanics of mine subsidence, in *Proceedings European Congress on Ground Movement*, Leeds, pp. 27–38.
24. Whetton, J. and King, J. (1959). Aspects of subsidence and related problems, *Transactions Institution of Mining Engineers*, **118**, 663–76.
25. Corden, C. and King, H. J. (1965). A field study of the development of surface subsidence, *International Journal of Rock Mechanics and Mining Sciences*, **2**(1), 43–55.
26. Gilluly, J. and Grant, U. S. (1949). Subsidence in the Long Beach Harbor area, *Bulletin Geological Society of America*, **60**, 461–530.
27. Hackett, P. (1964). *Prediction of Rock Movement by Elastic Theory Compared with In-Situ Measurement*, Rock Mechanics and Engineering Geology, Supplementum 1, pp. 88–102.
28. Berry, D. S. (1969). An elastic treatment of ground movement due to mining, *Journal of Mechanics and Physics of Solids*, Part 1, **8**(4), 280–92.
29. Nishida, T. and Goto, K. (1960). *On the Relation Between Geological Disturbance and Surface Movement Due to Mining Excavation*, Report No. 27, Research Institute of Science and Industry, Kyushu University, pp. 66–72.
30. Skempton, A. W. and MacDonald, D. H. (1965). Allowable settlements of buildings, *Proceedings Institution of Civil Engineers*, Part III, **5**, 727.
31. *Partial Extraction as a Means of Reducing Subsidence Damage* (1961). Information Bulletin 61/231, National Coal Board Production Department, London.
32. Johnson, G. and Soule, J. H. (1963). *Measurements of Surface Subsidence, San Manuel Mine, Pinal County, Arizona*, Report of Investigations 6204, U.S. Bureau of Mines, U.S. Department of the Interior.
33. Salamon, M. and Munro, A. (1967). A study of the strength of coal pillars, *J. Sth African Inst. Min. Metall.*, **68**, 55–67.
34. Cook, N. (1967). The design of underground excavation, in *Proceedings Eighth Symposium on Rock Mechanics*, American Institute of Mining, Metallurgical and Petroleum Engineers, pp. 167–93.
35. Hiramatsu, Y. and Oka, Y. (1968). Precalculation of ground movements caused by mining, *International Journal of Mechanics and Mining Sciences*, **5**(5), 399–414.
36. Orchard, R. J. (1964). Partial extraction and subsidence, *The Mining Engineer*, **123**(43), 417–30.
37. Wardell, K. and Eynon, P. (1968). Structural concept of strata control and mine design, *Transactions Institution of Mining Engineers*, **127**(95), 633–56.
38. Crane, W. R. (1929). *Subsidence and Ground Movement in the Copper and Iron Mines of the Upper Peninsula, Michigan*, Bulletin 295, U.S. Bureau of Mines, U.S. Department of the Interior.

39. Hakala, W. W. (1970). Subsidence caused by an underground nuclear explosion, in *Symposium on Engineering with Nuclear Explosives*, U.S. Atomic Energy Commission, **2** (Conf—700101), pp. 1428–55.
40. Houser, F. N. (1970). *Sequence of Surface Movement and Fracturing during Sink Subsidence, Nevada Test Site*, U.S. Geological Survey, Report USGS-474-56.
41. Kaneshige, O. and Okamura, H. (1970). The rock pressure phenomenon and deformation around a cavity in the upper rock layer owing to lower mining work, *Rock Mechanics in Japan*, **1**, 106–8.
42. Morrison, R. G. (1970). *A Philosophy of Ground Control*, Ontario Department of Mines, Toronto.
43. Gray, R. E. and Meyers, J. F. (1970). Mine subsidence and support methods in Pittsburgh area, *Journal of the Soil Mechanics and Foundation Division*, **96** (SM4, Paper 7407), 1267–87.
44. Orchard, R. J. and Allen, W. S. (1970). Longwall partial extraction systems, *Mining Engineer*, **129**, 523–35.
45. Grard, C. (1969). Mining subsidence. Flow to limit its effect on surface, *Revue de l'Industrie Minerale*, **51**(1), 35–70.
46. Kumar, R. and Singh, B. (1969). Investigations into movement of surface rocks due to underground opening, *Journal of Institute of Engineers, India*, **49**(7), 57–61.
47. Mansur, C. I. and Skouby, M. C. (1970). Mine grouting to control building settlement, *Journal of the Soil Mechanics and Foundations Division*, **96** (SM2, Paper 7166), 511–22.
48. Mather, J. D., Gray, D. A. and Jenkins, D. G. (1969). The use of tracers to investigate the relationship between mining subsidence and groundwater, *Journal of Hydrology*, **9**(2), 136–54.
49. Lee, K. L. and Shen, C. K. (1969). Horizontal movements related to subsidence, *Journal of the Soil Mechanics and Foundations Division*, **95**(SM1), 139–66.
50. Price, D. G., Malkin, A. B. and Knill, J. L. (1969). Foundations of multi-storey blocks on the coal measures with special reference to old mine workings, *Quarterly Journal of Engineering Geology*, **1**(4), 271–322.
51. Shelton, J. W. (1968). Role of contemporaneous faulting during basinal subsidence, *Bulletin American Association of Petroleum and Geology*, **52**(3), 399–413.
52. Parker, J. M. (1967). Salt solution and subsidence structures, Wyoming, North Dakota and Montana, *Bulletin American Association of Petroleum and Geology*, **51**(10), 1929–47.
53. Meade, R. H. (1967). *Petrology of Sediments Underlying Areas of Land Subsidence in Central California*, U.S. Geological Survey, Professional Paper N 497-C.
54. Jenike, A. W. and Leser, T. (1962). Caving and underground subsidence, *Transactions Society of Mining Engineers of AIME*, **223**(1), 67–73.
55. Potts, E. L. J. (1974). Mining subsidence and the environment, in *Minerals and the Environment Symposium*, Institution of Mining and Metallurgy, London, pp. 661–83.
56. *The Master Plan—Report* (1952). Peterlee Development Corporation.

57. King, H. J., Whittaker, B. N. and Shadbolt, C. H. (1974). Effects of mining subsidence on surface structures, in *Minerals and the Environment Symposium*, Institution of Mining and Metallurgy, London, pp. 617–42.
58. Orchard, R. J. and Allen, W. S. (1974). Time-dependence in mining subsidence, in *Minerals and the Environment Symposium*, Institution of Mining and Metallurgy, London, pp. 643–59.

13

Legislation and Control

13.1. INTRODUCTION

Public concern over the effects of mining upon the natural environment has become prominent, largely in the last 10–15 years. One effect of this upsurge in public interest has been a proliferation of legislation and control measures designed to reduce the impact of mining activity upon the environment. A wide range of environmental legislation has either been enacted or is under consideration in many countries and, as a measure of expediency, in some countries long-established statutes are being pressed into service to control mining activity. In this fluid situation, a comprehensive account of laws affecting mining and the environment is impracticable and such a listing would, in any case, rapidly become obsolescent as new laws and control regulations replace existing legislation.

However, despite the disparity in detail, it is apparent that most legislation is based on one or more of a small number of fundamental concepts. In this chapter these concepts are discussed and illustrated, where appropriate, with existing or proposed statutes.

Legislation and control measures may conveniently be divided into three categories:

   (i) Pre-operational—at this stage the decision is taken whether or not to permit mining operations to take place and, if so, under what conditions.

  (ii) Operational—if mining is permitted to proceed, the manner in which operations are conducted is normally controlled by public authorities.

 (iii) Post-operational—when mineral production ceases, legislation may enforce rehabilitation and long term care and maintenance of the site.

13.2. LAND USE PLANNING

Virtually all developed, and many developing countries exercise some form of land use planning. It is self-evident that any modern society uses land for a very wide variety of purposes which can be classified under a number of general headings such as agricultural, residential, urban, industrial, recreation and amenity, etc. Furthermore it is clear that some land uses can exist relatively harmoniously in close juxtaposition, for instance agriculture and amenity, whilst others, such as housing and heavy industry, are much less compatible.

Because of these considerations, the concept of zoning is fundamental to land use planning. In essence, zoning means assigning one or more priority land uses to particular areas of land. The development of a regional or national land use plan using a zoning concept is a complex operation. Most major land uses are subject to many constraints. Mineral operators frequently state, with obvious truth, that minerals can only be worked where they are found. But equally many types of agriculture are subject to topographical, hydrological, climatic and soil constraints; industry requires good communications, an adequate labour force living within easy travel distance, water and other services; many leisure and amenity pursuits depend upon a scenic landscape; and so on.

The development of a land use plan is thus dependent upon:

(a) A detailed knowledge of the land within the area under consideration, including existing land uses, geography, geology, climate,
communications and many similar factors.
(b) An appreciation of the land area required for the various land uses
and the constraints upon the suitability of land for each purpose.

When (a) and (b) have been adequately surveyed, the planner is faced with a matrix of considerable complexity. Individual land uses must be located in areas to which they are suited and as far as possible must avoid incompatible adjacent activities.

The process of land use planning as described above is an over-simplification, for two main reasons. First, the plan must be dynamic, not static, taking account not only of existing requirements but also future development. Secondly no region, or country, is a completely self-contained unit. Any given region is likely to be well suited to some land uses and unsuited to others. For instance there may be an abundance or lack of mineral deposits; a good or bad communications network; high grade productive farming land or large tracts unsuitable for agriculture. It is therefore probable that any region will tend to specialise in particular land uses to which it is best suited. Thus, for instance, about 40% of the world's production of china clay comes from one small area of South West England. Rational planning is therefore facilitated by the development of national policies and this is particularly so for mineral deposits because of their spasmodic occurrence.

Low value, relatively ubiquitous, minerals such as concrete aggregate can normally be planned on a national scale (although few countries have so far planned a rational extraction programme), but the production and distribution of less common, high value, minerals such as tin depends upon international trade and agreement.

Land use planning considerations, national minerals policies and international trade agreements thus play a most significant role in minerals development. They affect every aspect of mineral working from the decision whether or not to permit a particular deposit to be exploited, to the controls exercised during the life of the mine and the type of rehabilitation required.

13.3. PRE-OPERATIONAL PHASE

The first stage in the life of any minerals operation is the location of mineral resources and investigation to determine the quantity and quality of the deposit. Initial exploration by geological mapping and geophysical or geochemical techniques normally has little or no environmental impact and can usually be undertaken without any disruption of existing land uses. The second stage of exploration, once a resource has been located, involves detailed investigation of quantity and quality to determine the viability of mining. This often requires a drilling programme which has a greater, albeit temporary, environmental impact.

The impact of exploration activity is seldom, of itself, sufficient to cause public concern although ancillary activities, such as construction of access roads, exploration trenches, pits or adits, can be more disruptive. Conflicts generally arise in areas where existing land uses are regarded as incompatible with mining, and these areas are commonly National Parks or other regions of special amenity value. Opposing viewpoints are well summarised by Stevens et al.[1]:

(a) 'most planning authorities . . . take the view that . . . any discussion or consideration of a theoretical full-scale mineral working is irrelevant to the question whether planning permission should be given for investigatory activities designed to search for, and prove, mineral deposits;

(b) most conservationists . . . argue that permission should not be given for relatively inoffensive but expensive investigatory activities if, at the end of the day, it is likely that permission to exploit any mineral deposits which are found will be, or ought to be, withheld. To grant permission for investigatory activities . . . makes it difficult subsequently to refuse permission for exploitation.'

Not all governments require consent for exploration activity but, where such consent is necessary, it may in sensitive areas be dependent upon which of the above philosophies is applied. Stevens concludes 'that it cannot be right to determine any planning application (*i.e.* for exploration activity) on a consideration of some other hypothetical development for which permission has not been, and may never be, sought'. This view is not universally held even in government circles, as evidenced by the recently reported[2] closure of 200 million ha (500 million acres) of federal land in the U.S.A. to all mining and exploration activity. Even where exploration is not itself prohibited, other restrictions (for instance on road construction or bush clearance) may effectively limit full evaluation of mineral resources. Thus the U.S. Wilderness Act 1964 prohibits (in designated Wilderness Areas) temporary or permanent roads; aircraft landing strips and heliports or helispots; use of motor vehicles or boats or other forms of mechanical transport; aircraft landing, or dropping persons or materials from aircraft; structures or installations and tree felling for non-wilderness purposes.[22] Thus although the Act permits prospecting and gathering of information on minerals, it effectively precludes adequate evaluation of any resources discovered.

Opposition to, and control of, exploration activity is usually limited to areas of special sensitivity. There is commonly much more widespread objection to proposals to develop a mining operation. The extent of this opposition depends principally upon the amount of land use competition, the nature of existing land uses and their compatibility with mining.

Until recently, proposals to mine in remote or wilderness areas were unlikely to encounter significant opposition because of a lack of competing land uses and the absence of adjacent human activity which might suffer disruption. There is now, however, a growing feeling amongst conservationists that areas of land relatively untouched by the hand of man have a special ecological value (the 'wilderness' concept) and ought, therefore, to be preserved. Thus development proposals in the remotest regions will probably meet opposition and this has been well illustrated recently by the resistance, largely on ecological grounds, to the proposed oil pipeline through the Alaskan tundra.

Many countries designate areas of land of particular value for reasons of landscape, natural history, scientific interest, etc., and mining may be prohibited or permitted only under special circumstances. The National Parks of the U.S.A. and the U.K. are typical examples of this type of designation. In Britain the Hobhouse Committee Report,[3] which formed the basis of the National Parks and Access to the Countryside Act of 1949 stated that in National Parks 'new mineral workings should be permitted only on grounds of proven national necessity'. This certainly does not totally exclude mining, as was made plain by the Minister during the second reading debate in the House of Commons:[4]

'It may be necessary to utilise the mineral wealth which lies in those areas (*i.e.* National Parks) for the purpose of ensuring the economic life of our people. I do not think anybody would seriously suggest that we should ignore the existence of this mineral wealth and fail to use it, subject to a number of conditions. The first condition is that it must be demonstrated quite clearly that the exploitation of those minerals is absolutely necessary in the public interest. It must be clear beyond all doubt that there is no possible alternative source of supply and if those two conditions are satisfied then the permission must be subject to the condition that restoration takes place at the earliest possible opportunity'.

The stringency of the criteria used to determine the acceptability of mining in designated areas varies from one country to another, but the British approach typifies the kind of policy often adopted.

Elsewhere, in non-designated areas, the reaction to mining proposals depends upon a number of factors of which the most important are usually the amount of land use competition, the nature of existing land uses, the anticipated environmental impact of the mine, and the state of the local economy.

As the first stage in assessing an application for mineral workings, it is increasingly common for the regulating authority to require proof of the quantity and quality of the deposit. Some authorities also require marketing information to determine that a sufficient demand exists for the mineral. Information on reserve tonnage and grade is normally obtained by the prudent operator at the early stages of development and is frequently disclosed publicly in order to attract investment in the project. However, this is not always the case, particularly for low value industrial minerals such as limestone, where operators sometimes assume geological continuity without detailed site investigation. Most mineral producers accept it as reasonable that a regulatory body should request assurance that the proposed mine has adequate reserves to be viable, but a requirement to prove market demand in relation to a particular deposit at a particular site is frequently difficult.

Having received assurance that the proposal is viable and would generate sufficient cash to allow adequate environmental protection measures, some assessment of the environmental impact of the operation is almost universally required.

Statutory requirements vary widely and this is one of the most active fields for new legislation. The U.S.A. probably has the most advanced specific laws on environmental impact. The National Environmental Policy Act of 1969 requires that before taking any 'major Federal action having a significant impact on the quality of the human environment' each Federal agency prepare an environmental impact statement (E.I.S.) and take into account the environmental impact of its action. The effect of this legislation is reviewed by Bullock *et al.*[5] The compilation of an E.I.S. requires information of two types:

(i) A baseline study of the area likely to be affected by the proposed development. This study may encompass the numbers and varieties of flora and fauna; water and air quality and other ecosystem details; and other pertinent factors such as ambient noise levels.
(ii) A complete and accurate description of the proposed development.

With this information an E.I.S. can be prepared assessing the impact of the factors in (ii) upon those in (i). It is the responsibility of industry to supply information and, in the case of mineral developments, this includes descriptions of the site, the mineral deposit, the processes involved and proposed effluent and dust controls; a plan of the operation; assessment of the life of the project, workforce employed, transport and other relevant factors; information and comment on meteorological, ecological, economic, social and related matters.

Although few countries have a requirement as specific as the U.S. Environmental Impact Statement, many require the developer to supply information roughly analogous to the E.I.S. Burford[20] lists environmental impact study requirements in New South Wales and these are summarised in Table 78.

TABLE 78

ENVIRONMENTAL IMPACT STUDY REQUIREMENTS IN
NEW SOUTH WALES[20]

1. Statement of the major objective of the project.
2. Analysis of technological possibilities of achieving objective.
3. Statement of alternative plans considered to reach objective.
4. Statement of characteristics and conditions of existing environment prior to implementing project.
5. Separate report on each alternative engineering plan considered.
6. For each plan, an assessment of probable impacts on existing environment.
7. Summary or recommendation.

In Britain, Stevens[1] has proposed the use of a standardised minerals application form requiring sufficiently detailed information to obviate the need for an E.I.S. It is generally an objective that the environmental impact of a proposed mine should receive wide publicity in order that the general public may participate in the debate on whether the development should be allowed to proceed.

Bullock *et al.*[5] list the practical difficulties encountered in the U.S.A.

with the E.I.S. system. Apart from problems of legal interpretation, these are principally:

(a) Enforcement administration—some agency is needed to ensure that the E.I.S. is submitted when required.

(b) Public participation—probably because of cost, there have been difficulties in informing all interested sections of the general public.

(c) Cost and lack of experts—the cost of the E.I.S. to the developer may be up to $750 000 for a major mining proposal, and its preparation requires the involvement of experts. Likewise, the federal or state agency is involved in some expense and needs expert help. Governor Rockefeller vetoed a bill requiring E.I.S. procedure in New York because 'this legislation would require the state to add an indeterminable number of new positions and other costs' and the head of the environmental advisory council of Pennsylvania, which has no E.I.S. requirement, considers the 'availability of expertise' to be one of the major problems.

The importance attached to environmental impact assessment is in part dependent upon local economic and social factors. In regions of high unemployment or low *per capita* income new mining developments are welcomed, at least by some sections of the public, irrespective of environmental impact because of the jobs and wealth which will be created. Thus the possible development of a major open pit copper mine in the Snowdonia National Park of North Wales was met with widespread opposition on ecological and amenity grounds, but received support from the three local district Councils because of the jobs and prosperity which would have been created. This attitude is common in developing countries where national priorities are often to raise the standard of living by developing primary and secondary industry, and environmental factors, whilst not totally discounted, play a subordinate role to fiscal considerations.

If permission to proceed with a mineral development is forthcoming, it is normal to impose constraints designed to reduce the environmental impact. These may include limitations on output; effluent, dust, noise and vibration standards; landscaping, screening and restoration works; transport restrictions. Some authorities stipulate in detail how the deposit must be worked and within what timescale, whilst others merely require conformity with general existing standards and do not impose conditions specific to individual sites.

Commonly there is no provision for revision of the conditions attached to a permit. This can be a problem with many forms of mineral working. The life of a mine is frequently measured in decades. Over this period of time there can be significant technological and economic changes which affect methods of working and pollution control. Social attitudes may also change, particularly with increasing affluence. The problem is exemplified in Britain where, under present legislation, mineral consents may not be revised or revoked without compensation. Many amenity organisations claim that some consents which predate present concern for the environment are no longer appropriate. Mineral operators understandably resist suggestions that consents should be subject to review, claiming that this would increase the risk element in mining and hence compound the problem of attracting investment. The Stevens Committee[1] has recommended that mineral consents should have a fixed life not exceeding 60 years and that conditions should be subject to quinquennial review, provided that review powers do not extend to:

(a) size of the permitted area;
(b) depth of extraction;
(c) any period specified within which the development must start or finish;
(d) specified maximum rate of extraction.

This proposal effectively represents a compromise between the operator's need for security and the authority's desire to change conditions in the light of technical and socio-economic developments.

Even more stringent than these proposals is the Ontario Bill 120 of 1971 'to regulate pits and quarries and to provide for their rehabilitation'. This Act subjects some types of mineral working to annual review with possible revocation or suspension of licences without compensation.

Although the details of legislation may vary from one country to another, there is a well established trend that operators are required to supply detailed information on the proposed development and its environmental impact, submit proposals to public scrutiny, accept environmental constraints before consent is forthcoming, and be prepared to modify the conditions of the consent during the life of the operation.

13.4. OPERATIONAL PHASE

Some of the controls applied during the life of the operation are frequently specified as conditions of the permit. This is particularly so for landscaping and similar measures designed to reduce visual intrusion. Many of these are 'one off' operations undertaken early in mining development. For instance, amenity banks are often built from overburden and are graded and vegetated in the first year or two of production. Other factors affecting visual intrusion, such as the design and siting of buildings and waste disposal facilities, require a single decision rather than regular monitoring and hence feature little in ongoing control.

Regulations relating to environmental impact during the life of a mine concentrate principally on pollution control, particularly of liquid and gaseous effluents. Two very distinct philosophies have developed in the control of pollution. The first of these concentrates on what standards are technologically and/or economically practicable, and the second concentrates on the nature and capacity of the receiving environment.

Regulations of the first type are often categorised as 'best practicable means' or 'best available means'. They are inward looking, considering the mining operation in detail, rather than outward looking, concentrating on the surrounding environment. The basic approach is to consider the processes which give rise to pollution, the control technology available and to set standards based upon the expected performance of control equipment. Standards are thus set for effluents at the point of discharge into the surrounding environment.

As generally interpreted, 'best practicable means' takes account of individual site factors such as the age and size of the operation, the sophistication and expertise of the company, and the level of profitability. This approach can lead to different effluent standards for individual operations in the same industry. For instance a new, highly profitable, mine would be expected to meet more stringent standards than an older and less profitable similar concern.

'Best available means' takes no account of variations in individual sites but considers only the best available control technology applicable to the industry as a whole. Standards are based on this criterion and universally applied, and those unable to meet the standards, because of old plant or lack of funds to buy new control equipment, are obliged to cease working.

This 'best technology' concept is typified by the U.S. Federal Water Pollution Act of 1972 which has an objective of setting progressively more exacting standards in order to encourage the development of pollution control technology. The Act requires the establishment of effluent discharge limitations which conform by 1st July 1977 with 'the best practicable control technology currently available' and by 11th July 1983 with 'the best available technology economically achievable.' In discussing this legislation T. E. Carroll[17] of the U.S. Environmental Protection Agency states 'It is interesting though to keep in mind that these effluent standards now passed by Congress . . . have no direct relationship to the quality of the receiving body of water'.

In Britain the Alkali Inspectorate, which controls some gaseous and particulate emissions from mineral workings, likewise adopts a 'best practicable means' approach.

The philosophy underlying the alternative approach of defining ambient standards is that it is ultimately the capacity of the receiving environment to absorb effluents which is of importance. Thus ambient standards recognise that effluents, whether gaseous or liquid, are diluted when discharged and it is the level of pollution after dilution which determines the effect upon the natural environment.

This approach is apparent in the U.S. Clean Air Act of 1970 which required the Environmental Protection Agency to set primary ambient standards for health protection and secondary standards for environmental or other considerations. The primary target was that health standards must be met no later than 1975 unless required technology is not demonstrated in which case the deadline specified was 1977 for an industry and 1978 for a few specific facilities. The secondary standards were to be met as soon as practicable after 1975 and 1977.

J. B. Purves[6] describes a similar approach towards water pollution control in the South West of England, one of the oldest mining areas in the world.

One interesting problem of ambient standards is that air and water pollution levels in any region are seldom the responsibility of a single operator, or even a single industry. It is therefore insufficient merely to impose regulations requiring conformity with ambient standards since, in the event of non-compliance, it may be impracticable to assign responsibility. Therefore even where ambient standards are in force, it is normally necessary to apply effluent control standards. The difference between these and apparently similar 'best technology' standards is that they are set with regard to existing pollution levels in the environment and dilution factors.

In comparing the 'best technology' and 'capacity of receiving environment' approaches, the main advantages and disadvantages of the former compared to the latter are:

(a) The 'best technology' concept in conjunction with target dates for achievement ensures a gradual and continuing improvement in pollution control without imposing intolerable burdens on industry which could have widespread effects on profitability and employment; each individual operator within a sector is likely to have to bear proportionately similar costs.

(b) 'Best technology' standards relate to point emission sources which are easily monitored.

(c) It is probably inherently less difficult to determine the best control techniques than to determine the capacity of the receiving environment.

(d) Some important sources of pollution which are difficult to control, such as dust from blasting or water leaching of waste dumps, may be overlooked by the 'best technology' approach.

(e) Ambient standards are directly related to environmental damage whereas 'best technology' standards may still yield widespread damage or even, occasionally, waste money by being unnecessarily stringent.

(f) Ambient standards result in the operators working in the most environmentally sensitive localities bearing the highest costs; this may conflict with wider community interests such as continuing employment.

Control regulations to date have generally concentrated on measurable phenomena which can be related to ecosystem damage. Nuisance aspects of mining, which require subjective assessment, not only pose more difficult problems in deriving standards but also, in the opinion of many conservationists and government authorities, are less urgent considerations.

In the formulation of control standards, some compromise between the desirable and the practical is usually necessary. This is in part because it would be socially unacceptable to close down whole industries on environmental grounds, and in part because the relationship between pollution levels and the environment is complex. The latter consideration suggests intricate and sophisticated standards monitored carefully over wide areas. However, practical limitations favour simple standards with limited monitoring in order to keep costs within acceptable bounds and ease the problems of enforcement.

The field of pollution control has been in a state of flux in the last decade and this seems likely to continue into the foreseeable future. There is at present vast research effort into control technology and the environmental effects of pollution and this will certainly lead to new, and probably increasingly stringent, control regulations in most countries.

13.5. POST-OPERATIONAL PHASE

Regulations concerning the mine site after the cessation of production have principally related to restoration or rehabilitation and, to a lesser extent, the control of long term pollution.

Understandably the earliest schemes to become effective concerned those mineral workings which pose fewest long term problems. In general these are shallow thin stratified deposits which are readily amenable to rolling restoration techniques. True restoration is impossible since the removal of mineral precludes any chance of leaving the land in an identical condition to that existing prior to mining. However, restoration in the sense of returning the land to its original use is easiest for thin shallow deposits. Amongst the earliest schemes of this type was the British Ironstone Restoration Fund established under the Minerals Working Act, 1951. A levy per ton of ironstone is imposed on the producer, with a smaller contribution from central government, and the fund created is used to restore worked out areas. The success of this scheme may be related to the following factors:

(a) there is only a short time between mineral production and restoration;

(b) there are few technical problems in returning the land to a productive use;

(c) costs can be accurately predicted and hence the levy set at a realistic rate;

(d) different operators face similar problems and costs.

A scheme similar to the Ironstone Restoration Fund applies in the German brown coalfields where a levy per tonne likewise ensures restoration. Some States of the U.S.A. have restoration legislation applicable to strip mining of coal. For instance Pennsylvania's All Surface Mining Conservation and Reclamation Act, 1971 operates a licensing system which requires detailed plans of working and restoration to be submitted and the operator must post a performance bond based on the estimated cost to complete the approved reclamation plan. The U.S. Surface Mining Bill S.425 which would have regulated restoration or rehabilitation of stratified and other deposits was vigorously opposed by all sectors of the industry,[9] mainly on the grounds that the provisions of the Bill were financially untenable and that it is impossible to legislate effectively for the very wide range of conditions met throughout the industry.

The essential problems in dealing with rehabilitation of many types of quarry and non-stratified open pit mine are:

(i) the timescale of working is measured in decades rather than years;

(ii) the landform is completely altered, making restoration impossible and rehabilitation to productive new land uses very difficult;

(iii) because of (i) and (ii) the costs of rehabilitation cannot be established with any accuracy, which complicates the posting of performance bonds or setting the value of a levy.

The Stevens Committee[1] found that the normal commercial market could not meet the needs of long-term performance bonds for this type of operation and concluded that such a scheme would require government sponsorship. Because of problems of this nature, permits for non-stratified deposits have seldom specified rehabilitation other than general site clearance.

There has for some time been a growing feeling in conservationist and governmental circles that the widespread dereliction resulting from mining non-stratified deposits is socially unacceptable and that, whatever the practical difficulties, the problem should be tackled by legislative action. Ontario, the most populous province of Canada, produced one of the first statutes to attempt a solution of the rehabilitation problem. Bill 120,

'An Act to regulate Pits and Quarries and to provide for their Rehabilitation', was enacted in 1971 and was directed mainly at aggregate producers. The Act calls for detailed rehabilitation plans secured by a bond and an annual licensing system. The scheme is supported by landscape plans showing:

> 'as far as possible, ultimate pit development, progressive and ultimate road plan, any water diversion or storage, location of stockpiles for stripping and products, tree screening and berming, progressive and ultimate rehabilitation and, where possible, intended use and ownership of the land after extraction operations have ceased'.

The bond has been fixed at 2 cents per ton subject to a maximum of $100 000 or $500 per acre ($1200/ha). The scheme seems to have worked well so far, facilitated by flexibility and pragmatism on the part of operators and government.

A similar, but less comprehensive scheme operates in France where all aggregate producers pay a levy of 5 centimes per tonne to finance rehabilitation. In Britain the Stevens Committee,[1] whilst rejecting bond or levy systems as impractical, recommends that ultimate rehabilitation plans should form a part of the application to work a mineral.

Whatever the differences in legislative approach, the concept of design for abandonment is finding growing favour in many countries. It has been recognised that mines can cause environmental impact long after working ceases at a time when there is no income available for remedial measures and, indeed, the operating company may have gone out of business. If these problems are to be avoided, control measures must be taken during the life of the mine and there must be an obligation to leave the mined out area in a condition which will minimise long term impact.

Although most legislation to date has concentrated on dereliction and, to a lesser extent, continuing pollution, it is apparent that other aspects of mining can create long term problems. The Tribunal investigating the appalling Aberfan disaster in South Wales in 1966, when a colliery tip slide took 144 lives, discovered that no legislation dealing with the safety of tips was in force anywhere except for some limited provisions in West Germany and South Africa.[10] The Mines and Quarries (Tips) Act 1969 remedied the situation in Britain by stipulating measures to ensure the security of tips, and similar legislation has been enacted or is under consideration in many countries.

Design for abandonment seems certain to find increasing support in many countries and will result in statutes to control dereliction, long term pollution, the safety of waste disposal areas including tailponds, subsidence, and necessary maintenance of landscaping or other schemes after the mine closes down.

13.6. CONCLUSIONS

Environmental concern in the general public shows every sign of increasing in the future, and seems to have been little affected by the economic pressures associated with the depression of the last two years. This concern will undoubtedly manifest itself in a continuing proliferation of legislation and control measures aimed at reducing the impact of industry.

Some balance must be struck between the needs of the built and the natural environment. This point is well expressed by Fish:[11]

'the amenity of a person alongside a quarry may be adversely affected, but the person using the house, office, school, factory or hospital built with the quarry products enjoys an improvement in amenity'.

Conservationists fear that the short term financial and social benefit of industry and the community at large is being achieved only at the expense of irreversible damage to the natural environment. Industry fears that it may be subjected to unnecessary or unreasonable constraints which cannot, for technical or economic reasons, be met and that the resultant collapse of industry will threaten the whole fabric of society. The answer must lie in research and investigation to increase knowledge of the nature of environmental impact, and consultation with all persons expressing a coherent point of view, with government as the ultimate arbiter of action. The overall objectives are well stated by Burd:[14]

'The direction of change in environmental protection is toward more and stronger laws and toward more frequent enforcement of those laws. The goal of laws and enforcement is not punishment, assignment of blame, or payment of damages. The goal is the assurance that man and his social and economic activities will blend and conform to the laws of nature. This is society's problem'.

REFERENCES

1. Sir Roger Stevens, Fleming, M. G., Nardecchia, T. J. and Taylor, J. C. (1976). *Planning Control over Mineral Working*, Report of the Committee on Mineral Planning Control, H.M.S.O., London, 448 pp.
2. Withdrawal of Federal lands from exploration is detailed in Interior study (1975). *Engineering & Mining Journal*, **176**(11), 25.
3. *Report of the Hobhouse Committee* (1974). H.M.S.O., London.
4. Hansard, 31st March, 1949 Col. 1492; Silkin (the Minister), during second reading debate in the House of Commons.

5. Bullock, W. D., Brittain, R. L. and Place, G. A. (1975). Environmental aspects of the North American mining industry, in *Minerals and the Environment* (ed. M. J. Jones), Institution of Mining & Metallurgy, London, pp. 27–44.

6. Purves, J. B. (1975). Aspects of mining and pollution control, *ibid.*, pp. 159–79.

7. Wixson, B. G. (1975). Development of a co-operative programme for environmental protection between the lead-mining industry, Government and the University of Missouri, *ibid.*, pp. 3–12.

8. Kysel, P. (1972). Protection of the environment: research, legislation and government structure, *Trans Instn Min. Metall.*, **81**(782), A38–A42.

9. Beukena, C. F., MacGregor, I. and Place, J. B. M. (1973). Mining industry faces total shutdown under S. 425, *Mining Congress Journal*, **59**(4), 105–36.

10. Townshend-Rose, F. H. E. and Thompson, G. M. (1971). Security of tips—statutory and technical requirements, *Mining Engineer*, **130**(125), 293–309.

11. Fish, B. G. (1973). Towards a strategy for quarrying, *Quarry Managers' Journal*, **57**(8), 275–80.

12. Swan, D. (1973). Relationship of the environment and public policy, *Mining Congress Journal*, **59**(2), 74–6.

13. Herbert, C. F. (1971). A mining impact statement for Alaska, *Mining Congress Journal*, **57**(12), 44–6.

14. Burd, R. S. (1971). Mining and environmental protection, *Mining Congress Journal*, **57**(12), 46–8.

15. Ellsaesser, H. W. (1973). Air pollution control: does our program fit the problem? *Mining Congress Journal*, **59**(6), 36–7.

16. Beukena, C. F., MacGregor, I. and Place, J. B. M. (1973). Mining industry seeks realistic surface mining legislation, *Mining Congress Journal*, **59**(6), 88–111.

17. Carroll, T. E. (1972). Updating the policies and activities of the Environmental Protection Agency as related to the mining industry, *Mining Congress Journal*, **58**(12), 61–3.

18. Grove-White, R. (1975). The framework of law: some observations, in *The Politics of Physical Resources* (ed. P. J. Smith), Penguin Education/Open University Press, Harmondsworth, pp. 1–21.

19. Williams, R. E. (1973). Federal Water Pollution Control Act: what mining industry must do to comply, *World Mining*, **26**(3), 64–9.

20. Burford, J. H. (1973). Quarrying and the environment, *Quarry Managers' Journal*, **57**(1), 26–8.

21. Surface mining legislation moving (1972). *Mining Congress Journal*, **58**(4), 76–81.

22. La Grange, J. H. (1971). Effect of wilderness policy on exploration activities, *Mining Congress Journal*, **57**(3), 23–7.

23. *The Control of Mineral Working* (1960). Ministry of Housing & Local Government, H.M.S.O., London, 71 pp.

24. *The Monitoring of the Environment in the United Kingdom* (1974). Department of the Environment, H.M.S.O., London.

25. Jacobs, C. A. J. (1974). Structure plans and the quarrying industry, *Quarry Management & Products*, **1**(7), 243–6.

26. Hastings, R. H. (1970). Federal-state relationships in mineral waste control, in *Proc. 2nd Mineral Waste Utilization Symposium* (ed. M. A. Schwartz), Illinois Institute of Technology Research Institute, Chicago, Illinois, pp. 131–8.

27. Abdnor, J. S. (1968). Environmental controls, *Mining Congress Journal*, **54**(12), 54–61.

28. Excerpts from report of the Public Law Review Commission (1970). *Mining Congress Journal*, **56**(7), 76–87.

29. Hellier, M. J. (1966). Surface minerals and planning powers, *Quarry Managers' Journal*, **50**(2), 57–66.

30. Northcutt, E. (1971). *Summary of Mining and Petroleum Laws of the World*, U.S. Bureau of Mines, information circular 8514, Washington, D.C., 104 pp.

14
Future Trends

14.1. INTRODUCTION

As one of the two primary industries supplying all of the raw products to satisfy man's material needs, it is certain that mineral production will continue in some form as long as the human race flourishes. The nature and extent of that production will depend principally on three factors. These are the state of future technology, the population of the world, and individual affluence expressed as *per capita* consumption of goods. The future impact of mining upon the environment can only be assessed quantitatively if these three factors are first quantified. Although there has been widespread speculation in the last decade, few would claim that any of these factors can be predicted accurately for more than a very limited period of time ahead. Future sociological and political events will significantly affect all three, and chance may also play its part.

A discussion of the future impact of mining upon the environment is thus necessarily full of the uncertainties which result from extrapolating from the known into the unknown. Nonetheless such qualitative assessments do serve a purpose, since they highlight the logical conclusion to which particular trends will lead and the future effects of decisions taken today.

14.2. FUTURE DEMAND

The first stage in assessing future demand is to estimate future population. The difficulties of this are emphasised by Ehrlich and Ehrlich:[2]

'Projections of population growth, even more than estimates of population sizes and growth rates, are subject to a substantial margin of error. Besides the problem of scanty or inadequate data to start with, predicting the future reproductive behaviour of any society, even one whose population structure may be known in detail, is notoriously difficult. In the past demographers have erred fairly consistently on the low side'.

Historical analysis of world population shows that the rate of increase has been exponential rather than linear, *i.e.* growth has occurred at a percentage rate per annum, not by a fixed amount per annum. This has important implications for the timescale within which severe problems may occur. In a finite world no growth rate, even linear, can be sustained indefinitely. However, it is a characteristic of exponential growth that large numbers can be generated very rapidly. Meadows[1] cites an example of exponential growth to illustrate this point:

'There is an old Persian legend about a clever courtier who presented a beautiful chessboard to his king and requested in return 1 grain of rice for the first square on the board, 2 grains for the second square, 4 grains for the third, and so forth . . . the tenth square took 512 grains . . . the twenty-first gave more than a million grains. By the fortieth square a million million grains . . . and the king's entire rice supply was exhausted long before he reached the sixty-fourth square. Exponential increase is deceptive because it generates immense numbers very quickly'.

It is convenient to think of exponential growth in terms of the doubling time, *i.e.* the time it takes the growing quantity to double in size. The world population has not only experienced exponential increase in the past, but the growth rate has also increased, thus leading to a diminution of the doubling time. This is illustrated in Table 79. It is self-evident that this

TABLE 79
DOUBLING TIME OF WORLD POPULATION
(after Ehrlich and Ehrlich[2])

Date	Estimated world population	Time for population to double
8000 B.C.	5 million	1500 y
1650 A.D.	500 million	200 y
1850 A.D.	1000 million	80 y
1930 A.D.	2000 million	45 y
1975 A.D.	4000 million	predicted 35 y

exponential increase in population cannot be sustained indefinitely, or even for many decades at the present growth rate. There has recently been much speculation on whether world population will level off at a stable figure within the capacity of the earth to support life, or whether exponential growth will continue beyond the earth's capacity and the population be decimated by war, famine and disease. Such speculation is beyond the scope of this book, and the answer probably lies in the nature and effectiveness of international effort to reduce population growth rates in the next two or three decades. However, it has been pointed out by Meadows[1] that, even with the most optimistic assumption of decreasing fertility, there is little chance of the population growth curve levelling off before the year 2000 A.D. This is because most of the prospective parents of 2000 A.D. have already been born, and hence a world population of about 7 billion in 30 years time seems certain unless some cataclysmic event intrudes to cause a sharp upsurge in the death rate.

Thus, although the long term future is probably unpredictable, it is fairly certain that the mineral industry will have to cater to the needs of about double the present world population shortly after the turn of this century. Even assuming no general increase in material standards of living, this implies a doubling of mineral production, for, as Fish[9] says, 'every child born into the world brings with him a sizeable order for . . . minerals'. However, this assertion is an over-simplification since both population growth and consumption of materials vary widely in different countries, as illustrated in Table 80.

In essence, the picture that emerges from Table 80 is that most of the population growth is occurring in the poorer countries and most of the increase in individual affluence is taking place in the richer countries. If these trends are extrapolated into the future, the disparity in the material affluence between developed and developing countries will become even greater and, for instance, the G.N.P. *per capita* in the U.S.A. in 30 years' time will exceed that of Indonesia by a factor of almost 100 compared to a

TABLE 80

ECONOMIC AND POPULATION GROWTH RATES
(after Meadows[1])

Country	1968 population (million)	Average annual population growth rate (1961–1968) (% p.a.)	1968 G.N.P. per capita (U.S. $)	Average annual growth rate of G.N.P. per capita (1961–1968) (% p.a.)
China	730	1·5	90	0·3
India	524	2·5	100	1·0
U.S.S.R.	238	1·3	1100	5·8
U.S.A.	201	1·4	3980	3·4
Pakistan	123	2·6	100	3·1
Indonesia	113	2·4	100	0·8
Japan	101	1·0	1190	9·9
Brazil	88	3·0	250	1·6
Nigeria	63	2·4	70	−0·3
Federal German Republic	60	1·0	1970	3·4

factor of 40 in 1968. It seems unlikely that the poorer countries will continue indefinitely to accept such a situation, particularly since it will be increasingly necessary to exploit the mineral wealth of developing countries if supply is to match demand.

It is therefore likely that demand for minerals will grow exponentially at a rate exceeding the population growth rate. This is first because of the universal desire amongst parents that their children should have an improved material standard of living, and secondly because there will be increasing pressure to close the 'affluence gap' between rich and poor countries. Relative rates of growth of population and material requirements have been estimated by the U.S. National Commission on Materials Policy.[3] World population is expected to increase from 2·7 billion in 1953 to 6·4 billion in 2000 whilst, during the same period, gross world domestic product (which is roughly proportional to material requirements) grows from 1·7 to 11·5 trillion U.S. dollars (at 1971 value).

Detailed estimates of future demand for minerals are discussed by several authorities.[1,3] The validity of the specific figures quoted may be open to discussion, but it does appear reasonably certain that world population will double in the next 30 years and that during this period demand for minerals will increase by a factor exceeding two and perhaps even as high as four.

14.3. FUTURE SUPPLY

The effect of rapid increase in demand for minerals upon mining technology has already been mentioned in Chapter 1. There is a natural tendency to work the highest grade, most accessible mineral deposits. This means that as time goes by the trend is always towards lower grade, less accessible deposits. Technological developments of the last 20 years have increasingly favoured large scale surface mining in which the economies of scale associated with very large equipment have permitted large volumes of material to be moved at low cost. Presently about 70% of the world's minerals are won by surface methods.

There has recently been much discussion on the adequacy of the resources of the earth's crust to support continuing exponential growth in demand for minerals. It is obvious that such growth could not be supported indefinitely and hence most discussion centres on the limits to growth and the levels of mineral consumption which can be maintained indefinitely.

Economically mineable mineral reserves are dynamic rather than static. With the passage of time there is a tendency for technological innovation and economic changes to lower the grade of economically workable minerals and thus to transfer deposits from the category of resources to exploitable reserves. The assessment of future reserves is thus complicated by two factors. First, the grade of mineral which will be of future economic interest depends upon unknown future technical and economic developments. Secondly, exploration effort has largely concentrated on the upper few hundred metres of the earth's crust and the more developed countries of the world; thus the full potential of the earth's crust is inadequately determined. The existence of these unknown factors has led to varying degrees of optimism and pessimism in predicting future mineral supply.

Meadows,[1] after studying known global reserves of most major minerals, and after considering various assumptions about future demand and new reserves, concluded that:

'Given present resource consumption rates and the projected increase in these rates, the great majority of the currently important non-renewable resources will be extremely costly 100 years from now. The above statement remains true regardless of the most optimistic assumptions about undiscovered reserves, technological advances, substitution or recycling, as long as the demand for resources continues to grow exponentially'.

The U.S. Council on Environmental Quality stated in 1970:[6]

'Even taking into account such economic factors as increased prices with decreasing availability, it would appear at present that the quantities of platinum, gold, zinc and lead are not sufficient to meet demands. At the present rate of expansion . . . silver, tin and uranium may be in short supply even at higher prices by the turn of the century. By the year 2050, several more minerals may be exhausted if the current rate of consumption continues.

Despite spectacular recent discoveries, there are only a limited number of places left to search for most minerals. Geologists disagree about the prospects for finding large, new, rich ore deposits. Reliance on such discoveries would seem unwise in the long term'.

It is Ehrlich's opinion[2] that:

'certain economists think that only economic considerations determine the availability of mineral resources. They have the idea that as demand increases, mining will simply move to poorer and poorer ores, which are assumed to be progressively more abundant. These economists have misinterpreted the 'arithmetic–geometric ratio' (A/G ratio) . . . that as the grade of ore decreases arithmetically, its abundance will increase geometrically until the average abundance in the earth's crust is reached . . . the geological facts of mineral distribution do not support the simplistic views of these cornucopians . . . although some ores approximate a distribution where the A/G ratio may be applied, most do not'.

However, on a more optimistic note, the U.S. National Commission on Materials Policy[3] states that:

'Those unaware of the vast potential of our earth's finite crusts and seas fear increased consumption levels will deplete the earth's resources. Faulty comparisons of currently known mineral reserves with future demand, predicated upon exponential projections, serve to accentuate those fears . . . Materials should be viewed primarily not as commodities but as substances that provide properties that serve the functions and needs of man. Our search must be directed not only for commodities but also for properties that might be provided by more than one substance'.

Whatever the precise accuracy of these viewpoints may be, it seems certain that there will be growing difficulty in meeting the increased demand for minerals from traditional sources. It will be necessary to mine lower grade deposits and to widen the search for minerals to the remoter parts of the globe and deeper into the earth's crust.

One source of minerals hitherto almost untapped, but of growing commercial interest, is the oceans that cover two thirds of the world's surface. There are 1450 km^3 (350 million cubic miles) of ocean water, each cubic mile containing about 165 million tonnes of dissolved solids.[7] Average concentration of various elements ranges from 21 million tonnes/km^3 (89·5 million tonnes/cubic mile) for chlorine to 11 tonnes/km^3 (47 tons/cubic mile) for zinc, iron, aluminium and molybdenum, and 0·02 tonne/km^3 (0·1 ton/cubic mile) for lead and mercury. The apparent magnitude of these dissolved resources should not obscure the fact that current consumption is also of great magnitude. The oceans may contain 35 million tons of dissolved lead, but we already consume 3·5 million tons of lead per annum so that even with 'no growth' the ocean resource would suffice only for a decade. Unconsolidated minerals on the sea bed range from the sand and gravel deposits of continental shelves, extensively worked by dredging in some areas, to an estimated 1·5 trillion tonnes of ferromanganese nodules thought to occur throughout the world's ocean floors at depths from 30 to 3000 m (100–10 000 ft). Economic interest centres principally on the copper, nickel and cobalt content of nodules, although other metals are also found in significant amounts, and commercial consortia have already spent over $200 million in exploration and development. A further potential source of minerals is the deposits lying in the earth's crust below the ocean floor.

These marine sources of minerals exemplify the changes in exploration and mining technology which will occur as man strives to find supplies of minerals to meet growing demand.

In addition to the problems of assessing future technology, economics and resources, the demand pattern for individual minerals is complicated by the possibilities of substitution and recycling.

Traditional thinking expresses demand in terms of specific minerals—iron, copper, aggregates, etc. In reality the need is not for these particular minerals, but rather for materials with particular properties such as electrical conductivity, strength, resistance to corrosion, etc. When expressed in this way, it is clear that consumption trends for individual minerals may vary quite widely in the future. For instance, if tin were to be in very short supply and thus costly, its use in traditional tin cans might be discontinued in favour of a cheaper more abundant mineral such as aluminium. In this case the 'demand' for tin might grow at a rate considerably less than the average increase in mineral consumption whilst that of aluminium might be higher. Thus overall trends of the future can contain within them significant variations in the production pattern of individual minerals.

Very few minerals (except fuels) are consumed in the sense that usage changes their composition in such a way that they could not be re-used. Nearly all major materials are to some extent recycled. Table 81 shows the recycling of some major materials in the U.S.A. in 1967.[8]

TABLE 81

RECYCLING OF MAJOR MATERIALS FROM WASTES IN THE U.S.A.
(1967)

Material	Total consumption (million tonnes)	Total recycled (million tonnes)	Recycling as a percentage of consumption
Iron and steel	105·900	33·100	31·2
Aluminium	4·009	0·733	18·3
Copper	2·913	1·447	49·7
Lead	1·261	0·625	49·6
Zinc	1·592	0·201	12·6

Decisions to recycle are usually taken on the basis of short term economic considerations, by comparing the cost of collection and reprocessing with the cost of new mineral. There is evidence in the U.S.A.[3] that recycling is declining. Purchased iron and steel scrap consumption in proportion to total ferrous metallics consumption declined from 36·1% in 1951 to 28·2% in 1970, and similar figures were found for other minerals. Recycling by itself cannot fully meet demand when consumption is growing. However, even if action is not taken by governments on environmental grounds to reduce resource depletion, it seems likely that economic interest in recycling will grow to meet shortfalls between supply and demand.

In summary, the major trends in future supply of minerals seem to be:

(a) Exponential growth in the next few decades leading to total consumption of minerals in the year 2000 of between 2 and 4 times the 1970 levels. Thereafter the future is less certain.
(b) Ever-widening search for new mineral deposits in remote and inaccessible areas of the world.
(c) Steadily decreasing average grade of deposit.
(d) Increasing average cost (in real terms) of mineral production.
(e) Development of new technology to exploit new mineral resources (such as those of the sea).
(f) Alteration of the present pattern of minerals consumption by substitution from scarce and costly minerals to those which are more abundant and cheaper.
(g) Growing interest in recycling of minerals.

14.4. FUTURE ENVIRONMENTAL IMPACT OF MINING

The National Commission on Materials Policy[3] cites three widely held beliefs which need examination. These are:

(i) That natural resources can be obtained without limit to meet all public demands for goods and services;
(ii) that the well-being of society is adequately measured by the aggregate volume of the production of goods, *per capita* use of goods, or aggregate consumption of materials and energy;
(iii) that technological development should and will continue indefinitely to contribute to and accelerate the increased consumption of materials per person as it has in the past.

The implications of these assumptions on future demand for minerals and supply problems have already been discussed. It is clear that continued exponential growth also has implications on the future environmental impact of mining.

In relation to the total global land area, the amount of land disturbed directly by mineral production is very small. Even in the U.S.A., which has consistently been amongst the world's major mineral producers this century, only 0.20% of the nation's land area had been disturbed by all surface mining activity up to 1972. It is therefore apparent that considerable expansion of mineral production could take place without major disruption of existing land uses by the land directly required for mining, except in areas of intense land use competition. Conflict is more likely to arise because of the nature and location of mining activity, and because mining can cause environmental effects over a much wider area than that actually needed for production purposes.

As deposits in the more accessible, and socially acceptable, areas are worked out there will be increasing pressure to explore and mine in less accessible regions including wilderness areas, National Parks and other designated lands. This seems certain to cause increasing conflicts between the demands of the built environment and the desire of many to preserve the natural environment in areas of special significance. Such conflicts will be resolved by political decision.

The trend from higher to lower grade deposits has important implications for environmental impact. A lower grade of ore means that, in order to produce a tonne of product, there must be:

(a) more material removed from the earth's crust and hence a greater land area used, more potential for noise and vibration problems and visual intrusion;

(b) more material processed, and probably more sophisticated processing, yielding greater quantities of effluents and solid waste for disposal.

Thus it is probable that most of the environmental problems associated with mining will grow at a faster rate than the world output of minerals because of the steadily decreasing average grade of orebodies. It is implicit in this statement that the pattern of minerals production continues approximately as at present, *i.e.* conventional open pit mining supplying the preponderance of mineral output. Changes in the production pattern could also affect the pattern of environmental impact. It is generally accepted that underground mining causes fewer environmental problems than surface extraction and thus a major swing from open pits to deep-seated orebodies worked by underground methods could diminish environmental disruption. Similarly the widespread use of new techniques such as *in situ* leaching of base metal orebodies and marine mining would alter, and could reduce, environmental disruption trends.

Mining effluents disseminate pollutants into the ecosphere. There is insufficient evidence at present to determine the long term effects of these pollutants from current and future minerals production. The world wide distribution of pollutants has been inadequately studied and in any case many arise from the whole spectrum of industrial and agricultural activity. It is almost impossible at present to specify the pollution levels attributable to mining, except in the immediate environs of the mine, and thus there is little point in speculating on future pollution from mining until adequate data is available. Meadows[1] makes the following important points in relation to pollution:

(a) the few kinds of pollution that actually have been measured over time seem to be increasing exponentially;
(b) we have almost no knowledge about where the upper limits to these pollution growth curves might be;
(c) the presence of natural delays in ecological processes increases the probability of underestimating the control measures necessary, and therefore of inadvertently reaching those upper limits;
(d) many pollutants are globally distributed; their harmful effects appear long distances from their points of generation.

From the above it is clear that intense research into the levels of pollutant emission and the absorbing capacity of the natural environment is needed.

14.5. CONCLUSIONS

The future impact of mining upon the environment is a matter for speculation. Political decisions taken nationally and internationally will control whether man develops a society whose material requirements can be satisfied indefinitely. If present trends are extrapolated into the future it is clear that sooner or later they will be disrupted by forces outside man's control because of the finite capacity of the earth to supply raw materials and absorb waste products. The subjects of principal environmental interest to the mining industry in the future will thus be the means to avoid depletion of resources and to control the dissemination of pollution.

REFERENCES

1. Meadows, D. H. (1972). *The Limits to Growth*, Earth Island Ltd, London.
2. Ehrlich, P. R. and Ehrlich, A. H. (1972). *Population, Resources, Environment*, W. H. Freeman, San Francisco.
3. *Material Needs and the Environment Today and Tomorrow* (1973). Final Report of the National Commission on Materials Policy, U.S. Government Printing Office, Washington, D.C.
4. Boyd, J. (1974). Mineral Resources, in *Minerals and the Environment Symposium*, Institution of Mining and Metallurgy, London, pp. 187–93.
5. Goldsmith, E., Allen, R., Allaby, M., Davull, J. and Lawrence, S. (1972). *A Blueprint for Survival*, Tom Stacey Ltd, London.
6. *First Annual Report of the Council on Environmental Quality* (1970). U.S. Government Printing Office, Washington, D.C.
7. Wenk, E. (1969). The physical resources of the ocean, *Scientific American*, **221**(3), 167–76.
8. *Salvage Markets for Materials in Solid Wastes* (1967). Environmental Protection Agency, U.S. Government Printing Office, Washington, D.C.
9. Fish, B. G. (1975). Living with quarrying, *Quarry Management and Products*, **2**(4), 87–8.

APPENDIX
Conversion Factors

Below are listed the most commonly required conversions from imperial to metric units. Conversion graphs are provided for four hybrid units (ft^3/min—m^3/min, lb/acre—kg/ha, ton-miles/gallon—tonne-kilometres/litre and lb/in^2—g/cm^2).

Length

in to mm	× 25·4	mm to in	× 0·039
ft to m	× 0·305	m to ft	× 3·281
yd to m	× 0·914	m to yd	× 1·094
mile to km	× 1·609	km to mile	× 0·621

Area

in^2 to cm^2 × 6·452 cm^2 to in^2 × 0·155
ft^2 to m^2 × 0·093 m^2 to ft^2 × 10·764
yd^2 to m^2 × 0·836 m^2 to yd^2 × 1·196
$mile^2$ to km^2 × 2·59 km^2 to $mile^2$ × 0·386
acre to ha × 0·405   ha to acre × 2·471
1 acre = 4840 yd^2
1 ha = 10 000 m^2

Volume

in^3 to cm^3 × 16·387 cm^3 to in^3 × 0·061
ft^3 to m^3 × 0·028 m^3 to ft^3 × 35·315
yd^3 to m^3 × 0·765 m^3 to yd^3 × 1·308
U.K. gallon to litres × 4·546   litres to U.K. gallon × 0·220
U.S. gallon to litres × 3·785   litres to U.S. gallon × 0·264

Weight

g to oz × 0·035   oz to g × 28·350
lb to kg × 0·454   kg to lb × 2·205
long ton to tonne × 1·016   tonne to long ton × 0·984
long ton to short ton × 1·12   short ton to long ton × 0·89

Pressure
lbf/in² to   bar × 0·069   bar to lbf/in² × 14·504
lbf/in² to kgf/cm² × 0·070   kgf/cm² to lbf/in² × 14·223
lbf/in² to   kPa × 6·895   kPa to lbf/in² × 0·145

Flow
ft³/min to litre/min × 28·314 litre/min to ft³/min × 0·035
ft³/min to m³/min × 0·028   m³/min to ft³/min × 35·315

Velocity
 ft/s to m/s   × 0·305 m/s   to ft/s   × 3·281
mi/h to km/h × 1·609 km/h to mi/h × 0·621

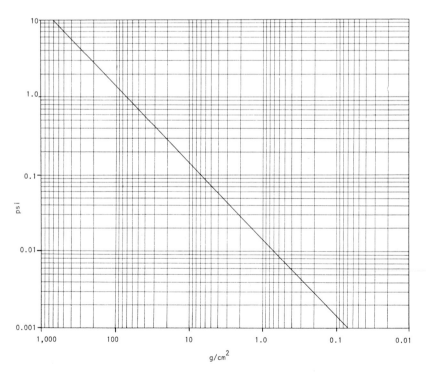

FIG. 105. Conversion graph for lb/in² to g/cm².

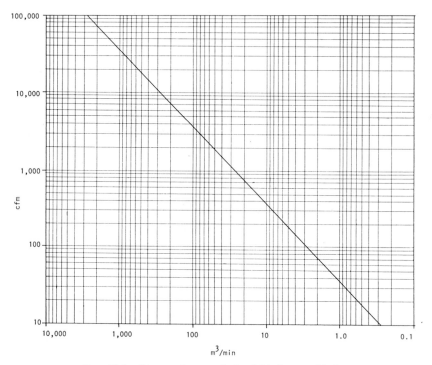

FIG. 106.　Conversion graph for ft^3/min to m^3/min.

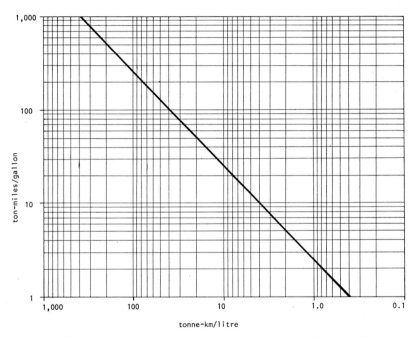

FIG. 107. Conversion graph for ton-miles/gallon to tonne-kilometres/litre.

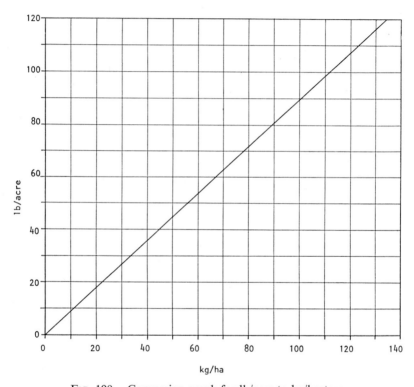

Fig. 108.   Conversion graph for lb/acre to kg/hectare.

Index